OCEANIC PROCESSES IN MARINE POLLUTION
VOLUME 2

OCEANIC PROCESS IN MARINE POLLUTION

Editors: Iver W. Duedall, Florida Institute of Technology, Melbourne, Florida
Dana R. Kester, University of Rhode Island, Kingston, Rhode Island
P. Kilho Park, U.S. National Oceanic and Atmospheric Administration, Rockville, Maryland

Volume 1:   Biological Processes and Wastes in the Ocean
Editors: Judith M. Capuzzo and Dana R. Kester

Volume 2:   Physicochemical Processes and Wastes in the Ocean
Editors: Thomas P. O'Connor, Wayne V. Burt, and Iver W. Duedall

Volume 3:   Marine Waste Management: Science and Policy
Editors: Michael A. Champ and P. Kilho Park

Volume 4:   Scientific Monitoring Strategies for Ocean Waste Disposal
Editors: Donald W. Hood, Amy Schoener, and P. Kilho Park

Volume 5:   Urban Wastes in Coastal Marine Environments
Editors: Douglas A. Wolfe and Thomas P. O'Connor

Volume 6:   Physical and Chemical Processes: Transport and Transformation
Editors: Donald J. Baumgartner and Iver W. Duedall

OCEANIC PROCESSES IN MARINE POLLUTION
Volume 2

PHYSICOCHEMICAL PROCESSES AND WASTES IN THE OCEAN

Edited By

THOMAS P. O'CONNOR

Ocean Assessments Division
U.S. National Oceanic and Atmospheric Administration
Rockville, Maryland

WAYNE V. BURT

School of Oceanography
Oregon State University
Corvallis, Oregon

IVER W. DUEDALL

Department of Oceanography and Ocean Engineering
Florida Institute of Technology
Melbourne, Florida

ROBERT E. KRIEGER PUBLISHING COMPANY
MALABAR, FLORIDA   1987

Original Edition 1987

Printed and Published by
ROBERT E. KRIEGER PUBLISHING COMPANY, INC.
KRIEGER DRIVE
MALABAR, FL 32950

Library of Congress Cataloging in Publication Data

Main entry under title:
Physicochemical processes and wastes in the ocean.
   (Oceanic processes in marine pollution; v. 2)

   1. Waste disposal in the ocean—Addresses, essays, lectures. 2. Waste disposal in the ocean—Case studies—Addresses, essays, lectures. 3. Marine pollution—Addresses, essays, lectures. 4. Marine pollution—Case studies—Addresses, essays, lectures. I. O'Connor, Thomas P. II. Burt, Wayne V. III. Duedall, Iver W. IV. Series.
TD763.P49 1987   628.4'459   84-29746
ISBN 0-89874-811-9

10 9 8 7 6 5 4 3 2

CONTRIBUTORS

ADAM, YVES, Management Unit of the North Sea and Scheldt Estuary Mathematical Models, Ministry of Public Health and Environment, Rue J. Wytsman 14, B-1050 Brussels, Belgium

ALTMANN, R. SCOTT, Environmental Engineering and Science, Department of Civil Engineering, Stanford University, Stanford, California 94305

ARNOLD, ROBERT G., Environmental Quality Laboratory, W. M. Keck Laboratories 138-78, California Institute of Technology, Pasadena, California 91125

ASTON, SIMON R., International Laboratory of Marine Radioactivity, Musée Océanographique, International Atomic Energy Agency, Principality of Monaco

BEWERS, J. MICHAEL, Atlantic Oceanographic Laboratory, Bedford Institute of Oceanography, Dartmouth, Nova Scotia, B2Y 4A2, Canada

BROOKS, NORMAN H., Environmental Quality Laboratory, W. M. Keck Laboratories 138-78, California Institute of Technology, Pasadena, California 91125

BRUNK, JAMES L., Environmental Sciences Division, Lawrence Livermore National Laboratory, University of California, P.O. Box 5507, Livermore, California 94550

BURT, WAYNE V., School of Oceanography, Oregon State University, Corvallis, Oregon 97331

CHOU, CHIU L., Halifax Fisheries Laboratory, P.O. Box 550, Halifax, Nova Scotia, B3J 2S7, Canada

CHURCHILL, JAMES H., Woods Hole Oceanographic Institution, Woods Hole, Massachusetts 02543

CROSBY, STUART A., Department of Marine Science, Plymouth Polytechnic, Plymouth, Devon, PL4 8AA, United Kingdom

CSANADY, GABRIEL T., Woods Hole Oceanographic Institution, Woods Hole, Massachusetts 02543

DELVIGNE, GERARD A. L., Delft Hydraulics Laboratory, P.O. Box 177, 2600 mh Delft, The Netherlands

DEVINE, MICHAEL, Ocean Assessments Division, U.S. National Oceanic and Atmospheric Administration, N/OMS32, Rockville, Maryland 20852

D'HONDT, PHILIPPE, Management Unit of the North Sea and Scheldt Estuary Mathematical Models, Ministry of Public Health and Environment, Rue J. Wytsman 14, B-1050 Brussels, Belgium

DUEDALL, IVER W., Department of Oceanography and Ocean Engineering, Florida Institute of Technology, Melbourne, Florida 32901

EAGLE, RODNEY J., Environmental Sciences Division, Lawrence Livermore National Laboratory, University of California, P.O. Box 5507, Livermore, California 94550

FAISST, WILLIAM K., Brown and Caldwell, 1501 N. Broadway, Walnut Creek, California 94596

FISHER, NICHOLAS S., Oceanographic Sciences Division, Brookhaven National Laboratory, Upton, New York 11973.

FOWLER, SCOTT W., International Laboratory of Marine Radioactivity, Musée Océanographique, International Atomic Energy Agency, Principality of Monaco

GONCHAROV, ANDREY A., State Committee for Hydrometeorology and Control of the Natural Environment of the U.S.S.R., U.S.S.R. State Institute of Oceanography, Kropotkinsky per 6, Moscow, Union of the Soviet Socialist Republics

GURBUTT, PAUL A., Fisheries Laboratory, Ministry of Agriculture, Fisheries and Food, Lowestoft, Suffolk, NR33 0HT, United Kingdom

HILL, MARION D., National Radiological Protection Board, Chilton Didcot, Oxfordshire, OX11 0RQ, United Kingdom

JACKSON, GEORGE A., Institute of Marine Resources, Scripps Institute of Oceanography, La Jolla, California 92093

JACQUES, THIERRY G., Management Unit of the North Sea and Scheldt Estuary Mathematical Models, Ministry of Public Health and Environment, Rue J. Wytsman 14, B-1050 Brussels, Belgium

JOKELA, TERRENCE A., Environmental Sciences Division, Lawrence Livermore National Laboratory, University of California, P.O. Box 5507, Livermore, California 94550

KILLWORTH, PETER D., Robert Hooke Institute, Department of Atmospheric Physics, Oxford, Oxfordshire, United Kingdom

KOH, ROBERT C. Y., Environmental Quality Laboratory, W. M. Keck Laboratories 138-78, California Institute of Technology, Pasadena, California 91125

KRANCK, KATE, Atlantic Oceanographic Laboratory, Bedford Institute of Oceanography, Dartmouth, Nova Scotia, B2Y 4A2, Canada

LANGSTON, WILLIAM J., The Laboratory, Marine Biological Association, The Hoe, Plymouth, Devon, PL1 2PB, United Kingdom

LECKIE, JAMES O., Environmental Engineering and Science, Department of Civil Engineering, Stanford University, Stanford, California 94305

LEVAQUE CHARRON, RENE, Brunswick Mining and Smelting Corporation Ltd., Belledune, New Brunswick, E0B 1G0, Canada

LIVINGSTON, HUGH D., Department of Chemistry, Woods Hole Oceanographic Institution, Woods Hole, Massachusetts 02543

LORING, DOUGLAS H., Atlantic Oceanographic Laboratory, Bedford Institute of Oceanography, Dartmouth, Nova Scotia, B2Y 4A2, Canada

LYASHENKO, ANDREY F., State Committee for Hydrometeorology and Control of the Natural Environment of the U.S.S.R., U.S.S.R. State Institute of Oceanography, Kropotkinsky per 6, Moscow, Union of the Soviet Socialist Republics

MARSH, JOHN G., Department of Marine Science, Plymouth Polytechnic, Plymouth, Devon, PL4 8AA, United Kingdom

MEARNS, ALAN J., Ocean Assessments Division, U.S. National Oceanic and Atmospheric Administration, N/OMA 32x2, Seattle, Washington 98115

MILLWARD, GEOFFREY E., Department of Marine Science, Plymouth Polytechnic, Plymouth, Devon, PL4 8AA, United Kingdom

MOBBS, SHELLY F., National Radiological Protection Board, Chilton Didcot, Oxfordshire, OX11 0RQ, United Kingdom

MOMMAERTS, JEAN-PAUL, Management Unit of the North Sea and Scheldt Estuary Mathematical Models, Ministry of Public Health and Environment, Rue J. Wytsman 14, B-1050 Brussels, Belgium

NEEDLER, GEORGE T., Institute of Oceanographic Sciences, Brook Road, Wormley, Godalming, Surrey, GU8 5UB, United Kingdom

NIELSEN, PALLE BO, Danish Isotope Centre, Skelbaekgade 2, DK-1717 Copenhagen V, Denmark

NORTON, MICHAEL G., British Embassy, Washington, D.C. 20008

NOSHKIN, VICTOR E., Environmental Sciences Division, Lawrence Livermore National Laboratory, University of California, P.O. Box 5507, Livermore, California 94550

O'CONNOR, THOMAS P., Ocean Assessments Division, U.S. National Oceanic and Atmospheric Administration, N/OMS32, Rockville, Maryland 20852

O'NIELL, PETER, Department of Environmental Science, Plymouth Polytechnic, Plymouth, Devon, PL4 8AA, United Kingdom

ROBINSON, DOUGLAS G., Halifax Fisheries Laboratory, P.O. Box 550, Halifax, Nova Scotia, B3J 2S7, Canada

SAYLES, F. L., Department of Chemistry, Woods Hole Oceanographic Institution, Woods Hole, Massachusetts 02543

SEIBERT, GERALD H., Atlantic Oceanographic Laboratory, Bedford Institute of Oceanography, Dartmouth, Nova Scotia, B2Y 4A2, Canada

SHEPHERD, JOHN G., Fisheries Laboratory, Ministry of Agriculture, Fisheries and Food, Lowestoft, Suffolk, NR33 0HT, United Kingdom

UTHE, JOHN F., Halifax Fisheries Laboratory, P.O. Box 550, Halifax, Nova Scotia, B3J 2S7, Canada

WHITFIELD, MICHAEL, The Laboratory, Marine Biological Association, The Hoe, Plymouth, Devon, PL1 2PB, United Kingdom

WOLFE, DOUGLAS A., Ocean Assessments Division, U.S. National Oceanic and Atmospheric Administration, N/OMA3, Rockville, Maryland 20852

WONG, KAI M., Environmental Sciences Division, Lawrence Livermore National Laboratory, University of California, P.O. Box 5507, Livermore, California 94550

Preface To
Oceanic Processes In Marine Pollution

It is becoming increasingly evident that the marine environment—oceans, seas, and estuaries comprising over 70% of the entire earth's surface—are not separate entities but are, instead, parts of a continuum. As such, the impacts dealt to one part of the marine environment must be expected to affect eventually the whole. As the ocean is used increasingly as a reservoir for wastes, it is imperative to ensure that the marine resources be used wisely and efficiently. One important approach to such a global perspective is the exchange of information vital to the health and careful management of the ocean. Many people associated with the marine scientific and technical community worldwide are now working to find solutions to problems of mutual interest. The results of their endeavors can be efficiently used through a common medium, such as *Oceanic Processes in Marine Pollution*.

The *Oceanic Processes in Marine Pollution* series is organized into individual reference books that focus on specifically related topics or themes in order to provide unity throughout. To date, six volumes are scheduled for publication:

Volume 1: *"Biological Processes and Wastes in the Ocean,"*
Volume 2: *"Physicochemical Processes and Wastes in the Ocean,"*
Volume 3: *"Marine Waste Management: Science and Policy,"*
Volume 4: *"Scientific Monitoring Strategies for Ocean Waste Disposal,"*
Volume 5: *"Urban Wastes in Coastal Marine Environments,"* and
Volume 6: *"Physical and Chemical Processes: Transport and Transformation."*

The objective of this series is to bring together selected scientific and technical information about fundamental oceanic processes, marine pollution, and marine waste management. The series will cover a broad range of topics, be interdisciplinary in scope, and be of use to scientists and researchers, environmental managers and engineers, administrative officials and legislative staffers, oceanographers, and students.

The driving force behind this series is our belief that some human activities, such as industrial and wastewater discharges to sea, ocean dumping, and inadvertent discharges from non-point sources, could lead to serious problems of marine pollution. A key element in this thought is the recognition that the combination of many dynamic biological, chemical, and physical processes in the sea sometimes reduces the effects of marine pollution; however, serious problems of marine pollution can and do occur even in the most dynamic system if it is overloaded with contaminants. The marine scientist, aware of the dynamics of this system, must, when called upon, be able to translate available technical findings into something that will be of use to the policy maker, the environmental manager, or the engineer who is given the task of regulating or setting policy for the use of the ocean. With this series we hope to provide a cohesive overview of the state-of-the-art technology and thought in a wide range of marine pollution topics.

The chapters included in this and succeeding volumes of this series have been subjected to both external and editorial reviews. We are especially grateful to the reviewers of these chapters for the time and effort that they devoted to the development of the final manuscripts.

Oceanic Processes in Marine Pollution could not have been produced without the willingness and support of the Ocean Assessments Division of the U.S. National Oceanic and Atmospheric Administration.

The Series Editors
Iver W. Duedall
Dana R. Kester
P. Kilho Park

PREFACE

This volume emphasizes physical and chemical processes associated with wastes disposed of into the ocean. Physical processes include dilution, dispersion, and mixing of wastes. Chemical processes include adsorption kinetics, binding of metals, and bioavailability of radioactive materials. This book also examines vertical transport of wastes and the role of marine organisms in the transport of waste materials. Several chapters examine the physical and chemical processes that can be modelled.

Many people helped to make this book a reality. We acknowledge the scientific community for their help in the peer-review process. We thank the staff at the Center for Academic Publications, Florida Institute of Technology, for their help; in particular, we thank Nancy Hatchett, Barbara Hutchko, and Dorothy Scofield for their efforts in manuscript preparation; Mary Ann Nelson for providing editorial assistance and coordinating the production of this volume; Elizabeth Schafer for organizing the author index; Barbara Struttmann for drafting most of the illustrations; and Diane Quimby for her care and patience in proofreading. Finally, we are especially grateful to the authors themselves, without whom this body of information could never have been possible, for their cooperation and patience. This work was supported by the U.S. National Oceanic and Atmospheric Administration.

THOMAS P. O'CONNOR
WAYNE V. BURT
IVER W. DUEDALL

Rockville, Maryland

CONTENTS

PART III. CHEMICAL PROCESSES AND TRANSPORT

GLOSSARY OF ACRONYMS AND INITIALISMS

BOD	Biological oxygen demand
CHC	Chlorinated hydrocarbons
CSDOC	County Sanitation Districts of Orange County
DDT	Dichlorodiphenyltrichloroethane
DHL	Delft Hydraulics Laboratory
GEOSECS	Geochemical Oceans Sections Study
GESAMP	United Nations Joint Group of Experts on Scientific Aspects of Marine Pollution
IAEA	International Atomic Energy Agency
ICES	International Council for the Exploration of the Sea
ICRP	International Commission on Radiological Protection
IMO	International Maritime Organization
NEA	Nuclear Energy Agency
NOAA	U.S. National Oceanographic and Atmospheric Administration
PCBs	Polychlorinated biphenyls
SVD	Singular value decomposition
VCF	Volume concentration factors
WCF	Wet-weight concentration factors
WHIMP	Woods Hole Interstitial Marine Probe

PART I: INTRODUCTION

Chapter 1

Scales of Biological Effects

Thomas P. O'Connor

Ocean Assessments Division
U.S. National Oceanic and Atmospheric Administration
Rockville, Maryland

Michael G. Norton

British Embassy
Washington, D.C.

Alan J. Mearns

Ocean Assessments Division
U.S. National Oceanic and Atmospheric Administration
Seattle, Washington

Douglas A. Wolfe

Ocean Assessments Division
U.S. National Oceanic and Atmospheric Administration
Rockville, Maryland

Iver W. Duedall

Department of Oceanography and Ocean Engineering
Florida Institute of Technology
Melbourne, Florida

ABSTRACT

As an introduction to subsequent chapters on physical and chemical factors controlling the distributions of contaminants in the ocean, we have briefly reviewed data on the scales over which biological effects have been shown or strongly suggested to have been induced by waste inputs. These scales of biological impact provide insights on scales that are appropriate for physical modelling of distributions for different categories of contaminants. Observed effects extend up to 10^5 km² (for hypoxia), but most known effects have been much more localized, a fact that emphasizes the importance of small-scale models in predicting the effects of waste inputs. Larger-scale models remain necessary when predicting overall distributions of persistent contaminants and when considering large amounts of very hazardous materials.

1.1. INTRODUCTION

The following sixteen chapters in this book deal with processes that determine the distribution of contamination caused by the introduction of wastes into the ocean. Chapters 2–11 are primarily concerned with quantifying physical mixing processes. The scale of interest in these chapters varies from the less than 1 km² important in initial mixing (Chapter 2) to ocean-wide mixing (Chapter 9). Between these extremes, a wide range of scales is examined by various authors (Chapters 3–8 and 10), usually based on considerations that are specific to the geography and current regime in an area used for waste disposal. Chapters 11–17 deal with processes that control the partitioning of contaminants between dissolved and particulate forms or between aqueous and biological phases. These are intensive, as opposed to extensive, processes and are basically independent of spatial scale. Nevertheless, the characteristics of oceanic areas under consideration are important because they determine the availability of particles and biota for interaction with contaminants.

The models described in the subsequent chapters are valuable for estimating the contaminant distributions resulting from existing or proposed disposal operations, and they are an important prerequisite to managing the use of the ocean for that purpose. It is equally important to know what concentrations of contaminants can give rise to biological effects. These two categories of information provide management with the ability to predict the types of change, their location, and the scales over which they may occur. For effective and economic management, modelling efforts should be scaled appropriately for the contaminants and the effects of concern. We do not delve into the management issues here, but, as a prelude to chapters that span a wide range of scales, we consider it useful to examine the spatial scales over which effects have been observed and attributed to waste disposal. In doing so, we have not distinguished between regulated and unregulated discharges nor between point sources and diffuse runoff. We have, however, tried to identify the largest scales associated with a particular type of change.

As a point of departure let us consider two extremes of

scale each of which has relevance in different circumstances. The United Nations Joint Group of Experts on the Scientific Aspects of Marine Pollution (GESAMP) (1983) report describes processes and models for quantifying oceanic distributions of contaminants (see Chapter 8). Within the GESAMP report, an area with a 30-km radius (i.e., approximately 1000 km²) was considered the "extremely near field." The report centered on waste disposal on the deep seafloor, and a detailed consideration of so small a scale was considered inappropriate. Essentially, the authors concerned themselves with near- and farfield distributions given a point source of 1000 km². This is, of course, reasonable when that 1000 km² is a remote area on the deep seafloor, which itself spans 2×10^8 km².

At the other extreme are the zones of initial dilution used in the United States for regulating the dumping of wastes from barges and the discharging of wastes through pipelines. In the case of dumping, wastes must be diluted below 1% of their 96-h LC_{50} within a zone that is 200 m wide, 20 m deep, and as long as the dumping track (U.S. Environmental Protection Agency [EPA], 1977). While this dilution factor is subject to case-specific modification, a discharge over a distance of 5 km would thus define an initial mixing zone of 1 km² (see Chapter 2 for a discussion of the initial mixing process). For discharges from pipes, the zone of initial dilution is again subject to case-specific factors but can be estimated as a circle with a radius equal to the depth of the discharge (EPA, 1982). A discharge at a depth of 30 m would therefore define about a 0.003-km² zone of initial dilution.

It would be misleading to leave the impression that the authors of the GESAMP (1983) report are unconcerned about effects over less than 1000 km² or that oceanic disposal regulations in the United States do not consider effects beyond zones of initial dilution. The dichotomy between these extremes, however, emphasizes the importance of disposal location and the importance of the possible effects being considered. The GESAMP (1983) report is a theoretical consideration of very hazardous wastes and is primarily involved with the transfer of chemicals from wastes to humans via seafood. Large-scale models are necessary because it is only on such a scale that chemicals, initially placed on the deep seafloor, could enter the human food chain. The considerations of initial mixing, on the other hand, emphasize effects on marine life and are based on the philosophy that wastes should become dilute enough over small scales so that large-scale effects are avoided. The very small initial-mixing zone defined around outfalls reflects the fact that outfalls are usually close to shore and therefore in valued locations. Actual effects attributed to waste disposal generally cover spatial scales that are larger than initial-mixing zones but smaller than or of the same order as what GE-SAMP (1983) refers to as "extremely near-field."

The persistence of a contaminant in the environment is an important consideration in modelling efforts, because it influences the scale over which effects may occur. The GE-SAMP (1983) report is primarily concerned with radioactive waste and, therefore, incorporates first-order decay rates. Other materials may be degraded through photochemical or biological oxidation, which are also usually first-order processes but with rates dependent on environmental conditions. The scale of effects from waste inputs is ultimately dependent on the balance between the rates of waste input and the rates of physical processes and contaminant decay.

1.2. SCALES OF BIOLOGICAL EFFECTS

Among the largest scales over which waste-induced biological change has been observed are in estuaries, in coastal areas, and in seas where excessive inputs of nutrients have enhanced production of organic matter whose subsequent decay has led to hypoxic conditions in the lower layer of stratified water columns. Some such instances are noted in Table 1.1. We make no claim to have included all possible entries to this table, but we believe that we have incorporated the largest-scale examples. We have emphasized large scales by listing the areas over which oxygen concentrations in bottom water were ≤ 2 mg liter^{-1} in the years of greatest effect. In most areas for which it has become regarded as a waste-induced problem, hypoxia has historically occurred on a smaller scale. For example, Sindermann and Swanson (1979) in the case of the New York Bight and Gerlach (1981) in the case of the Baltic Sea discuss the relative influences of nutrient inputs and the hydrographic and climatological factors in determining the scale of hypoxic areas. An exception is the recently define hypoxia in the Gulf of Mexico (Rabalais and Boesch, 1986). In that case there is no historical information with which to compare the present. In all of the instances of oxygen depletion cited in Table 1.1, there was either direct or anecdotal data on the effects of hypoxia on marine organisms.

The New York Bight represents the largest scale over which benthic communities are known to have been altered by inputs of particulate organic matter or simply by the dumping of solid wastes (Table 1.1). For the New York Bight and southern California outfalls we have emphasized scale by listing the largest areas over which biomass has increased. In these instances and others (see Pearson and Rosenberg, 1978), the most dramatic effects are generally over smaller scales for which the input of organic matter is highest, and the benthic biomass tends to increase as a result of high densities of opportunistic species and to the detriment of sensitive organisms. The example from the U.K.

coastal dumpsites (Table 1.1) represents the depletion of the benthos by direct burial and by physical changes to the seafloor as a result of dumping fly ash from coal-fired electric power stations.

Oxygen depletion and benthic alterations can result from inputs of nutrients and organic matter or solids and are not generally due to contamination by particular chemicals that have reached toxic concentrations. The remaining effects listed in Table 1.1 may be attributed to specific contaminants although their identity is not always known.

Information on the scale of water-column effects is provided by scope for growth in mussels and oyster larvae bioassays (Table 1.1). One of the most sensitive indicators of biological stress has been a decreased scope for growth among mussels (*Mytilis edulis*) transplanted to different parts of estuaries. In southern San Francisco Bay and in Narrangansett Bay, three or four measurements at discrete points along the length of part of the estuary have defined a gradient in scope for growth. In order to place areal dimensions on those essentially linear observations, we have assumed that the entire water-body widths along those gradients enclose areas within which chemical contamination has exerted an effect. The other water-column effect for which we have information on the scale is that from oyster larvae bioassays in Puget Sound. The effect was attributed to discharges of sulfite waste liquor from pulp mills, and its extent decreased dramatically after such discharges were controlled (Cardwell et al., 1977). The negative effects on oyster larvae were not simply due to the oxygen demand of such wastes because the assay procedures compensated for that possible effect.

Effects have also been attributed to sediment toxicity. Long (1985) has summarized the results of numerous assays in which test organisms were exposed to sediment samples taken from various locations within Puget Sound. The results indicate very discrete zones of toxicity and clearly show that continuous toxicity cannot be assumed between zones. In total, the results indicate that approximately 10 km^2 of the sediments within Puget Sound are toxic to at least one of the tested organisms. Given the discrete nature of locations within which bottom fish have been found to exhibit hepatic neoplasia in Puget Sound (Malins et al., 1984) and Boston Harbor (Murchelano and Wolke, 1985), it would seem that the cumulative area affected in each case would also be in the range of 1 to 10 km^2; however, since neither author considered areal extent, we have not included such estimates on Table 1.1.

Observations of another anomalous condition in bottom fish, fin rot (or fin erosion), can be put into a spatial context. O'Connor et al. (1986) conclude that, over and above annual variations in the frequency with which fin

Table 1.1   Scales over Which Biological Effects Have Occurred in Marine Waters and Been Attributed to Waste inputs

Effect (Example[s]):	Areal Extent (km^2)[b]	Reference
Hypoxia		
Baltic Sea	100,000	Gerlach, 1981
Gulf of Mexico	10,000	Rabalais and Boesch, 1986
New York Bight	8,600	Sidermann and Swanson, 1979
Kattegat and Belt Sea	7,000	Rosenberg, 1985
Chesapeake Bay	3,000	Officer et al., 1984
Altered Benthic Biomass		
New York Bight	200	Boesch, 1982
Coastal outfalls, California	<1–100	Mearns and Word, 1982
Coastal dumpsites, U.K.	6–21	Norton, 1985
Scope for Growth in *Mytilis edulis*		
South San Francisco Bay	400	Martin and Severeid, 1984
Narragansett Bay	100	Widdows, 1985
Oyster Larval Toxicity		
Puget Sound	30–50	Cardwell et al., 1977
Sediment Toxicity		
Puget Sound	10	Long, 1985
Fin Rot in Bottom Fish		
Inner New York Bight	3,000	O'Connor et al., 1986
Southern California outfall	100	Cross, 1985
Epidermal Papilloma		
German Bight	1,000	Dethlefson, 1984
Closed Fisheries		
New York marine waters	9,000	Faber, 1986
New Bedford Harbor	45	Weaver, 1984
Belledune Harbor	1	Chapter 11, this volume
New York Bight	800	Stanford et al., 1981
Marine Bird Reproductive Losses		
Brown pelicans, Southern California	10,000	Anderson et al., 1975
Ospreys, Long Island Sound	1,000	Spitzer et al., 1978

[a]No attempt has been made to include all possible examples of any particular type of change. We do think that we have included the largest-scale examples.

[b]Where only one significant figure is given, the area has been estimated from data provided in the cited source, usually a map.

rot is observed, there is a 3000-km^2 area within the Hudson Estuary and inner New York Bight where the condition is more frequent relative to reference areas. Cross (1985) analyzed fin-erosion data collected between 1971 and 1982 near the wastewater discharge from Los Angeles County. He found that the frequency of occurrence increases with proximity to the discharges and that, beyond an area of sediment contamination (approximately 100 km^2 from the map in the text), the frequency decreases to what could be background levels.

Epidermal papilloma in German Bight dab is another type of abnormality that has been attributed to waste inputs (Dethlefsen, 1984). The author is careful to indicate that the condition to which he refers by that name is not the same as the type of epidermal papilloma due to a parasitic infection that has been found worldwide in bottom fish. He indicates that, during the spring of some years, dab taken over an area of about 1000 km^2 within the German Bight showed a statistically significant increased frequency of epidermal papilloma over that among dab collected in nearby areas. On the scale of the North Sea, other areas also showed apparently higher prevalences of the disease. Since the prevalence and its distribution were seasonally variable and since it was highest in areas where fishing intensity was high and where the density of fish was high, there is reason to suspect a natural cause for the condition. Nevertheless, after taking all these types of influences into consideration, Dethlefsen (1984) concludes that increased prevalence of epidermal papilloma in the German Bight is due to the dumping of waste from titanium dioxide production.

The next category of entries on Table 1.1 refers to areas over which commercial fishing has been prohibited due to the public health hazard associated with seafood contamination. In these cases there has not necessarily been a biological effect in the sense that organisms harboring the chemicals are suffering; however, since resources have been effectively lost, we do include this as a biological effect. The first three of the four examples are closures due to chemical contamination. The particular chemicals causing the problems are polychlorinated biphenyls (PCBs) in the marine waters of New York, PCBs in New Bedford Harbor, and cadmium in Belledune Harbor. The last example is the closure of commercial shellfishing in the New York Bight because of bacterial contamination. While that is the largest single example, shellfish area closures are quite common in coastal areas. In the United States in 1985, about 970 separate prohibitions amounted to a total of 10,710 km^2 of shellfish beds closures, representing about 16% of the total harvestable area (U.S. National Oceanic and Atmospheric Administration–Food and Drug Administration, 1985).

The final entry in Table 1.1 concerns reproductive failure attributed to dichlorodiphenyltrichloroethane (DDT) and DDT-related compounds among marine birds. Anderson et al. (1975) found that the reproductive success of California brown pelicans was compromised by DDT contamination, which caused egg-shell thinning, and that hatching success improved when DDT levels declined. Spitzer et al. (1978), observing ospreys in Connecticut and Long Island, also found egg-shell thinning to diminish and hatching success to increase as DDT levels decreased. It is difficult to put spatial dimensions on these effects because the nesting sites are discrete locales within the overall ranges of the birds. In the case of California brown pelicans, we have estimated scale by noting that the major effect was concentrated along southern California to

northwest Baja California (Anderson et al., 1975), a distance of about 250 km, and offshore to the breeding islands within about 20 km of the mainland. In the case of ospreys, the observations center on eastern Long Island Sound, which has an area of about 1000 km^2.

1.3. CONCLUSION

This brief review suggests that scales of biological effects that have been attributed to waste inputs range from those that are small relative to oceanic scales to effects over a sea or region. The overall problem of marine pollution may be viewed as a collection of local problems. For the most part, observed effects imply that only over small scales are contaminant concentrations large enough to cause effects and, in that context, the appropriate physical models are those yielding detailed distributions over small scales. Some of the more wide-ranging effects such as hypoxia may, however, require modelling on the scale of regional seas.

In addition to their climatological and geochemical significance, large-scale models have important applications in considerations of waste disposal. First, there are some contaminants that are distributed over oceanic scales. Some of these chemicals are lead (Schaule and Patterson, 1981), whose input into surface waters from human activity has overwhelmed natural inputs worldwide, and PCBs (Harvey and Steinhauer, 1976) and fallout nuclides (Bowen et al., 1980), whose actual existence is due to people. Knowledge of the oceanic distributions of these chemicals is, in fact, used to improve large-scale oceanic models. Second, as already mentioned, in a consideration of the disposal of very hazardous wastes in remote locations (e.g., GESAMP, 1983) it is the large-scale distributions that have to be predicted. Last, we cannot ignore the possibility that there are large-scale biological changes that have been due to waste disposal but which have not been recognized. Stegeman et al. (1986) have found that deep-sea fish have elevated hepatic enzyme activities due to their accumulation of PCBs and other chlorinated compounds. Whether there is any physiological consequence to this biochemical change remains to be determined for coastal as well as deep-sea fish, but this finding emphasizes the importance of large-scale physical models in predicting effects of waste inputs.

ACKNOWLEDGMENT

We thank Joel S. O'Connor, Charles A. Parker, John Paul Tolson, and Dorothy L. Leonard of the U.S. National Oceanic and Atmospheric Administration Ocean, Assessments Division, and Edward D. Santoro of Region II of the U.S. Environmental Protection Agency for helpful discussion and assistance with finding data on biological effects. Mary Ann Nelson of the Florida Institute of Technology provided valuable editorial assistance.

REFERENCES

Anderson, D. W., J. R. Jehl, Jr., R. W. Risebrough, L. A. Woods, Jr., L. R. Deweese, and W. G. Edgecomb. 1975. Brown pelicans: improved reproduction off the Southern California coast. *Science*, **190**, 806–808.

Boesch, D. F. 1982. Ecosystem consequences of alterations of benthic community structure and function in the New York Bight Region. *In*: Ecological Stress and the New York Bight: Science and Management, G. F. Mayer (Ed.). Estuarine Research Federation, Columbia, South Carolina, pp. 543–568.

Bowen, V. T., V. E. Noshkin, H. D. Livingston, and H. L. Volchok. 1980. Fallout radionuclides in the Pacific Ocean: vertical and horizontal distributions, largely from GEOSECS stations. *Earth and Planetary Science Letters*, **49**, 411–434.

Cardwell, R. D., C. E. Woelke, M. I. Carr, and E. W. Sandborn. 1977. Evaluation of the efficiency of sulfite pulp mill pollution abatement using oyster larvae. *In*: Aquatic Toxicology and Hazard Evaluation, F. L. Mayer and J. L. Hamelink (Eds.). ASTM STP 634, American Society of Testing and Materials, Philadelphia, Pennsylvania, pp. 281–295.

Cross, J. N. 1985. Fin erosion among fishes collected near a Southern California municipal wastewater outfall (1971–1982). *Fishery Bulletin*, **83**, 195–206.

Dethlefsen, V. 1984. Diseases in North Sea fishes. *Helgolander Meeresunters*, **37**, 353–374.

Faber, H. 1986. State to ban all commercial striped bass fishing. *The New York Times*, April 22, p. 1.

Gerlach, S. A. 1981. Marine Pollution—Diagnosis and Therapy. Springer-Verlag, Berlin, 218 pp.

Harvey, G. R., and W. G. Steinhauer. 1976. Transport pathways of polychlorinated biphenyls in Atlantic water. *Journal of Marine Research*, **34**, 561–575.

Long, E. R. 1985. Status and trends in sediment toxicity in Puget Sound. *In*: Ocean Engineering and the Environment, Conference Record (12–14 November 1985) San Diego, California. Marine Technology Society and IEEE Ocean Engineering Society, Washington, D.C., pp. 919–925.

Malins, D. C., B. B. McCain, D. W. Brown, S-L. Chan, M. S. Myers, J. T. Landahl, P. G. Prohaska, A. J. Friedman, L. D. Rhodes, D. G. Burrows, W. D. Gronlund, and H. O. Hodgins. 1984. Chemical pollutants in sediments and diseases of bottom-dwelling fish in Puget Sound, Washington. *Environmental Science and Technology*, **18**, 705–713.

Martin, M., and R. Severeid. 1984. Mussel watch monitoring for the assessment of trace toxic constituents in California marine waters. *In*: Concepts in Marine Pollution Measurements, H. H. White (Ed.). University of Maryland Sea Grant, College Park, Maryland, pp. 291–324.

Mearns, A. J., and J. Q. Word. 1982. Forecasting the effects of sewage solids on marine benthic communities. *In*: Ecological Stress and the New York Bight: Science and Management, G. F. Mayer (Ed.). Estuarine Research Federation, Columbia, South Carolina, pp. 495–512.

Murchelano, R. A., and R. E. Wolke. 1985. Epizootic carcinoma in the winter flounder, *Pseudopleuronectes americanus*. *Science*, **228**, 587–589.

Norton, M. G. 1985. Colliery-waste and fly-ash dumping off the northeastern coast of England. *In*: Wastes in the Ocean, Vol. 4: Energy Wastes in the Ocean, I. W. Duedall, D. R. Kester, P. K. Park, and B. K. Ketchum (Eds.). Wiley-Interscience, New York, pp. 423–448.

O'Connor, J. M., R. A. Murchelano, and J. Ziskowski. 1986. Index of fish and shellfish disease. NOAA Special Report, U.S. National Oceanic and Atmospheric Administration, Ocean Assessments Division, Rockville, Maryland, in press.

Officer, C. B., R. B. Biggs, J. L. Taft, L. E. Cronin, M. A. Tyler, and W. R. Boynton. 1984. Chesapeake Bay anoxia: origin, development, and significance. *Science*, **223**, 22–27.

Pearson, T. H., and R. Rosenberg. 1978. Macrobenthic succession in relation to organic enrichment and pollution of the marine environment. *Oceanography and Marine Biology Annual Review*, **16**, 229–311.

Rabalais, N. N., and D. F. Boesch. 1986. Extensive depletion of oxygen in bottom waters of the Louisiana shelf during 1985. *Coastal Ocean Pollution Assessment News*, **3**, 45–47.

Rosenberg, R. 1985. Eutrophication—the future marine coastal nuisance? *Marine Pollution Bulletin*, **16**, 227–231.

Schaule, B. K., and C. C. Patterson. 1981. Lead concentrations in the northeast Pacific: evidence for global anthropogenic perturbations. *Earth and Planetary Science Letters*, **54**, 97–116.

Sindermann, C. J., and R. L. Swanson. 1979. Historical and regional perspective. *In*: Oxygen Depletion and Associated Benthic Mortalities in New York Bight, 1976, R. L. Swanson and C. J. Sindermann (Eds.). NOAA Professional Paper 11, U.S. National Oceanic and Atmospheric Administration, Rockville, Maryland, pp. 1–16.

Spitzer, P. R., R. W. Risebrough, W. Walter II, R. Hernandez, A. Poole, D. Pulestone, and I. C. T. Nisbet. 1978. Productivity of ospreys in Connecticut–Long Island increases as DDE residues decline. *Science*, **202**, 333–335.

Stanford, H. M., J. S. O'Connor, and R. L. Swanson. 1981. The effects of ocean dumping in the New York Bight ecosystem. *In*: Ocean Dumping of Industrial Wastes, B. H. Ketchum, D. R. Kester, and P. K. Park (Eds.). Plenum Press, New York, pp. 53–86.

Stegeman, J. J, P. J. Kloepper-Sams, and J. W. Farrington. 1986. Monooxygenase induction and chlorobiphenyls in the deep-sea fish *Coryphaenoides armatus*. *Science*, **231**, 1287–1289.

United Nations Joint Group of Experts on the Scientific Aspects of Marine Pollution. 1983. An Oceanographic Model for the Dispersion of Wastes Disposed of in the Deep Sea. Report No. 19 of the

Joint Group of Experts on the Scientific Aspects of Marine Pollution, International Atomic Energy Agency, Vienna, 182 pp.

U.S. Environmental Protection Agency (EPA). 1977. Ocean Dumping: Final Revisions of Regulations and Criteria, United States Environmental Protection Agency. *Federal Register*, **42**, 2462–2490.

U.S. EPA. 1982. Revised Section 301(h) Technical Support Document. Report 430/9-82-011, U.S. Environmental Protection Agency, Washington, D.C., 236 pp.

U.S. National Oceanic and Atmospheric Administration–Food and Drug Administration. 1985. 1985 National Shellfish Register of Classified Estuarine Waters. U.S. Food and Drug Administration (Shellfish Sanitation Branch), National Marine Fisheries Service (Industry Development Division), National Ocean Service (Ocean Assessments Division), NOAA/OAD, Rockville, Maryland, 19 pp.

Weaver, G. 1984. PCBs in and around New Bedford Harbor, Mass. *Environmental Science and Technology*, **18**, 22A–27A.

Widdows, J. 1985. Physiological responses to pollution. *Marine Pollution Bulletin*, **16**, 129–134.

PART II: PHYSICAL PROCESSES AND MODELS

Chapter 2

Experiments on the Dilution Capacity of Wakes from Tankers Dumping in the North Sea

Gerard A. L. Delvigne

Delft Hydraulics Laboratory
Delft, The Netherlands

ABSTRACT

Field experiments in the North Sea used two tankers to dump acid wastes into a ship's wake. The dilution capacity in the wake was examined in the near-field phase (at approximately less than 12 min) and in the mid-field phase (up to 40 min after the passage of the ship). The near-field results were compared with the International Maritime Organization formula predicting dilution after 5 min. Experiments conducted by the Delft Hydraulics Laboratory confirm measurements with one of the tankers in 1969, when dilution was found to be two or three times greater than was calculated with the formula. This discrepancy is explained in terms of the negative buoyancy of the wake due to the large density difference between the waste fluid (approximately 1300 kg m^{-3}) and the seawater (approximately 1020 kg m^{-3}). The mid-field diffusion is in agreement with tracer diffusion experiments in the ocean. No measurable difference occurred between discharge along the hull of the ship and in the propeller stream.

2.1. INTRODUCTION

For many years tankers have been dumping fluid wastes into the North Sea. The Dutch discharge licenses require that wastes receive a predetermined amount of initial dilution. Calculation of that initial dilution is based on a formula relating the dilution of effluent in the wake of a ship to the discharge rate, ship length, and ship speed. The formula, upheld by the International Maritime Organization (IMO) (1975), was based on full-scale experiments with a tank-cleaning effluent (Dahl and Tollan, 1972; Dahl et al., 1973, 1975) and laboratory experiments with discharges of fluids with both positive and negative buoyancies (Mercier et al., 1973; Delft Hydraulics Laboratory, 1975):

$$S = C_1(vL^2/Q)(vt/L)^{0.4} \qquad\qquad 1$$

where S is dilution, defined by $S = C_d/C_{max}$ (C_d is the concentration of wastes at discharge, and C_{max} is the maximum concentration of wastes in a cross section of the ship's wake); Q is the discharge rate measured in m^3 s^{-1}; t is the time after the ship's passage (approximately 300 s); L is length of the ship measured in meters; v is the ship speed measured in m s^{-1}; and C_1 is a numerical coefficient with C_1 equal to 0.003 for single dis-

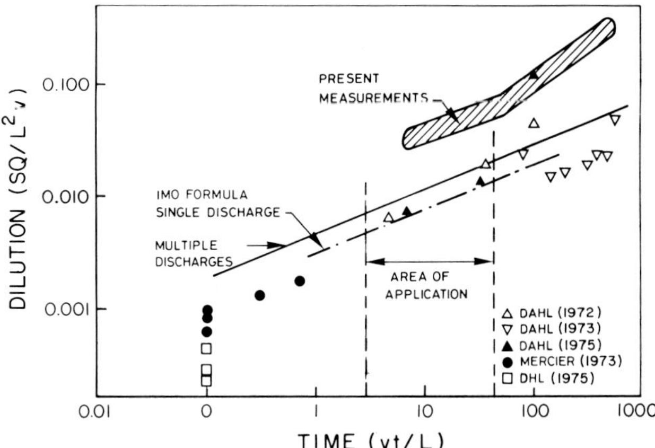

Figure 2.1. International Maritime Organization formula and supporting experimental data (IMO, 1975).

charge and C_1 equal to 0.0045 for multiple discharge with the discharges located symmetrically with respect to the centerline of the ship.

The range of applicability of the formula is $3 < vt/L < 40$, with only v and L as variables. Because of ecological considerations such as the exposure-reaction threshold time, the formula should be applied with t equal to 300 s. Figure 2.1 shows the IMO formula in terms of the non-dimensional dilution SQ/vL^2 versus the non-dimensional time vt/L. The range of application has been given together with the supporting experimental data.

Experiments in the North Sea in 1969 showed a dilution rate in the ship's wake to be nearly double that of the dilution as calculated by equation 1 (Delft Hydraulics Laboratory, 1970). Therefore, the Delft Hydraulics Laboratory (DHL) performed a second, similar experiment on the dilution of acid wastes in a ship's wake. The goals of the second experiment were: (1) to provide a general investigation of the dilution rates in the wakes of the tankers, including dilution as a function of time (t); (2) to provide adequate measurements of dilution at a t of 300 s for a comparison with the IMO formula; (3) to compare a hull discharge located 0.2 shiplengths forward of the stern, with a discharge into the propeller stream (the IMO formula is supported mainly by experimental data relating to a hull discharge, although it has also been accepted that a propeller-stream discharge gives a comparable dilution at t equal to 300 s); and (4) to compare the measured near-field dilution S with the time dependence $S \sim t^{0.4}$ of the IMO formula. Moreover, it is interesting to

investigate at what time the typical mixing process of the wake disappears and changes into the typical mixing process of the surrounding sea.

The measurements were carried out with two tankers dumping fluid wastes into the North Sea; one of the tankers [tanker B, (see Section 2.2)] was the same as was used in the 1969 measurements, and the discharges were the same type of acid wastes.

2.2. METHODS

2.2.1. The Tankers

Two discharge ports were installed symmetrically on either side of the hull of tanker A at a distance 0.2 shiplengths forward of the stern. A second set of ports, placed symmetrically on each side of the outside edges of the propeller stream, discharged directly into the propeller stream. Tanker B was equipped only for discharging into the propeller stream. Tanker A had a length of 72 m, a width of 11.5 m, and a draft of 3.5 m, whereas tanker B was 78 m long, 11.0 m wide, and had a draft of 5.0 m. During the experiments, the speed of both tankers was held at 5 m s^{-1}. Both tankers discharged acid wastes with a density (ρ_w) of approximately 1300 kg m^{-3}, whereas the density of the seawater (ρ_s) was approximately 1020 kg m^{-3}. The discharge took about 5 h to complete. Most measurements were carried out with tanker A; only one series of measurements were made from tanker B to compare the dilution capacity of the wakes from the two tankers.

2.2.2. Plume-Measuring System

The dumping events were designed so that the dumping tanker moved primarily in a direction normal to the prevailing current (Fig. 2.2). Immediately downstream of the initial dumping track, a current meter and sample-pumping array were moored to the seafloor (Fig. 2.3). The pumping array continually provided water from each of ten depths to the measuring ship (Fig. 2.3), where pH values were continuously recorded. The lateral transport of the wake past the pumping array enabled the determination of complete contours in a cross section of the wake.

The data set consisted of (1) pH curves of 10 sampling points between -1 and -14.5 m, (2) current velocity and direction profiles, (3) temperature and salinity profiles (no stratification was observed), and (4) the time and location for the dumping tanker. The minimum age of the wake as it crossed the pumping array was about 1.5 min; the maximum age of approximately 40.0 min was dependent on the ± 0.1 pH sensitivity of the electrodes.

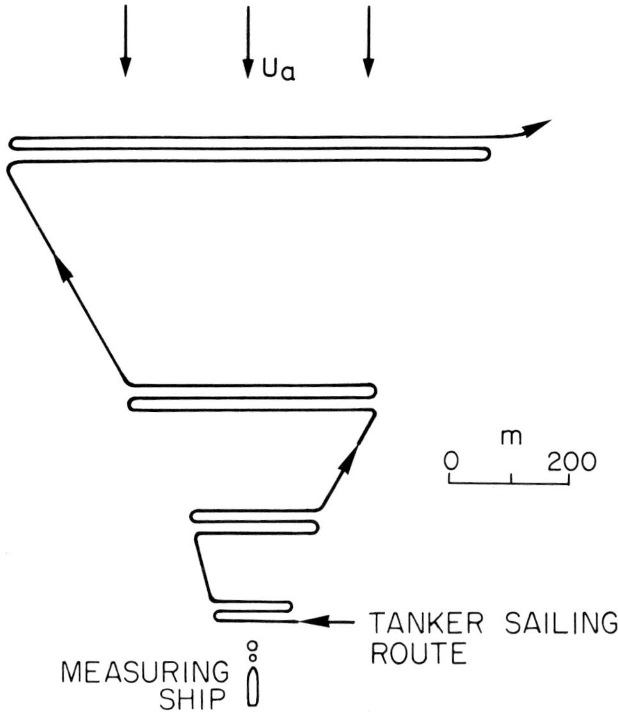

Figure 2.2. Sailing scheme of dumping tanker.

The pH measurements defined the lateral and vertical distribution of wastes in the wake at a particular time after the passage of the tanker. The pH data could not be transformed directly into dilution (S) data because of chemical reactions of the acids when they came into contact with the seawater. The actual pH-dilution curve (Fig. 2.4) was determined in the laboratory with three samples of acid wastes and with seawater. When comparing this data with the IMO formula predictions, only the minimum S at a given time is of interest.

2.2.3. Results

The measurements were made on 28 and 29 August 1981 during four periods of 4 h around the times of maximal tidal-current velocity ($U_a \simeq 0.6$ m s^{-1}) at a location about 70 km offshore and at a water depth of 25 m. There was little wind and the sea state was 2.

Four series of measurements with different discharge methods and discharge rates (Q) were carried out; these are summarized in Table 2.1. Each series consists of 10–12 measurements at cross sections to the wake at times between 1.5 and 40.0 min after a dumping event. Figures 2.5 and 2.6 show the results of the b and c series by ΔpH contours in the cross section of the wake at different times. Similar figures from the a and d series do not indicate other phenomena of any importance when compared with Fig. 2.5 (series b); therefore, these

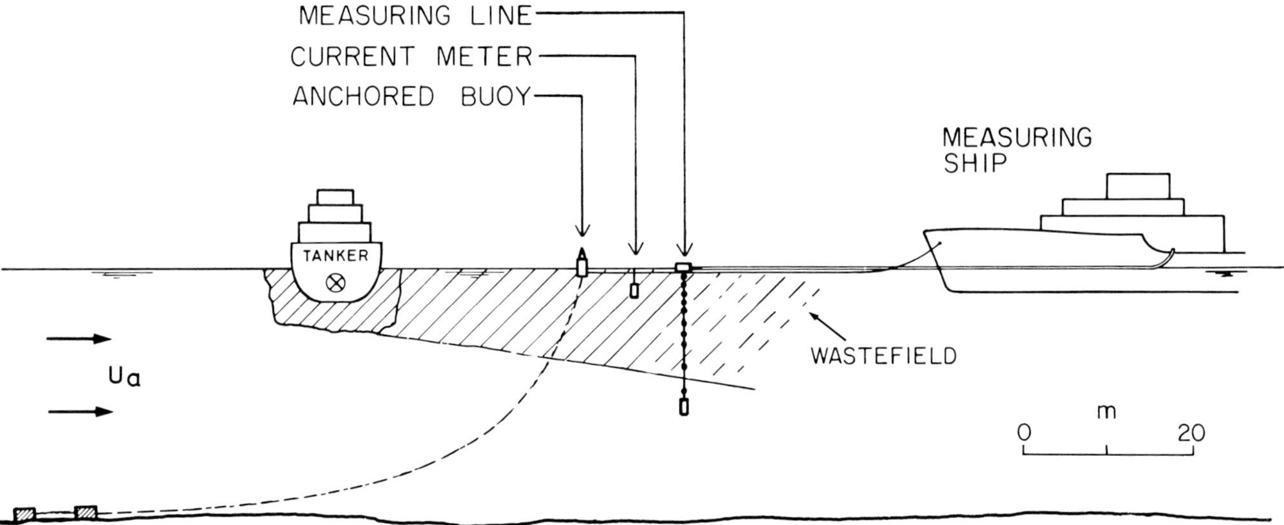

Figure 2.3. Side view of the measuring system during the experiments and the lateral transport of the tanker wake past the measuring line.

values have not been given here. The dilution versus time (S–t) measurements of the series are shown in Fig. 2.7. All data have been brought together in Fig. 2.8.

2.3. TURBULENCE AND MIXING IN THE SHIP'S WAKE AND SURROUNDING SEA

Three phases can be distinguished in the dilution process of wastes disposed of into a ship's wake.

1. Initial-dilution phase. The diposed wastes are distributed throughout the wake. Dilution takes place when the wastes mix with the water in the wake.

2. Near-field phase. Dilution takes place as the wake develops. The wake takes up water by entrainment from the surrounding sea, and dilution then occurs from the enlargement of the wake volume by entrainment.

3. Mid-field phase. Dilution takes place when turbulent diffusion mixes the wake with the surrounding sea.

From these three phases it is evident that dilution takes place by quite different mixing processes. The dilution in each phase, therefore, might be described by a distinct formula not applicable to the previous or to the following phase. The IMO formula should be applicable to the second, or near-field, phase. From observations of the wake cross-sections in Figs. 2.5 and 2.6, it is evident that the measurements discussed in this chapter do not cover the initial-dilution phase: the wastes had already been distributed throughout the wake by the time the first sample was taken.

2.3.1. Near-Field Phase: Comparison with IMO Formula

Due to the generation of turbulence by the passage of the ship, the turbulence in the wake during the near-field phase is much stronger than it is in the surrounding sea. Often the mass exchange between a highly turbulent area and an adjacent area of low turbulence can be approximated by the entrainment process, without taking into account the turbulence of the low-turbulent flow. The entrainment velocity is defined as the entrained volume of water per unit of time and per unit of interface area; it is usually proportional to the intensity of the turbulence. In the wake of a ship, the entrainment process causes the expansion of the wake as a function of time and, consequently, the decrease of the concentration of the wastes distributed throughout the wake. The expansion relation $b \sim t^{0.2}$ due to the entrainment for a momentumless wake (Birkhoff and

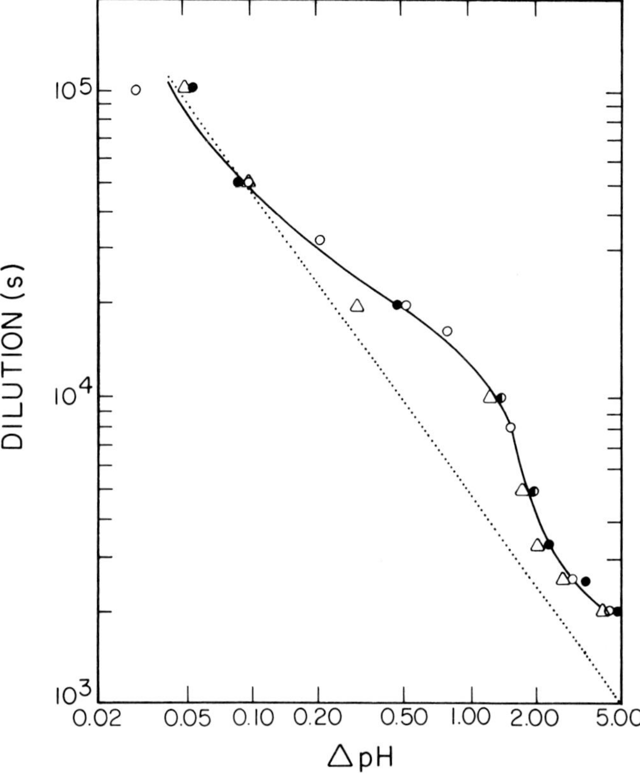

Figure 2.4. The ΔpH–S calibration curve for waste acid in seawater (from laboratory measurements). The dotted line indicates the concentration–dilution relation for a nonreactive system. The open circle, closed circle, and triangle denote waste samples from various ship loads.

Table 2.1. Discharge Methods

Experimental Series	Tanker	Type of Discharge	Q (m^3 s^{-1})
a	A	Hull	0.073
b	A	Propeller	0.073
c	A	Propeller	0.120
d	B	Propeller	0.073

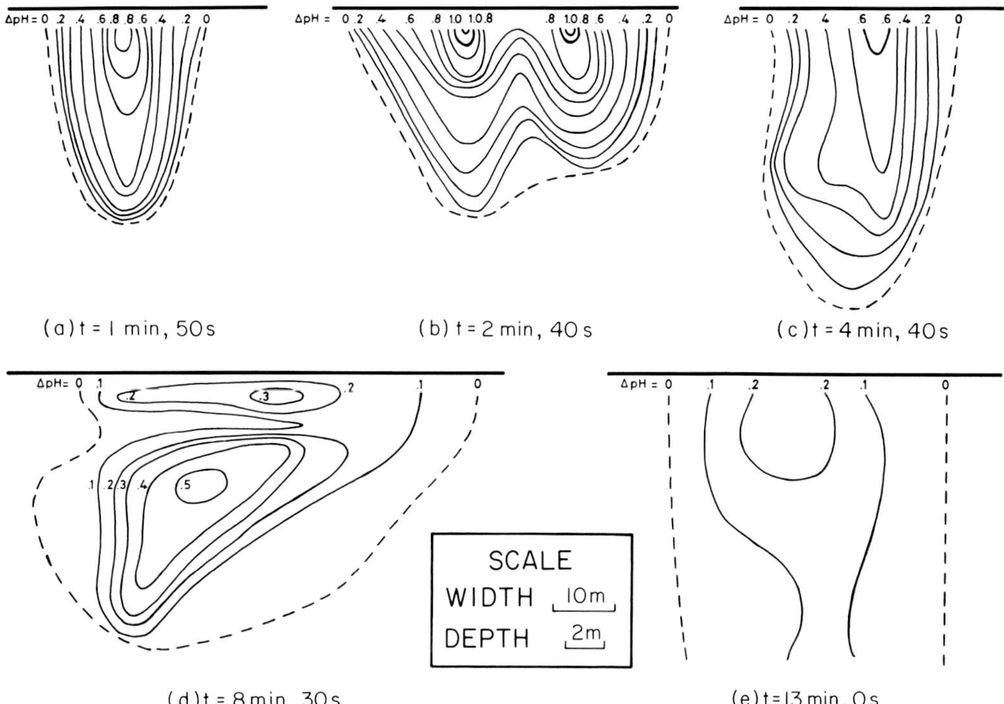

Figure 2.5. Contours of equal ΔpH in the cross sections of the wake in series b. Series a and d have similar contours.

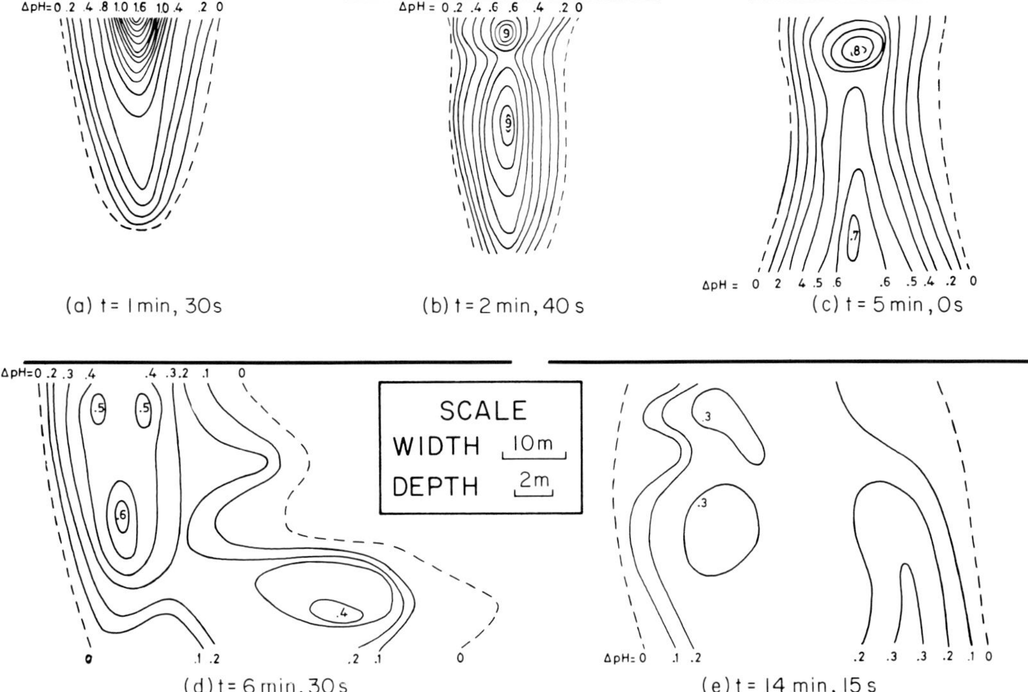

Figure 2.6. Contours of equal ΔpH in the cross sections of the wake in series c.

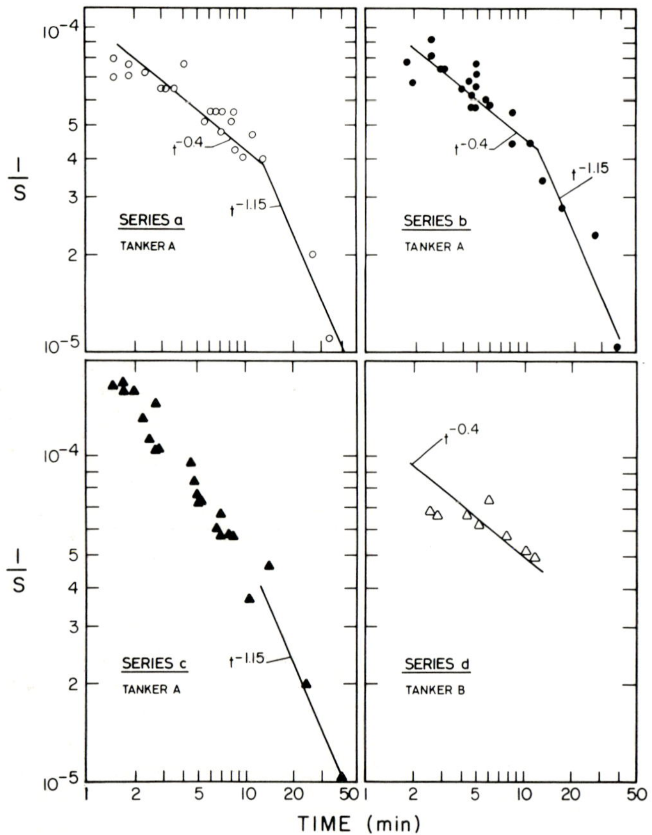

Figure 2.7. Data on dilution versus time of the wakes of series a, b, c, and d. The best fit with $t^{-0.4}$ and $t^{-1.15}$ curves has been given. Open circles denote series a, closed circles denote series b, open triangles denote series c, and closed triangles denote series d.

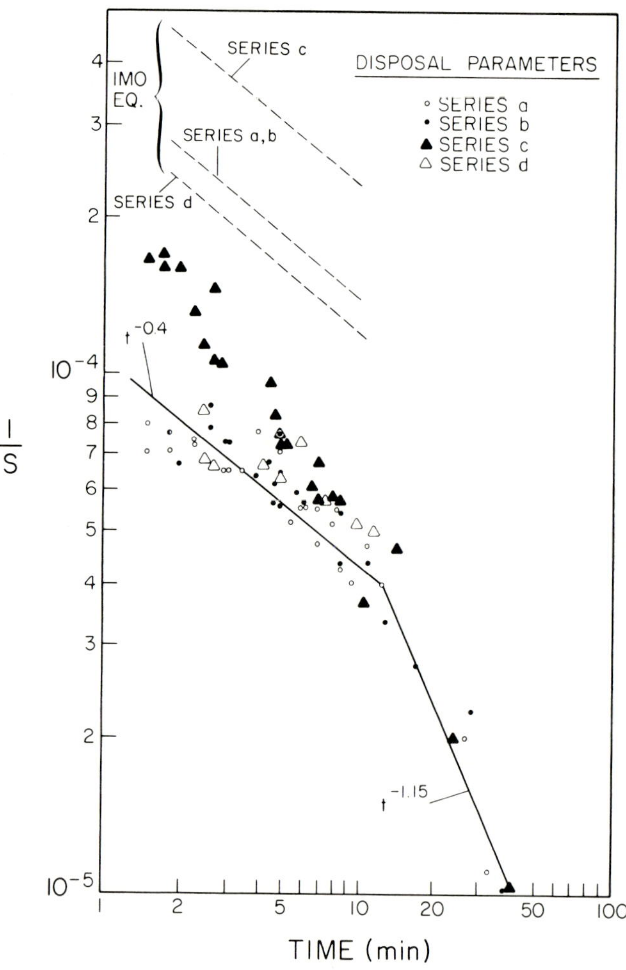

Figure 2.8. Data on dilution versus time for all series compared with calculations by the IMO formula.

Zarantonello, 1957), with b as the width of the wake, is the basis of the $S \sim t_{0.4}$ relation in the IMO formula.

In Fig. 2.7 (S versus t on log–log scale), two straight lines with different slopes may be drawn through the cluster of measured values. The a, b, and d series of measurements in the near-field phase can be fitted with an $S \sim t^{0.4}$ curve, as suggested by the IMO formula, but the c series, with an increased discharge rate, shows a remarkable deviation from that predicted by the formula. In Figs. 2.1 and 2.8 the data on the absolute dilution in the a, b, and d series have been compared with the IMO formula. It appears that the near-field measurements lay in the area of application of the formula but that the measured dilution was 2–3 times greater.

Three conclusions can be drawn from these figures:

1. No difference exists between discharge along the ship's hull and discharge into the propeller stream.

2. Except at the larger discharge rate (experiment c), the time dependence of the measured dilution $S \sim t^{0.4}$ is in agreement with the IMO formula.

3. The absolute wake dilution capacity of both tankers was 2.5 and 3.0 times more than the predictions of the IMO formula (especially at 5 min).

The discrepancies with the IMO formula will be discussed in Section 2.3.2.

2.3.2. Mid-Field Phase

After some time, which in these experiments was about 12 min, the typical high turbulence of the wake diminishes. The turbulence of the surrounding sea then penetrates the area of the wake, and further mixing of the wake with the sea may then be described by turbulent diffusion. The mid-field measurements can be fitted with the following equation

$$S \sim t^{1.15} \qquad\qquad 2$$

as is shown in Fig. 2.7. The exponent 1.15 was not derived so accurately from the measurements, but, instead, from a compilation of diffusion experiments.

Many diffusion experiments have been carried out in the North Sea over the past 20 y. Nearly all of these experiments were made with an instantaneous release of tracers, usually rhodamine. The results have been compiled in Fig. 2.9 (Delvigne and Karelse, 1978; Van Dam, 1982). For comparison, the "Okubo line," which indicates the best fit to data from similar experiments in other seas (Okubo, 1968), is also shown.

Figure 2.7 describes S–t data from instantaneous releases during series a, b, c, and d. The continuing dilution in the patches of acid wastes is due mainly to two-dimensional horizontal diffusion. These patches are either mixed throughout the whole water column (e.g., in the relatively shallow, unstratified North Sea), or the vertical growth is retarded by the often slow, vertical diffusion. The slope of the curves indicates that $S \sim t^{2.3}$. Assuming radially symmetrical horizontal diffusion (generally true in the open sea), then the dilution in a plume with one-dimensional lateral diffusion may be approximated from the patch data by

$$S \sim t^{2.3/2} = t^{1.15} \qquad\qquad 3$$

which is similar to equation 2.

2.4. DISCUSSION

The dilution data in this chapter, compared with other near-field and mid-field experiments, produce two agreements and two disagreements. On the one hand, the mid-field dilution data of all series can be fitted with the $S \sim t^{1.15}$ curve for $t \geqslant 12$ min. All four series

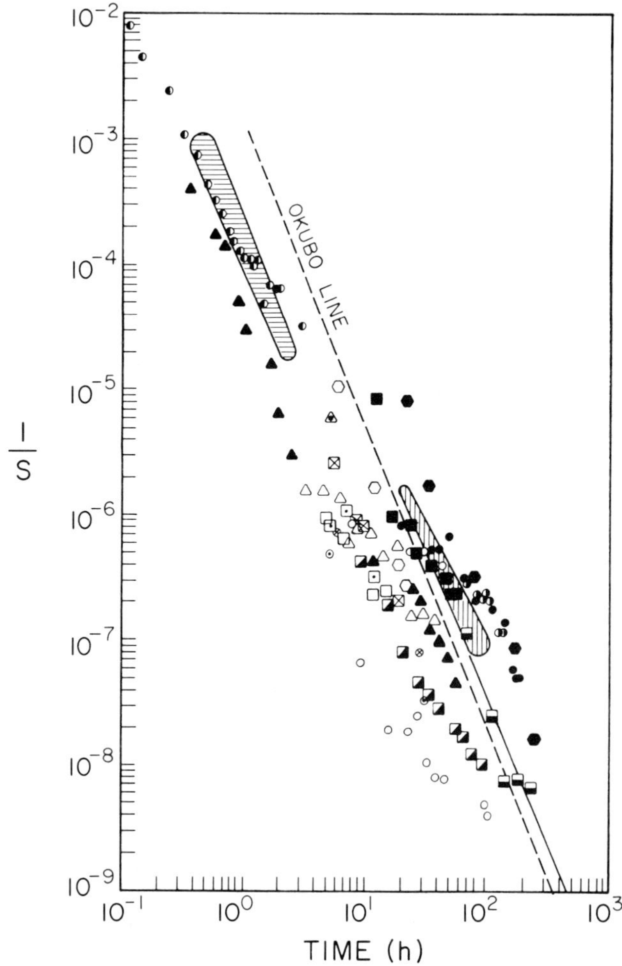

Figure 2.9. Data on dilution versus time for various instantaneous disposals in the North Sea.

suggest the junction of the near-field curves with the mid-field curves at a $t \simeq 12$ min, with a value of the dimensionless time $vt/L = 46$ to 50. Furthermore, the near-field dilution data of the a, b, and d series (with $Q = 0.073$ m³ s⁻¹) can be fitted with the relative dilution curve $S \sim t^{0.4}$, as suggested by the IMO formula. On the other hand, the absolute values of the near-field dilution of the a, b, and d series (with $Q = 0.073$ m³ s⁻¹) are two or three times larger than the IMO predictions. Also, the relative near-field dilution of the c series (with $Q = 0.120$ m³ s⁻¹) runs faster than the IMO predictions of $S \sim t^{0.4}$. The measured data satisfy $S \sim t^{\alpha}$, with $\alpha \simeq 0.7$.

Some experiments supporting the IMO formula refer to hull discharge and exhibit evidence of incomplete mixing within the wake in the initial dilution phase, especially in the case of the experiments described by Mercier (Mercier et al., 1973; Delft Hydraulics Laboratory, 1980). This phenomenon might explain qualitatively the disagreement on the absolute values of the near-field dilution; however, the ΔpH contours of the present propeller discharge measurements (Figs. 2.5 and 2.6) indicate complete mixing at $t \simeq 1.5$ min. Therefore, the reason for the discrepancies with the IMO predictions must lie elsewhere.

In the IMO experiments, the waste concentration in the wake extended to a depth only slightly greater than the drafts of the tankers, but the measurements in this study show depths two or three times greater than the draft of the tankers. Figures 2.5 and 2.6 give evidence of vertical stretching of the wake due to density differences between the waste and the seawater. Figure 2.5 shows that in the a, b, and d series the vertical stretching of the wake was almost completed in the time period before $t = 1.5$ min, but in the c series, with an increased discharge rate, the stretching continued after $t = 1.5$ min (Fig. 2.6). This phenomenon might explain the increased absolute dilution (shift of $S–t$ curves with a factor of 2 or 3 along the S scale) of the a, b, and d series in the near field after $t = 1.5$ min, and the increased dilution rate (indicated by the high level and steep slope of the $S–t$ curve) of the c series in the near field.

The stretching of the wake shown in series b apparently is related to density differences between the waste and the seawater. The stretching enlarges the interface between the wake and the sea and therefore increases the rates of entrainment and dilution. In the c series, with the large discharge rate of heavy fluid, the stretching continues into the near-field phase. In first approximation, the stretching should be proportional to the density difference and, therefore, to the discharge rate (Q). Near-field stretching with $Q_1 = 0.073$ m^3 s^{-1} of series a, b, and d, however, is not 1.6 times less than with $Q_2 = 0.120$ m^3 s^{-1} of series c. Careful observation of the ΔpH contours indicates a more pronounced maximum concentration in the central part of the wake with Q_2 than with Q_1. This might be caused by other discharge conditions, such as greater velocity, leading to a slightly reduced initial mixing.

2.5.   CONCLUSIONS

The analysis of the dilution of an acid waste dumped into the wake of a tanker as described in this chapter leads to the following five conclusions.

1. Experiments do not show any difference, either in near-field or in mid-field dilution, between discharge along the hull of a ship and discharge into the propeller stream.

2. For $Q = 0.073$ m^3 s^{-1}, the absolute dilution in the near field is 2.5 to 3 times larger than the predictions of the IMO formula. The depths of the wakes imply increased dilution through vertical stretching (due to negative buoyancy) of the wake in the initial dilution phase.

3. The experiments with $Q = 0.073$ m^3 s^{-1} result in a relative near-field dilution relation of $S \sim t^{0.4}$, which is in agreement with the IMO formula. For $Q = 0.120$ m^3 s^{-1}, the dilution is more rapid ($S \sim t^{\alpha}$, with $\alpha \simeq 0.7$) because the wake continues to stretch vertically during the near-field phase.

4. The mid-field dilution in all experiments satisfying $S \sim t^{1.15}$ is in agreement with other diffusion experiments in the North Sea.

5. In the experiments described in this chapter, the transition from near field ($S \sim t^{0.4}$) to mid-field ($S \sim t^{1.15}$) takes place about 12 min after the ship's passage ($vt/L \simeq 48$).

REFERENCES

Birkhoff, G., and E. H. Zarantonello. 1957. Jets, Wakes and Cavities. Academic Press, New York, 353 pp.

Dahl, J. B., and O. Tollan. 1972. Measurement of Dilution of Tank Waste Water in the Wake of M/T "Esso Bergen." Institute for Atomic Energy, Kjeller, Norway, 6 pp.

Dahl, J. B., U. Haagensen, C. Qvenild, and O. Tollan. 1973. Dilution in the Ship's Wake of Tank Washings Released from M/T "Esso Slagen." Institute for Atomic Energy, Kjeller, Norway, 6 pp.

Dahl, J. B., U. Haagensen, C. Qvenild, and O. Tollan. 1975. Measurement of the Dilution of Tank Washings in the Wake of M/T "Esso Slagen". Institute for Atomic Energy, Kjeller, Norway, 6 pp.

Delft Hydraulics Laboratory. 1970. Untersuchungen über die Vermischung von Dünnsäure mit Meerwasser im Schraubenstrahl eines Küstenmotorschiffes. Report M 939, Delft Hydraulics Laboratory, Delft, The Netherlands, 37 pp.

Delft Hydraulics Laboratory. 1975. Dilution of Liquids Discharged from a Ship. Report M 1312, Delft Hydraulics Laboratory, Delft, The Netherlands, 17 pp.

Delft Hydraulics Laboratory. 1980. Dilution of Liquids Discharged from a Ship: 1969 North Sea Experiments Compared with IMO Formula. Report R 1565, Delft Hydraulics Laboratory, Delft, The Netherlands, 16 pp.

Delvigne, G. A. L., and M. Karelse. 1978. Diffusion Mechanics in

Conducts, Channels, Rivers, Estuaria and Seas. Report R 895-2, Delft Hydraulics Laboratory, Delft, The Netherlands (Dutch text), 161 pp.

International Maritime Organization (IMO). 1975. Procedures and Arrangements for the Discharge of Noxious Liquid Substances: Method for Calculation of Dilution Capacity in the Ship's Wake. IMO document MEPC III/7, International Maritime Organization, London, 5 pp.

Mercier, J. R., R. I. Hires, and M. Wu. 1973. Model Study of the Dilution of Soluble Liquids Discharged from Tankers. Report No. CG-D-12-74, Department of Transportation, U.S. Coast Guard, Washington, D.C., 40 pp.

Okubo, A. 1968. A New Set of Oceanic Diffusion Diagrams. Chesapeake Bay Institute, Report No. 38, The Johns Hopkins University, Baltimore, Maryland, 52 pp.

Van Dam, G. C. 1982. Models of Dispersion. *In*: Pollutant Transfer and Transport in the Sea, Vol. I, G. Kullenberg (Ed.). CRC Press, Boca Raton, Florida, pp. 91–160.

Chapter 3

Environmental Engineering Analysis of Potential Dumpsites

Gabriel T. Csanady and James H. Churchill

Woods Hole Oceanographic Institution
Woods Hole, Massachusetts

ABSTRACT

In addition to the standard engineering calculation of initial dilution, a comprehensive design analysis of a proposed marine waste-disposal system should include projections of the frequency of visitation by relatively undiluted waste, of the background concentration field, and of deposition of particulate waste onto the seafloor. Such projections can be made where routine oceanographic observations, mainly long-term current-meter records, are available. Application of the suggested methods to a hypothetical dumpsite in the middle of a broad, flat continental shelf shows that neither the visitation nor the background concentration is likely to be a critical constraint. The only potential adverse consequences of dumping on the mid-continental shelf appear to be related to the concentration of waste in the sediment.

3.1. INTRODUCTION

In future designs of systems for marine waste disposal, much more careful attention will have to be paid to environmental impact than was deemed adequate in the past. Standard engineering practice in sewage outfall design has been to focus on initial dilution (the mixing of the waste with seawater through initial momentum and buoyancy) in a technologically controllable way. However, the various subtle ecological, sanitary, and esthetic consequences of waste release cannot be judged solely on the basis of initial dilution: a more comprehensive design analysis is needed. The rationale for marine waste disposal remains that possible adverse effects on the oceanic environment are to be controlled by the dilution of liquid waste and the dispersion of waste particles. Design projections, however, should be made not only of initial dilution, but also (1) of the dilution to be reached at increasingly longer periods after release, (2) of the parts of the water column affected, (3) of the exposure risk of certain sensitive regions to wastes of a given concentration, and (4) of the rate of deposition of waste particles near the release site or far from it. These and similar projections could be made on the basis of physical considerations alone. They would provide the foundation for further necessary investigations (e.g., of nutrient enrichment leading to algal blooms, of oxygen depletion below the pycnocline, or of the concentration of toxic substances in the sediment) involving other disciplines.

This chapter only discusses some physical aspects of such design projections. For the class of waste typified by sewage sludge, the following four physical measures of environmental impact, when taken together with appropriate biological and chemical considerations, may provide an adequate basis for the selection of design criteria.

1. Initial dilution. The dilution (total waste–seawater mixture divided by the released waste) achieved through the design of discharge ports, or controlled release rate from a moving barge, remains an important criterion. Generally speaking, initial mixing is relatively efficient and technologically controllable within certain limits, whereas subsequent dilution by natural turbulence and current shear is sluggish and controllable only by the choice of the disposal site. Marine organisms are exposed to the waste concentration reached at the end of the initial mixing phase (or something not much less) for a period of hours to days: hence the importance of initial dilution.

2. Visitation frequency. Oceanic currents vary irregularly and act in the long run as large-scale turbulence, dispersing batches of waste–seawater mixture over a region of ocean expanding slowly with the random walk of these batches. Immediately following release, however, the batches travel more purposefully, as it were, because the typical tidal and wind-driven currents of coastal waters have a persistence time of several hours to several days (Scott and Csanady, 1976). Thus, for example, in the first 12 h of travel, during which time the center concentration of waste in a large batch of waste–seawater mixture changes little, some batches might come to visit sensitive locations, such as a bathing beach, productive fishing grounds, or a marine sanctuary. Within the range of such short-term, direct excursions, an important measure of waste impact is the visitation frequency, defined as the fraction of time for which a given point is immersed in a relatively young (recently released) batch of waste–seawater mixture.

3. Background concentration. Outside young batches of waste–seawater mixture, the level of what may be called the background concentration of compounds found in the waste depends on the vigor of the large-scale random walk process, the presence of any physical barriers, and the intensity and distribution of sinks for the waste or waste constituent of interest. The background concentration is also an important measure of long-term exposure risk, and it enters into the calculation of secondary effects such as oxygen depletion (as a sink term, expressing the intensity of the biological oxygen demand).

4. Deposition rate of particles. The first three criteria make no distinction between the liquid and the particu-late constituents of wastes. The most significant adverse effects of marine waste disposal, however, have arisen as a result of deposition onto the seafloor (Swanson and Sindermann, 1979). Where the deposition rate of particles is high, both in absolute terms and also in comparison with the local deposition rate of particles naturally present in the sea, the character of the sediment changes, sometimes to the point of drastically altering benthic ecological relationships. The area thus affected and the maximum rate of waste-particle accumulation on the seafloor are important measures of pollution.

Design calculations of the aforementioned variables are based on idealized models for quantifying complex natural processes from a few bulk characteristics. These include a depth-averaged mean current velocity, an effective horizontal diffusivity, a settling velocity and a deposition velocity for a particle size-fraction, the density change across the pycnocline, or a more sophisticated quantity such as the frequency of currents of given speed and direction. The estimation of such bulk parameters for a potential dumpsite is made easier by the fact that they vary relatively little within coherent oceanographic zones as the mid- and outer continental shelf in the Mid-Atlantic Bight and the southern California shelf (Beardsley et al., 1976).

The analysis and modelling of initial dilution following waste release from a pipe or multiport diffuser lying on the seafloor is a well-explored subject and is discussed in standard texts such as that by Fischer et al. (1979). Methods of calculating initial diffusion in the wake of a barge have been formulated more recently (Csanady, 1981; Chapter 2; Chapter 7). Less well explored are suitable, simple models of visitation frequency, background concentration, and particle-deposition rate; these will be discussed further in this chapter.

3.2. OCEANOGRAPHIC CRITERIA

3.2.1. Visitation Frequency

Visitation frequency is a measure of exposure risk to relatively young batches of a waste–seawater mixture. It is defined as that fraction of time γ (nondimensional) for which a target point of interest is immersed in a waste–seawater mixture released a given period t_d (diffusion time) earlier. At target points close to a release site, the visitation frequency is a better measure of waste effects than is the long-term mean concentration of waste, because at such points brief episodes of high concentration (i.e., during visitation by young batches) alternate with much longer periods of lower background concentration. The mean concentration is thus low and does not hint at the intense effects likely to be encountered at times of visitation. The high concentrations found in young batches can be estimated from classical plume models, such as those used in calculating initial dilution. The visitation frequency gives a measure of their total duration as a fraction of time; for example, 10^{-3} would be about one day in three years.

Nearshore sewage outfalls have generally been designed so that the concentration of indicator bacteria is acceptably low at the end of the initial mixing phase, regardless of where the waste–seawater mixture may move; however, high environmental standards applied to stable toxic-waste constituents often cannot realistically be satisfied close to a release site. If the waste-release operation is to be moved further from the shore or from other sensitive locations, the question arises of how far is far enough. A reasonable answer is, where the visitation frequency of young waste–seawater batches is low enough (i.e., 10^{-3} or less) that, with very few exceptions, waste–seawater batches that do arrive at the location in question are much more dilute than they are at the end of the initial mixing phase.

A crude model of visitation frequency $\gamma(x \mid t_d)$ at points x (radius vector) near a release site, for batches released t_d earlier, may be constructed from an observed speed–direction frequency distribution of currents (Csanady, 1983). A long-term current-meter record at the prospective site yields, in addition to the frequency distribution, a typical persistence time of currents (i.e., an estimate of the Lagrangian time scale t_L). For time periods short compared to t_L, batches of waste–seawater mixture travel more or less in a straight line from the source. Given $t_d < t_L$, a current with a speed of v and a direction of ϕ causes target points to be visited by waste–seawater mixtures younger than t_d to an area with a radius r equal to vt_d. The visitation frequency is then approximately

$$\gamma(r, \phi \mid t_d) = \frac{b f_\phi}{r \Delta \phi} F_\phi(r/t_d) \qquad 1$$

where b is batch diameter (often not much greater than diameter at the end of the initial mixing phase), f_ϕ is the directional frequency of currents within sector $\phi \pm \Delta\phi/2$, and F_ϕ is the cumulative frequency of currents

faster than v, in the same sector:

$$F_\phi = \sum_{v=r/t_d}^{\infty} f_{v\phi} \qquad 2$$

where $f_{v\phi}$ is the frequency of currents in the sector-segment $\phi \pm \Delta\phi/2, v \pm \Delta v/2$.

For comparison, the long-term mean concentration caused by batches younger than t_d *only* is

$$C^{\mathrm{I}} = \frac{q f_\phi}{r h \Delta\phi} \sum_{v=r/t_d}^{\infty} \frac{f_{v\phi}}{v} \qquad 3$$

where the superscript I denotes the contribution of young puffs to the concentrations C, q is the release rate of waste (in mass per unit time), and h is the depth of the water column or layer over which the waste is distributed. The sum in equation 3 may be written

$$\sum_{v=r/t_d}^{\infty} \frac{f_{v\phi}}{v} = \frac{F_\phi}{v_m} \qquad 4$$

where v_m is a mean speed weighted toward low velocity v. Therefore the concentration C^{I} and the visitation frequency γ are related by

$$C^{\mathrm{I}} = \frac{q}{v_m b h} \gamma \equiv C_b \gamma \qquad 5$$

where $C_b = q/v_m b h$ is the typical value of concentration in a plume carrying the waste away from the release point. This is also the typical concentration experienced at target points near the source at times of visitation. The long-term mean concentration is less by a factor equal to the visitation frequency. Note that, because of the lower limit on the sum in equations 3 and 4, the typical velocity v_m increases with distance from the source so that C_b in equation 5 decreases, eventually to zero. Physically, distant points are only visited when currents are fast; in a fast current, the concentration of a waste is lower for the same release rate, as is well known from standard plume models (Fischer et al., 1979).

A complication in the oceanic application of these

ideas is that at times of pronounced stratification, top and bottom layers are effectively insulated from each other by a stable thermocline (and possibly from one or two intermediate layers as well) (Swanson and Sindermann, 1979). Depending on the initial conditions, the release might be confined to a single layer, or it might be distributed over more than one. The aforementioned calculations may be carried out separately for each, with the appropriate value of q/h, provided that representative current-meter data are available.

3.2.2. Background Concentration

The total long-term mean concentration of waste at any given point is the sum of the contributions from young and old puffs:

$$C = C^{I} + C^{II} \qquad\qquad 6$$

where C^{I} is as calculated for equations 3 and 5, and C^{II} is contributed by batches released more than t_d seconds before the observation time. A separation of the mean concentration field according to equation 6 is conceptually rigorous, but there is a certain arbitrariness in the choice of the period t_d in comparison with which batches are considered young or old. However, if t_d were to be about equal to the Lagrangian time scale, the asymptotic properties of turbulent diffusion for short and long time scales can be utilized for realistic estimation of the two spatial distributions C^{I} and C^{II}, respectively. At short times the asymptotic behavior results in a concentration field as described by equation 3, with the distribution of waste reflecting random advection. At long times, the random walk of waste batches results in a waste distribution that can be described by solutions of the diffusion equation. This is a satisfactory model of C^{II}, if $t_d \sim t_L$ or longer.

At a distance far enough from a release point, no waste batch arrives in less time than any chosen t_d, so that all of the concentration is due to old batches. This is in agreement with the result that C^{I} vanishes at a large enough r, because the visitation frequency γ tends to zero, while both v_m and b increase (equation 5). At such distant points the fluctuations of waste concentration are moderate, and exposure effects may be judged by the mean concentration C, which is equal at that point to C^{II}. This concentration is the result of an interplay between the waste release rate and the effectiveness of the variable currents in removing waste batches from the neighborhood of the release site.

When the random walk is restricted by physical barriers or by internal separation surfaces (fronts) acting as barriers to turbulent mixing, high concentrations may be reached.

Close to the release site, C^{II} represents the average concentration outside waste batches younger than t_d, and will therefore be called the background concentration. Where the large-scale random walk is restricted, C^{II} may also become high close to the source. The fluid used to dilute the released waste then already contains a high fraction of waste contributed by old batches that have returned to the release site. This is a serious problem, because the background concentration field is smooth, varying over a large spatial scale as compared to water-parcel movements in a tidal cycle or in a wind-driven flow episode. A high value of C^{II} at the source usually signals high values much farther away, possibly at the shore or at some other sensitive location. The near-source background concentration, C^{II}, is therefore an important measure of pollution on its own merits. It is also usually the maximum value of the spatial C^{II} distribution, which is a measure of far-field exposure risk.

The concentration field produced by the horizontal random walk of old waste batches is well described by the advection–diffusion equation:

$$\boldsymbol{u}\cdot\nabla C^{II} = K_H\nabla_1^2 C^{II} + s(x,y,t_d) \qquad\qquad 7$$

where C^{II} is the waste (or constituent) concentration, $\boldsymbol{u}$ is the long-term mean (Lagrangian) current velocity, K_H is an effective (horizontal) diffusivity characterizing the irregular motions, and $s(x, y)$ is an effective source strength per unit area. The latter parameterizes the effects of random advection near the release point and is proportional to the Lagrangian particle displacement probability distribution $P(x,t)$ familiar in turbulent diffusion theory:

$$s(x,y,t_d) = \frac{qP}{h}(x,y,t_d) \qquad\qquad 8$$

where the dimensions of $s(x, y, t_d)$ are mass/(length)^3, time.

The short-term asymptotic distribution $P(\ )$ may be used as an approximation. In terms of the current frequency distributions introduced before, this is

$$P(r, \phi, t_d) = \frac{f_\phi}{r \, \Delta r \, \Delta \phi} \, f_{v\phi}(r / t_d) \qquad 9$$

A better approximation to $P(\quad)$ may be obtained by integrating a long-term current-velocity record for segments of duration t_d.

The diffusion equation 7 has often been used to model the oceanic distribution of some tracer or pollutant (Csanady, 1984), but without the refinement represented by equation 8. Far enough from any sources or sinks, the precise distribution of $s(x, y, t_d)$ is immaterial, and only the total (space-integrated) source strength enters the results, so that conventional model calculations are as accurate as those based on equation 7, which is more complete. Close to sources or sinks, however, the diffusion equation is valid only if random advection is taken into account. Also, the calculated field represents background concentration only, not the total concentration. These are crucial differences for the assessment of exposure risk near a dumpsite.

3.2.3. Deposition of Waste Particles

Although most wastes are released into the ocean in liquid form, they generally contain solid particles, and often also form more particles upon entry into the ocean through flocculation and other physicochemical processes (Morel and Schiff, 1984). A key characteristic of such particles is their settling velocity w_s. A large fraction of certain wastes such as sewage sludge consists of particles with a settling velocity of 10^{-5} m s^{-1} (~ 1 m d^{-1}) or less (Brooks et at., 1982). These particles behave much the same as do fine organic particles naturally present in the sea and deposit permanently where bottom currents are quiescent. On the other hand, particles with a much greater settling velocity quickly reach the bottom after release and form at least a temporary accumulation on the seafloor. Documented adverse effects of ocean dumping have so far always been ascribed to such accumulations of waste (see, for example, Swanson and Sindermann, 1979).

The fine particulate constituents of the waste mingle with similar particles naturally present, and their concentration in the sediment, where they eventually settle out, can be no larger than their concentration as a fraction of the total particulate load of the water col-

umn. This may be taken to be the background concentration, as defined in Section 3.2.2, of waste particles divided by the total (waste as well as naturally occurring) fine-particle load. A desirable goal is, for example, to keep this fraction to 10% or less (Committee on Ocean Waste Transportation, 1984). One notes that this is more easily accomplished where naturally occurring particles are plentiful (e.g., over continental shelves) than where they are not (in the deep ocean, far from continental margins) (McCave, 1972). Waste particles are diluted, in a sense, by naturally occurring fine particles, not by seawater, a conceptual difference with far-reaching implications.

The ultimate fate of fine particles in the ocean depends on many complex processes (consolidation, chemical transformations, effects of living organisms, and resuspension) within or involving freshly formed sediment (McCave, 1972); a discussion of these processes is, however, outside the scope of this chapter. From the point of view of large-scale diffusion models, one should keep in mind that such sediments can act as secondary sources, as well as sinks, for certain waste constituents. The subject of fine-particle behavior in the sea is in need of further observational studies and new ideas of how to come to grips with it quantitatively.

Coarse particles (those settling fast enough to overcome the effects of bottom turbulence) deposit when they reach the seafloor; that is, after a period $t_s = h/w_s$, where h is the height of their release above the bottom. The quantitative criterion for permanent deposition to occur is not very well known. If u_* is the friction velocity at the bottom, particles characterized by $w_s/u_* = 1$ certainly have a predominantly vertical motion and may be expected to settle out; however, when $w_s/u_* = 0.1$, deposition is unlikely.

A simple model for coarse-particle deposition can be constructed on the hypothesis that horizontal motion and settling occur independently. In the specially simple case in which the water column is well enough mixed for horizontal motions to be coherent, those motions are described statistically by the Lagrangian particle displacement probability distribution $P(x, y, t)$ mentioned in equation 8. The mean pattern of deposition on the seafloor is proportional to this function, with $t = t_s = h/w_s$, for a size fraction of settling velocity w_s. If the mass-fraction of particles of settling velocity w_s to $w_s + d\,w_s$ is $\sigma(w_s)\,d\,w_s$, and if all particles settling faster than some critical w_c deposit while all other particles remain in resuspension, then the aggregate deposition rate, D, is

$$D = q_p \int_0^{w_c} P\left(x, y, \frac{h}{w_s}\right) d(w_s) \qquad 10$$

with q_p the heavy-particle release rate in $kg\ s^{-1}$. This can also be written as

$$D = q_p P(x, y, t_{sm}) \qquad 11$$

where t_{sm} is a weighted-mean value of the fallout time.

3.3. APPLICATION TO A HYPOTHETICAL DUMPSITE

To illustrate the application of the approach of this chapter to a system design for waste disposal, a hypothetical dumpsite located on a continental shelf similar to that on the Atlantic coast of the United States will be considered. Long-term current-meter data obtained in water at a 15-m depth at the 60-m isobath off Atlantic City, New Jersey, have been processed to yield visitation frequencies, particle displacement probabilities, and other statistics as outlined in the previous sections (Fig. 3.1). For a more thorough discussion of these parameters, see Csanady (1983).

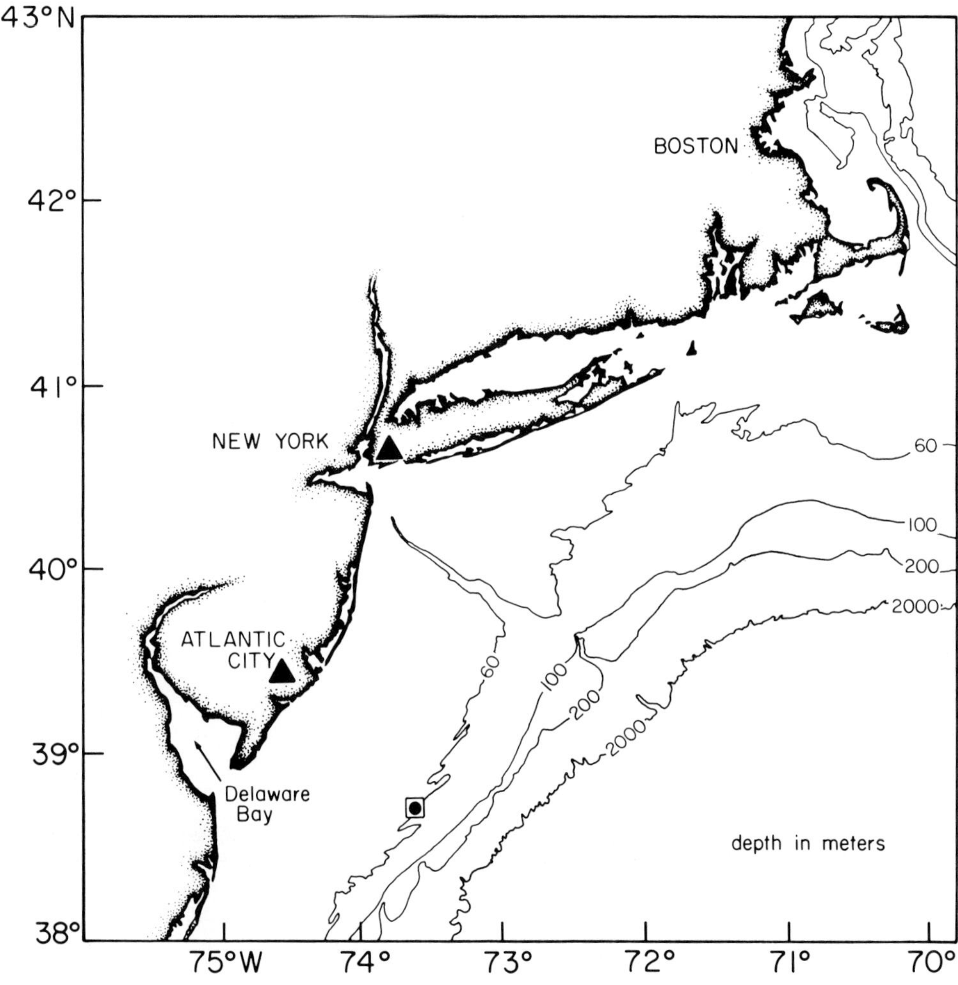

Figure 3.1. Location of current meter placed at the 60-m isobath off Atlantic City. Other isobaths shown (100-, 200-, and 2000-m) define the edge of the continental shelf, where fishing is intensive.

The autocorrelation function of hourly mean currents (Fig. 3.2) suggests a Lagrangian time scale (t_L) of at least 100 h. The time scale t_d of effective dilution by oceanic processes for a typical batch of dumped waste will be assumed to be 12 h, based on limited experience near Deepwater Dumpsite-106 (Csanady, 1981). In a period this long, the waste concentration should drop to 10 times below the value reached at the end of initial dilution, which would be reached within about 0.5 h after the dump. With $t_d \ll t_L$, the visitation frequencies may be calculated according to equation 1 and are shown in Fig. 3.3 (fraction of time, shown as parts per thousand). The 1 ‰ contour extends to about 10 km across the isobaths (the vertical direction in Fig. 3.3) and nearly 20 km southwest along the 60-m isobath. The distance to shore at this site is about 80 km and about 30 km to the shelf-edge front (an intensively fished region) (Fournier, 1978); thus, waste batches as young as 12 h probably never reach either of these sensitive locations.

The probability density of particle displacement at the end of a 100-h period (a crudely estimated Lagrangian time scale) is shown in Fig. 3.4. The mean cross-shore displacement is negligible, but alongshore there is a 40-km offset from the origin, yielding a mean velocity u equal to 0.11 m s^{-1}. The cross-shore standard deviation at this diffusion time (t_d) is $\sigma_y = 13.3$ km, and its growth with t_d seems to approach the growth curve calculated using $K_H = 300$ m^2 s^{-1} = const (Fig. 3.5). The latter value was estimated from evidence such as freshwater diffusion across the shelf (Ketchum and Keen, 1955). It is therefore appropriate to calculate the background-concentration field with the aid of the diffusion equation, using the aforementioned values of u and K_H. The source-strength distribution is taken to be $P(\ )$ for a t of 100 h, approximated by Gaussian distribution, of a standard deviation σ_y equal to 13.3 km in the cross-shore direction. The alongshore spread of the source is of no consequence, as is often the case in such plume-diffusion problems, because the nondimensional number measuring the importance of advection relative to diffusion is large:

$$A = \frac{\sigma_y u}{K_H} \gg 1 \qquad 12$$

In terms of traditional nondimensional numbers $A = LePe$, the product of the Lewis number and the

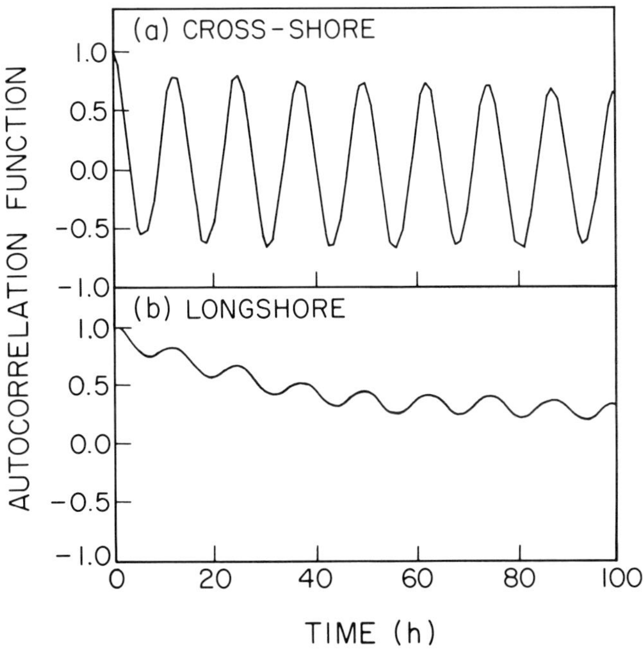

Figure 3.2. Autocorrelation function of velocity observed at a moored current meter. The tidal currents dominate cross-isobath motions, but the along-isobath currents have a time scale of approximately 100 h.

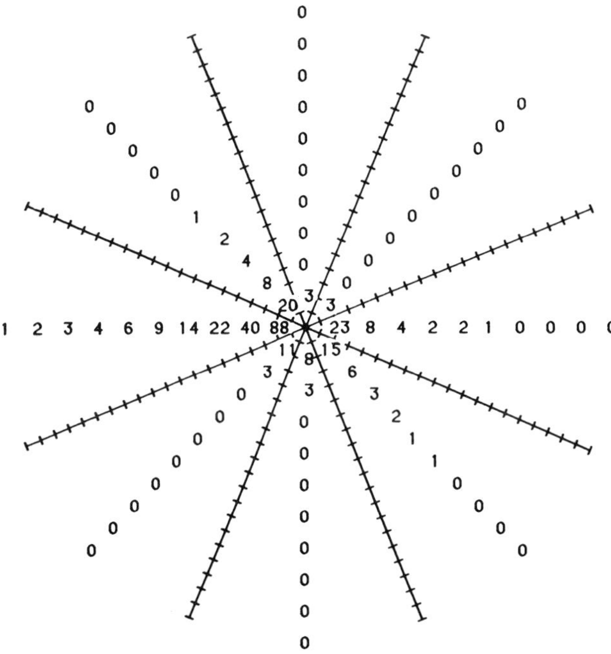

Figure 3.3. Visitation frequencies (parts per thousand) in 45° sectors, at 2-km intervals from the presumed source. Each tickmark denotes 1 km. The batch diameter was $b = 300$ m.

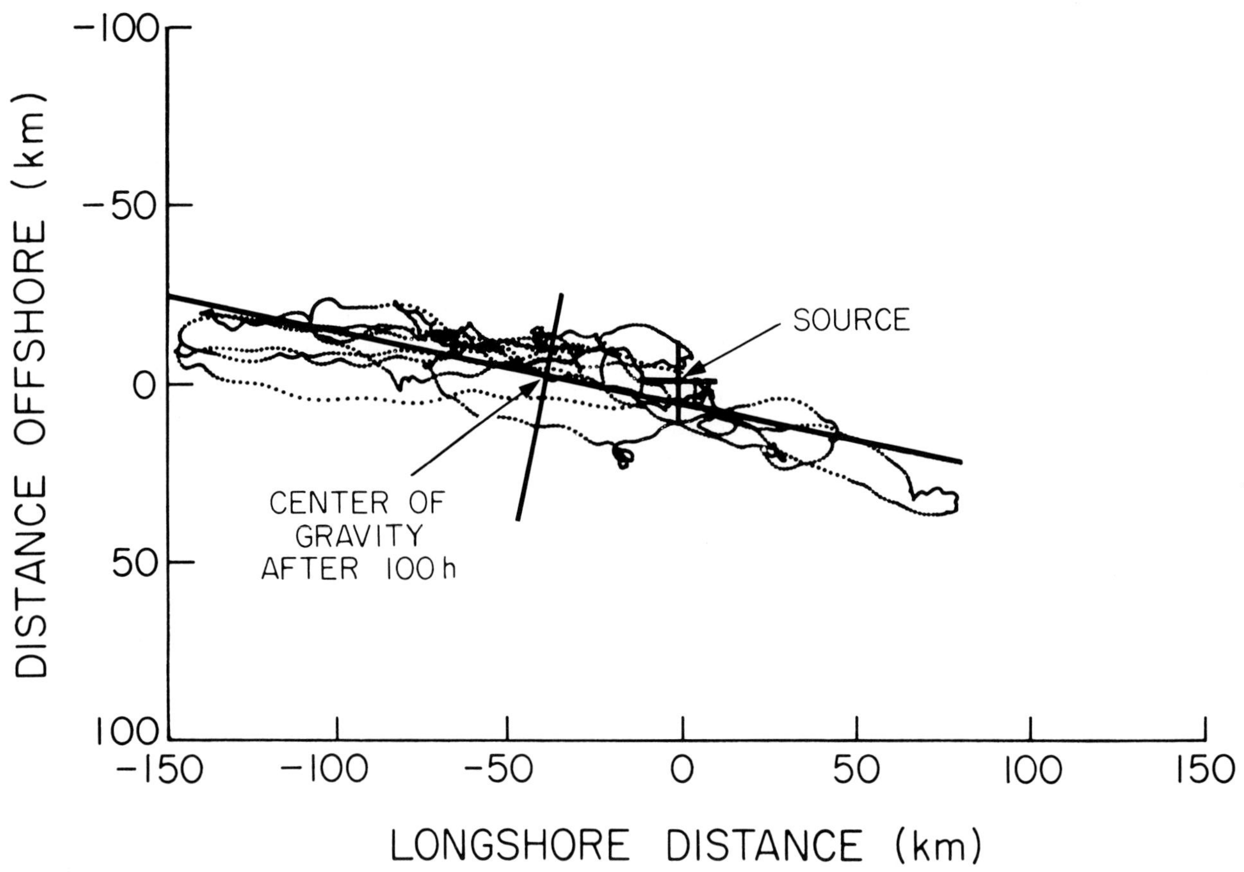

Figure 3.4. Distribution of particle positions after 100 h of travel, inferred by integrating hourly current-meter data. Irregular lines arise from coalescence of dots; principal axes of distribution are also shown. Note the different scales along and across isobaths.

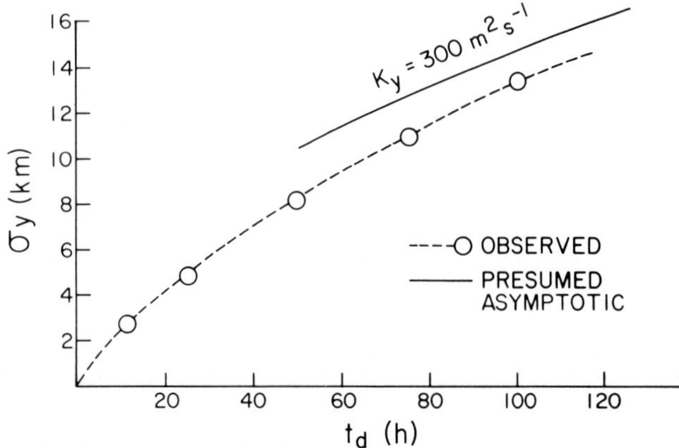

Figure 3.5. Cross-isobath standard deviation (σ_y) versus travel time, calculated from the inferred distribution of particle positions from Fig. 3.4.

Peclet number (Rohsenow and Choi, 1961). The simplified form of the diffusion equation may therefore be used:

$$u\frac{\partial \chi}{\partial x} = K_H \frac{\partial^2 \chi}{\partial y^2} \qquad 13$$

This also implies the simple mass balance

$$q = \int_{-\infty}^{\infty} u\chi h \, dy \cong \text{constant} \qquad 14$$

where q is the release rate, and χ is concentration.

The shore is at $-uy/2K_H = 15$, which is effectively infinity, and there is no need to consider its presence.

The near-source background concentration is then simply, as in a Gaussian plume:

$$\chi_m = q/\sqrt{2\pi}\ \sigma_y\ uh \qquad\qquad 15$$

The result shows that the background concentration is as if the release rate, q kg s^{-1}, were distributed over a stream of discharge $Q = \sqrt{2\pi}\ \sigma_y\ uh$, which is about 2.2×10^5 m^3 s^{-1} (the equivalent to 1 Amazon or 13 Mississippi rivers). The mechanism by which the release is distributed over such a wide cross section is random near-source advection through coherent motion, rather than through subsequent random walk. Note also that the source has so far been considered a point; however, where a barge distributes waste in the cross-shore direction in a line dump, the effective source dimension is increased further. The increase is significant when the cross-shore projection of the physical source size is comparable with or larger than σ_y. The effect can be calculated using a conventional plume model, so it will not be discussed further in this chapter.

For further appreciation of equation 15, consider the release of $q = 3$ kg s^{-1} of particulate matter, equal to the maximum projected release rate of dry sludge from the Orange County outfall (Brooks et al., 1982; Chapter 10). Equation 15 yields $\chi_m = 136 \times 10^{-5}$ kg m^{-3}, or about 0.013 mg liter^{-1} of fine particulate matter. This is one to two orders of magnitude less than the typical suspended-sediment concentration on an outer continental shelf (McCave, 1972). Furthermore, if the particulate matter were to be discharged in liquid form, in the typical 2% initial concentration of solids, then the calculated background concentration corresponds to a dilution of liquid waste by a factor of 1.5×10^6, or about 300 times more than at the end of initial mixing in an efficiently conducted dumping operation from a barge, with an initial dilution of 5000.

The background concentration in the water column is thus suitably low to eliminate any probability of toxic effects; however, the concentration of waste particles as a fraction of total fine particles present is seen to be as high as 1–10%. Thus, should these particles be deposited without significant further mingling with particles from other sources (a realistic possibility in view of the large spatial scale of the background concentration field), the deposit could contain the same, relatively high, although not dominant, fraction of waste particles. Deposition is clearly one mechanism concentrating particulate matter, including waste particles, and it may

well constitute the one critical aspect of oceanic dumping.

Turning now to the relatively heavy-particle fractions, from the observed *rms* velocity of 0.24 m s^{-1}, one may estimate the friction velocity at the bottom (u_*) to be be 0.02 m s^{-1}. It is reasonable to suppose, according to previous discussion, that particles with a settling velocity (w_s) greater than or equal to 1.5×10^{-3} m s^{-1} ($\equiv 5$ m h^{-1}) are deposited upon reaching the seafloor. At a 60-m depth, this corresponds to a settling time from the surface of 12 h. The particle-displacement probability distribution for this travel time (Fig. 3.6) encloses an ellipse about 40 km $\times$ 16 km, the standard deviations being 2.7 km cross-shore, 8.6 km alongshore. Using equation 11, and taking q_p as 0.1 kg s^{-1} for the release rate, one finds that the rate of deposition (D) is

$$D = 6.85 \times 10^{-10}\ \text{kg m}^{-2}\,\text{s}^{-1} \qquad\qquad 16$$

which is about 2 g cm^{-2} in 10^3 y, or 1 mm of consolidated deposit in the reasonably projected lifetime of a dumpsite (100 y). The long-term fine-particle sedimentation rate on the continental rise is of comparable magnitude (McCave, 1972).

3.4. CONCLUSION

One objective of the foregoing discussion has been to illustrate how to incorporate routine oceanographic observations into the design analysis of a marine waste-disposal system. By focusing on the simplest winter, or well-mixed, conditions on a broad, flat continental shelf assumed in the calculated examples, certain environmental influences have not been adequately emphasized, notably the influence of stratification. However, the extension of the ideas in this chapter to models involving several layers should be straightforward.

A major result of the analysis is the importance of random near-source advection (coherent motion) in distributing the dumped waste over what may be called an extended source region, from the point of view of long-term dispersal. An adverse aspect of near-source advection is to cause nearby points to be visited by relatively undiluted waste. This effect can be quantified, however, and its incidence kept low by a suitable placement of the dumpsite. For a dumpsite far enough from sensitive areas, near-source advection has the beneficial effect of rapidly dispersing the waste over a vast stream of water, which would reduce background concentra-

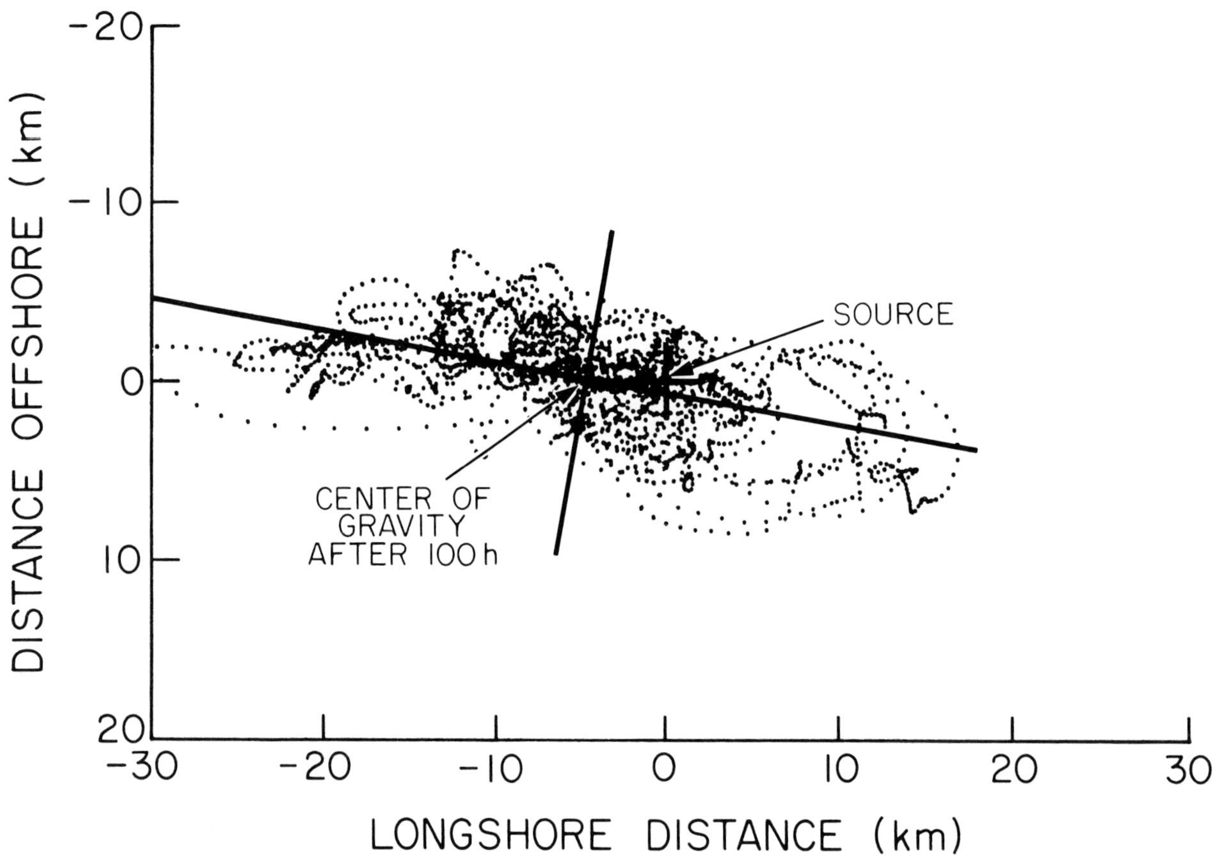

Figure 3.6. Distribution of particle displacement (with principal axes) for a 12-h travel time. Individual dots (one particle released each hour) are more readily visible than they are in Fig. 3.4.

tions (per unit mass of seawater) to negligible values on an open continental shelf. However, the concentration of particulate waste, expressed as a fraction of the total particulate matter present, might still be relatively high, typically 1–10% for a major sewage-sludge dumping operation. This also means that the waste fraction in the sediment may become important in places. At present, this appears to be the only potentially critical aspect of oceanic dumping.

ACKNOWLEDGMENTS

This work was supported by the Department of Energy, under a contract titled Coastal-Shelf Transport and Diffusion. Bradford Butman of the U.S. Geological Survey kindly placed the current-meter data at our disposal. Gabriel Csanady also benefited from many discussions with Norman H. Brooks of the California Institute of Technology.

REFERENCES

Beardsley, R. C., W. C. Boicourt, and D. B. Hansen. 1976. Physical oceanography of the Middle Atlantic Bight. *American Society for Limnology and Oceanography, Special Symposium*, **2**, 20–34.

Brooks, N. H., R. G. Arnold, R. C. Y. Koh, G. A. Jackson, and W. K. Faisst. 1982. Deep Ocean Disposal of Sewage Sludge off Orange County, California: A Research Plan. Environmental Quality Laboratory Report No. 21, California Institute of Technology, Pasadena, California, 101 pp.

Committee on Ocean Waste Transportation. 1984. Ocean Disposal Systems for Sewage Sludge and Effluent. National Academy Press, Washington, D.C., 126 pp.

Csanady, G. T. 1981. An analysis of dumpsite diffusion experiments. *In*: Ocean Dumping of Industrial Wastes, B. H. Ketchum, D. R. Kester, and P. K. Park (Eds.). Plenum Press, New York, pp. 109–129.

Csanady, G. T. 1983. Dispersal by randomly varying currents. *Journal of Fluid Mechanics*, **132**, 375–394.

Csanady, G. T. 1984. Advection, diffusion and particle settling. *In*: Ocean Disposal of Municipal Wastewater: Impacts on the Coastal Environment, E. P. Myers and E. T. Harding (Eds.). MIT Sea Grant College Program, Massachusetts Institute of Technology, Cambridge, Massachusetts, pp. 178–247.

Fischer, H. B., E. J. List, R. C. Y. Koh, J. Imberger, and N. H. Brooks. 1979. Mixing in Inland and Coastal Waters. Academic Press, New York, 483 pp.

Fournier, R. O. 1978. Biological aspects of the Nova Scotian shelfbreak fronts. *In*: Oceanic Fronts in Coastal Processes, M. J. Bowman and W. E. Esaias (Eds.). Springer Verlag, Berlin, pp. 69–77.

Ketchum, B. H., and D. J. Keen. 1955. The accumulation of river water over the continental shelf between Cape Cod and Chesapeake Bay. *Deep-Sea Research*, **3** (Supplement), 346–357.

McCave, I. N. 1972. Transport and escape of fine-grained sediment from shelf areas. *In*: Shelf Sediment Transport: Process and Pattern, D. J. P. Swift, D. B. Duane, and O. H. Pilkey (Eds.). Dowden, Hutchinson & Ross, Stroudsberg, Pennsylvania, pp. 225–248.

Morel, F. M. M., and S. L. Schiff. 1984. Geochemistry of municipal waste in coastal waters. *In*: Ocean Disposal of Municipal Wastewater: Impacts on the Coastal Environment, E. P. Myers and E. T. Harding (Eds.). MIT Sea Grant Program, Massachusetts Institute of Technology, Cambridge, Massachusetts, pp. 251–421.

Rohsenow, W. M., and H. Y. Choi. 1961. Heat, Mass and Momentum Transfer. Prentice Hall, Englewood Cliffs, New Jersey, 537 pp.

Scott, J. T., and G. T. Csanady. 1976. Nearshore currents off Long Island. *Journal of Geophysical Research*, **81**, 5401–5409.

Swanson, R. L., and C. J. Sindermann (Eds.). 1979. Oxygen Depletion and Associated Benthic Mortalities in New York Bight, 1976. National Oceanic and Atmospheric Administration Professional Paper 11, U.S. Department of Commerce, Washington, D.C., 345 pp.

Chapter 4

Deterministic and Stochastic Models of Dispersion of Released Effluents into Coastal Areas

Palle Bo Nielsen

Danish Isotope Centre
Copenhagen, Denmark

ABSTRACT

Mathematical dispersion models have been developed in order to assess the spreading and dilution of effluents released continuously in coastal areas. In order for the models to be run with a small desktop computer, only models with modest memory requirements and computing time can be considered. Deterministic models based on the superposition principle (puff models) and stochastic particle models have been considered appropriate in several instances. Recently, these types of models have been used to describe the effluent concentration field in the vicinity of an outfall 500 m off the western coast of Denmark. The problem of estimating the effluent flow into the Limfjord, which connects with the North Sea through a narrow channel 6.5 km north of the outfall site, has been particularly addressed. Input data for the models were derived mainly from a series of field measurements collected by the Danish Isotope Centre in 1981. The puff model output shows how spreading and dilution of effluents are related to environmental factors. A probability distribution of the environmental states has been set up and fed into the stochastic model to obtain pollution statistics (in terms of fractions of the released effluent) at significant locations.

4.1. INTRODUCTION

Mathematical models of the physical processes in bodies of water receiving effluents are a desirable or even necessary tool for description, prediction, and control of the receiving waters. Although an exhaustive understanding of the physics of the waters is impossible for practical (and theoretical) reasons, the mathematical modelling should be based primarily on a thorough knowledge of the governing processes. Such knowledge can be obtained through field measurements. Because input data to the models have to be derived from the measurements, the extent and kind of measurements determine which types of models that may come into play.

Detailed physical measurements are, for reasons of economy, typically restricted to cover only one or a few short periods. In this way, keeping in mind that measurements were perhaps not made during typical periods, a satisfactory coverage of processes within a wide spectrum of space scales can be obtained. The measurements can, for example, consist of three-dimensional mapping of natural or artificial tracers in the receiving waters to illustrate the transport and mixing conditions close to an outfall (existing or future). Equipment and techniques developed for such measurements have been described by Appelquist and Nielsen (1987).

Measurements of longer duration, on the other hand, are often accomplished at the expense of the level of details in space but imply the coverage of a wider spectrum of time scales. Examples are the continuous recording of currents, sea level, salinity, and temperature by means of moored, automatic recording instruments, giving rise to long time series for the measured parameters but usually at a small number of locations. Hence, it becomes possible to evaluate the effects of effluent releases taking place over a long period of time and, thus, to assess the risk of effluent accumulation at locations associated with special environmental concerns. Moreover, such data are important when setting up pollution statistics for particular locations.

In addition to the importance of available data, the models have to fit the computational facilities at hand. Computing time and memory requirements impose limits on the number of potential approaches and may also necessitate simplifications and approximations in the structure of the model.

In this chapter two mathematical models for dispersion in the sea are briefly described, and their use for prediction of effluent concentrations in the vicinity of an outlet off the western coast of Denmark is discussed. One model is based on the superposition principle (puff model), and the other is based on a Monte Carlo simulation technique (particle model). These models are applied by the Danish Isotope Centre in connection with waste-disposal investigations in coastal areas. Input data for the models were derived from results obtained during detailed field measurements accomplished by the Centre with the aid of artificial tracers (fluorescent rhodamine-B and radioactive ^{82}Br) and with previous data of the current as measured by a moored, automatic recording instrument. The example presented in this chapter illustrates how data derived from short, detailed measurements and a single long time series can be used for mathematical modelling of the transport and dilution of effluents in the receiving waters.

4.2. MODELS OF DISPERSION

In the following, the concept of model refers solely to mathematical models. Only the dispersion of dissolved matter with a density such that no buoyancy effects interfere after the initial mixing is considered; therefore, the effluent follows the motion of the water.

Dispersion models can be classified according to various points of view. Both models presented in this chapter can be classified as dynamic, since most of the factors involved are functions of time. Also, spatial inhomogeneities can be handled, so the spreading and dilution resulting from the interplay of the governing space- and time-dependent factors can be simulated. One of the models (puff model) is deterministic in that the simulations imply that the concentration field of the released effluent is uniquely determined by input data. The other model (particle model) is, however, stochastic, since the governing factors are assumed to be stochastically distributed with known probability functions. The output data are, consequently, also stochastic, and statistics for the concentration field can be set up from them.

4.2.1. The Puff Model

In the puff model, various factors such as current and turbulent diffusion are time-varying and spatially inhomogeneous. The model permits simulations of the spreading and dilution of effluents when the dynamics of the receiving waters are crucial. If, for instance, the prevailing currents are dominated by a periodic component generated by tidal forces, then released effluent may accumulate during reversals of the current. The principle of the puff model is that the continuously released effluent is considered as a series of puffs that successively leave the outlet site (Fig. 4.1). Each puff represents a given amount of effluent, M_{puff}, determined by the release rate of the effluent, Q, and the time interval, Δt, between each puff release:

$$M_{puff} = Q\,\Delta t \qquad\qquad 1$$

The horizontal current implies a motion of each puff as a whole, the amount being equal to the product of the Δt used and the current velocity:

$$x_{t+\Delta t} = x_t + u\,\Delta t \qquad\qquad 2a$$

$$y_{t+\Delta t} = y_t + v\,\Delta t \qquad\qquad 2b$$

where x and y are the coordinates for the position of the center of the puff at time t, and u and v are the horizontal velocities in a Cartesian coordinate system.

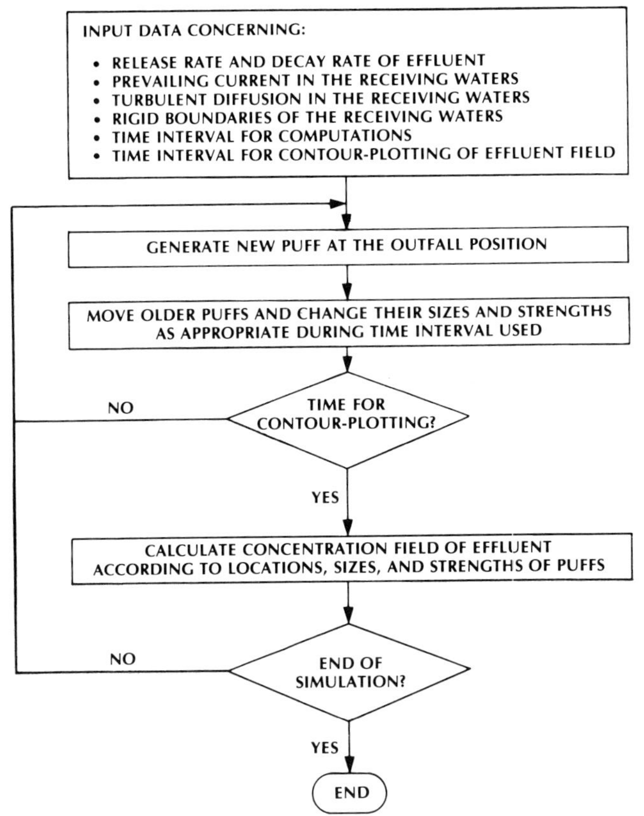

Figure 4.1.   Simplified flowchart for the puff model.

The effluent within a puff is assumed to fit a two-dimensional Gaussian distribution with variances increasing with time, and with turbulent diffusion determining the growth rate. The concentration is considered constant down to a given depth, d, which likewise increases with time, but with the depth to the bottom of the receiving waters as an upper limit. The effluent concentration, C_{puff}, within each puff is as follows:

$$C_{puff}(x, y) = M_{puff}\exp\left(- (x - x_0)^2/2\sigma_x^2\right)$$

$$\times \exp\left(-(y - y_0)^2/2\sigma_y^2\right)d^{-1} \qquad 3$$

where (x_0, y_0) is the position of the center of the puff and (σ_x^2, σ_y^2) are the variances in the two-dimensional Gaussian distribution. As the puffs grow, they will be deformed by the prevailing shears of the current velocity and, for modest shears, will move each puff according to the current averaged over the puff area. If the velocity shears are crucial or if long diffusion times are

considered (implying large puff areas), more sophisticated methods should be used. Such methods applied to atmospheric puff models have been reported (Sheih, 1978).

The effluent concentration at a particular location is calculated by summing the contributions from all puffs (superposition). When simulating the spreading and dilution of decaying components, zero-order decay is assumed and the amount of effluent in each puff is reduced after each Δt according to the decay rate, k:

$$\Delta M_{puff} = - k\Delta t \qquad 4$$

The presence of rigid horizontal boundaries (coastlines) necessitates modification of the puffs approaching a boundary. A reflection procedure has been adopted, implying that the part of a puff positioned behind a boundary is reflected before it contributes to the effluent concentration nearby. In this way, only boundaries resembling straight lines can be taken into account. Likewise, the boundaries act as reflecting barriers for the centers of the puffs when onshore currents occur.

Similar models have been worked out by several authors; a list of references can be found in a paper by Van Dam (1982). During the 1960s, Harremoës (1966) applied a puff model to simulate effluent plumes from distributions of radioactive isotopes released instantaneously in connection with field experiments carried out by the Danish Isotope Centre. One of the differences between Harremoës' model and the puff model presented in this chapter is that Harremoës used observed patches to determine the distribution of effluent in a puff, whereas Gaussian distributions are used for the puff model in this chapter.

The puff model is run on a desktop computer with 23 kbytes of read–write memory. Input data comprise information on the geometry of the receiving waters, the current conditions, turbulent diffusion, the release and the decay rate of the effluent, and the Δt used for the calculations. Output data are specified with regard to printing, plotting, and storing on magnetic tape.

Computing time and the necessary computer memory impose limits on the time interval that can feasibly be used between successive releases of puffs and, thus, the maximum number of operating puffs. On the other hand, when simulating continuous releases, it is necessary to use a small time interval in order to keep the distance between successive puffs small. The selection of the time interval between puffs is thus a compromise between several demands. The execution time of a simulation by the model is variable but typically amounts to 1–4 h. Until now, no special effort has been devoted to the reduction of computing time and memory requirements. Ludwig et al. (1977) have succeeded in making several simplifications in an atmospheric puff model to suit the model for real-time minicomputer use.

The puff model is typically used to simulate the spreading of an effluent from a continuous point source. When the variation of the current in terms of time and space is known from field measurements by automatic recording instruments or floats, and when the turbulent diffusion coefficients (constant or not) are derived from physical simulations with tracers, it is possible to evaluate the concentration field at particular states of the spreading process. The length of time accounted for by the simulations is limited by the capacity of the desktop computer. The Danish Isotope Centre has applied the puff model for simulations of effluent spreading taking place over a period of 1–5 d. Use of the model to simulate dilution of effluent released off the western coast of Denmark into the North Sea under various environmental conditions is described in Section 4.3.

4.2.2. The Particle Model

Particle models, like the puff models, can be used to simulate the spreading and dilution of released effluents under time-varying and inhomogeneous conditions. Examples are reported by Maier-Reimer (1975) and by Awagi et al. (1980). The particle model presented in this chapter was developed in order to address a more specific problem, as is described below.

If a given location is defined as a protected area, it may be desirable to set up statistics for the contamination of that location resulting from outfall(s) nearby. Simulations then have to yield the contamination in a large number of instances and sum the contributions after weighing them according to their probability of occurrence. The factors characterizing the potential contaminating situations must be selected in accordance with the problems at hand. If, for instance, the current in the receiving waters is primarily generated by the wind, statistics used for the wind would generate a set of statistics for the current. If the effluent spreading is simulated for a large number of situations with conditions for the currents selected in accordance with those statistics, then pollution statistics (e.g., excess probabilities of given concentration levels) can be computed.

The puff model can be used for this purpose; however, if only one or a few locations are of concern, it would be a waste of computing time to simulate the entire effluent plume associated with each environmental condition. The particle model was designed for pollution problems when the contamination at only one or a few locations is of concern.

The pollution statistics for a given location are often caused by the interplay of many factors (e.g., wind, currents, turbulent diffusion, or tides). Typically, the coupling between the factors is so complicated that only simple, approximate relations can be established. The residuals can be ascribed to an incomplete understanding of the processes, so stochastic descriptions become indispensable. In such cases, the pollution cannot be related uniquely to the governing factors, and simulations with the puff model would then represent an average picture that probably never occurs in nature. To cope with stochastic fluctuations, it seems appropriate to characterize the environmental states by probability distributions. To perform a simulation, an environmental state is constructed by random sampling (Monte Carlo method) from the various distributions (Fig. 4.2). Then a particle is released from the outfall,

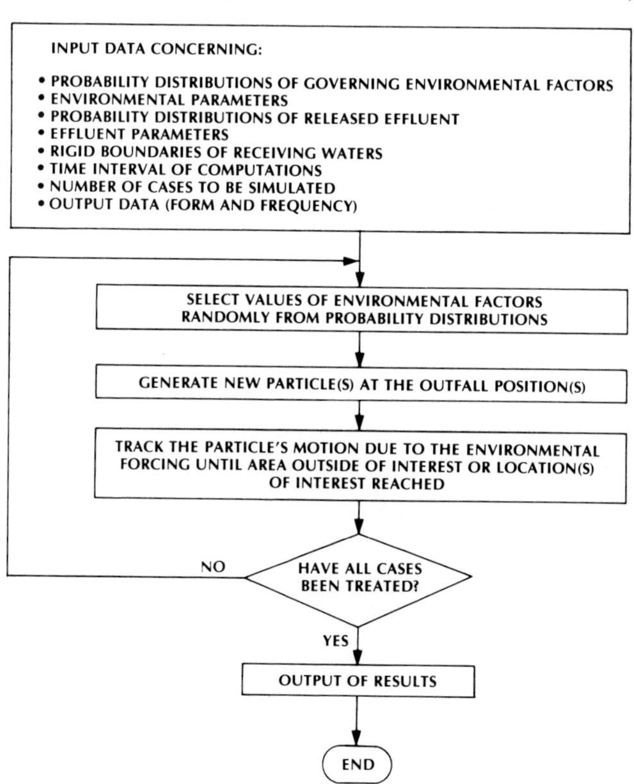

Figure 4.2.　Simplified flowchart for the particle model.

and its motion is tracked during a period of time until it is apparent whether it reaches the given location. When a large number of simulations has been performed, the percentage of particles that have reached the location of concern equals the fraction of effluent contributing to contamination at the location.

4.3. APPLICATION OF THE MODELS

4.3.1. Site and Measurements

In the autumn of 1981, the Danish Isotope Centre made field measurements off the western coast of Denmark in order to evaluate the spreading and dilution of effluents released from a chemical plant. The effluents (4000 m^3 d^{-1}) were discharged through two 400-m-long pipes, each supplied with a two-port diffusor, at a depth of 9 m. The distance between the diffusors was about 80 m, and the density of the effluents was close to the density of the receiving seawater, which is vertically homogeneous in most cases.

The current typically moves parallel to the coast (i.e., in a northerly or southerly direction). When the current is moving northward, the effluents are transported toward, and sometimes into, a channel (Thyborøn Channel) connecting the Limfjord to the North Sea (Fig. 4.3). The purpose of the field investigation was to obtain information on the dilution of the effluents into the coastal area between the outlet position and the channel, and also to evaluate the fraction of the released effluents being transported with the current into the channel, and possibly, further into the Limfjord.

The effluents were traced by means of the fluorescent dye rhodamine-B and by ^{82}Br, both of which were injected into the wastewater at the chemical plant. The measuring technique is described by Appelquist and Nielsen (1987). The data thus obtained have been used to estimate the conditions of the current and the turbulent diffusion in the outfall area.

The current along the western coast of Denmark can be considered as having a periodic component gener-

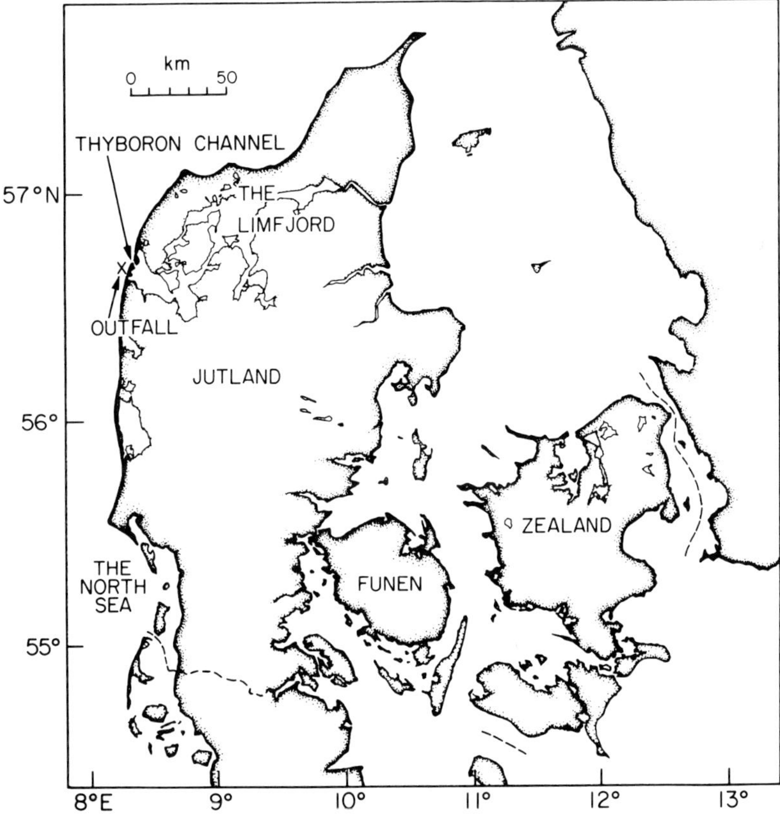

Figure 4.3. Map of Denmark, with the outfall indicated by an x.

ated by the semidiurnal astronomical tide and also as having a component generated by the wind. Both components are typically oriented parallel to the coast; thus, during stable weather conditions the current can be described as a constant net current (0–0.6 m s^{-1}) with a superimposed periodic tidal current with an amplitude of about 0.1 m s^{-1}.

The intensity of the turbulent diffusion was evaluated from the dilution of the tracer rhodamine-B. Reliable computations of the horizontal coefficient of diffusion presume that a tracer plume is formed under conditions of a steady current and that the plume is mapped in sections perpendicular to the current direction at different distances downstream of the outfall. In two instances these requirements were fulfilled. In the first, the width of the plume increased downstream corresponding to a constant coefficient of diffusion of 0.04 m^2 s^{-1}, which was valid for diffusion times up to 1 h. During the following hour, the diffusion coefficient increased by a factor of 10. The other instance revealed a diffusion coefficient that increased proportional with the diffusion time raised to the power of 1.63 during the first 1.5 h. Based on previous data and information from the literature, it was estimated that the diffusion coefficients did not exceed 1–10 m^2 s^{-1} for diffusion times shorter than 1 d.

Turbulent diffusion in the vertical direction was investigated by measuring the vertical distribution of the tracer at various distances from the outfall. In most cases the tracer was found to be mixed throughout the whole water column, indicating that the vertical diffusion was so intense that the physical limits of the receiving waters provided the limit of the vertical distribution of effluents, the immediate surroundings of the outlet position excluded.

The current in the Thyborøn Channel is dominated by a component generated by the semidiurnal astronomical tide. The residual current is coupled to the east–west component of the prevailing wind. These results were derived from a series of current observations made from 1968 to 1969 by an automatic recording instrument moored at a depth of 6 m. In calm weather, the current in the channel comes in toward the Limfjord when the tidal current along the western coast of Denmark is moving north (flood tide), and vice versa. This means that the effluent entering the channel typically arrives from the south and is drawn into the channel during flood tide. The tidal current (with an amplitude of about 0.5 m s^{-1}) will, when the residual current is minor, cause part of the incoming effluent to be returned after the current reverses. The field measurements have not shed light on this phenomenon

because the tracer substance was already very dilute at the time it entered the channel.

4.3.2. Simulations with the Puff Model

The puff model has been used to evaluate the dilution of the effluent during its motion along the western coast of Denmark. This dilution depends on the current velocity and the turbulent diffusion. As the field measurements have shown, these factors show temporal variations that cause the dilution to become variable.

In order to evaluate the dilution between the outfall and the entrance into Limfjord, simulations have been performed corresponding with realistic current and diffusion conditions. Based on the computed dilution field isodilution curves have been drawn. Figure 4.4 shows smoothed isodilution curves corresponding to a continuous release of 1 m^3 s^{-1} of nondecaying effluent for net currents ranging from 0.1 to 0.6 m s^{-1}, superimposed by a tidal current with an amplitude of 0.1 m s^{-1}. The turbulent diffusion is characterized either by a constant coefficient of 0.04 m^2 s^{-1} (Fig. 4.5 a–c) or by a coefficient increasing with the diffusion time with an upper limit of 1 m^2 s^{-1} (Fig. 4.5 d–f). When the net current is comparable with the amplitude of the tidal current (Fig. 4.5 a, b), the effluent accumulates every time that slack tide occurs. The locations of accumulation migrate downstream with the net current and are visible at constant intervals. When the net current is large compared to the tidal current the accumulation effect is negligible.

The computed isodilution curves are not to be considered realistic in that they should compare well with observed dilutions of the effluent. The simplifications inherent in the puff modelling lead to results that may apply only to averaged dilution patterns, so all small-scale irregularities may be smoothed out. For this reason, no definite verification of the model can be made. Nevertheless, the simulations provide a simple means of evaluating the dilution under conditions not covered by the field measurements, as well as of estimating the relative importance of the governing factors.

4.3.3. Simulations with the Particle Model

Northerly currents near the outfall transport the released effluents along the western coast of Denmark toward the entrance of the Thyborøn Channel. If the current in the channel is toward the Limfjord when the effluent approaches, it will be transported into the chan-

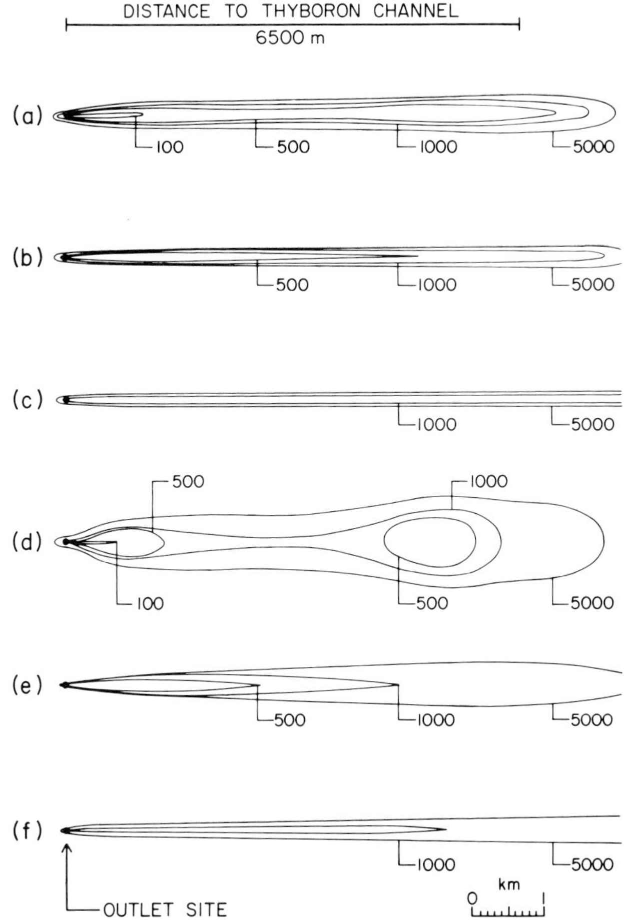

Figure 4.4. Isodilution curves calculated by means of a puff model (a release rate of 1 m^3 s^{-1} is assumed). The amplitude of the tidal current is 0.1 m s^{-1}; the net current is 0.1 m s^{-1} in (a) and (d), 0.3 m s^{-1} in (b) and (e), and 0.6 m s^{-1} in (c) and (f); and the coefficient of turbulent diffusion is constant at 0.04 m^2 s^{-1} in (a), (b), and (c), and increases with diffusion time raised to the power of 1.63 in (d), (e), and (f), with 1 m^2 s^{-1} as an upper limit.

nel. The field measurements showed examples of this phenomenon. One of the main objectives of the investigation was to find out the amount of released effluent entering the channel. A version of the particle model has been set up to solve this problem.

The current along the coast is assumed to be produced by the wind and the tidal forces. A probability distribution of the wind-generated portion can be established by statistics of wind patterns in the area and the empirical relations for the wind–current coupling. A probability distribution of the current in the channel can be set up in a similar way. A simulation consists of the following steps. A wind speed and a wind direction are selected randomly from probability distributions as they are provided by statistics on the wind in the outfall area. The wind-generated component of the currents along the coast and in the channel are calculated by using the empirical wind–current relations. The temporal variations in the currents are ascribed merely to the semidiurnal tides and are assumed to follow a cosine function. An effluent particle is released from the outfall after one randomly selects the phase-angle of the tide. The particle follows the current along the coast, and the effect of the turbulent diffusion is considered by adding a random velocity component. The tracking of the particle is carried out in time intervals, Δt, and can be described as

$$x_{t + \Delta t} = x_t + (u_w + u_t\cos(2\pi t/T + \phi) + u')\Delta t \quad \text{5a}$$

$$y_{t + \Delta t} = y_t + (v_w + v_t\cos(2\pi t/T + \phi) + v')\Delta t \quad \text{5b}$$

where x_t and y_t are the position of the particle at time t, u_w and v_w are the wind-generated velocities, and u_t and v_t are the amplitudes of the tidal current. The period of the tidal current is termed T, ϕ is the phase angle, and u' and v' are the turbulent components of the current related to a diffusion coefficient, D:

$$u' = \sqrt{6D/\Delta t}\,\text{rnd} \quad \text{6a}$$

$$v' = \sqrt{6D/\Delta t}\,\text{rnd} \quad \text{6b}$$

where rnd is a random number from a simple rectangular distribution and ranges from -1 to 1.

If the mean current is southbound, then the tracking of the particle is omitted because the particle will never enter the channel. A northbound particle will eventually reach the mouth of the channel, thus being affected by the channel flow. Depending on the direction of the current, the particle will approach and eventually enter the channel, or it will move away from the coast. The tracking of the particle terminates when the particle enters the channel, or when it is obvious that it never will. This procedure was repeated many times (more than 1000), and the relative number of particles entering the channel was computed to indicate the fraction of released effluent transported into the channel. The computed fraction is related to the input statistics and,

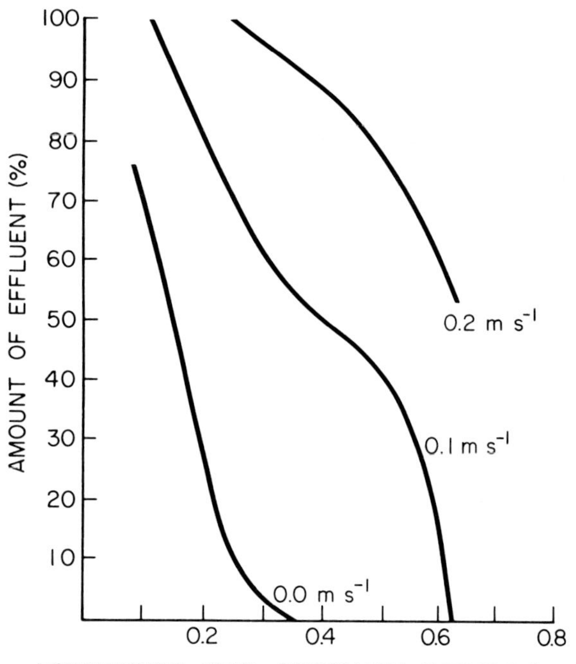

Figure 4.5. Percentage of effluent entering Thyborøn Channel. The percentages vary as functions of the intensity of wind-driven, northbound currents (*x* axis) and as functions of the intensity of wind-driven, inbound currents at the Thyborøn Channel (the three curves in the body of the figure).

since the wind statistics apply to a typical year, the computed fraction does the same.

The prevailing winds were treated separately in the particle model, and no allowance was made for phenomena arising from unstable weather or for changes from one wind condition to another. Only periodic variations according to the tides were incorporated, indicating that the model is semistationary. In the actual situation it was found that about 10% of the released effluent will, on the average, enter the channel. This fraction, however, is the result of a weighted sum of fractions, each associated with a particular wind condition. Figure 4.5 shows the fraction of released effluent entering the channel as a function of the wind-generated currents along the coast and at the entrance of Thyborøn Channel. The fraction appears to vary greatly, ranging from 0 to 100%. When the tidal current predominates the wind-generated component, however, a part of the effluent entering the channel should return to the North Sea after the next reversal in currents.

The particle model is able to cope with data input that is more detailed and complex than that presented in this chapter. This means that whenever such data become available, the estimates of the amount of effluent entering an area can be improved. The model nevertheless allows testing of whether estimates are sensitive to particular input data; this may be convenient in planning further measuring programs in the field.

4.4. CONCLUSIONS

In the experience of the Danish Isotope Centre, puff models and particle models are suitable tools for dealing with waste-disposal problems. These models can be adjusted to serve specific purposes; depending on the available computer facilities the models can be simplified or expanded, but they are capable of operating even with small desktop computers. It should be stressed, however, that all kinds of modelling involve simplifications that force the users to be cautious when interpreting the results. Only detailed field measurements in the receiving waters can indicate where the effluent actually goes. On the other hand, without models it is difficult to extrapolate the results of field measurements beyond the environmental conditions under which they have been carried out.

REFERENCES

Appelquist, H., and P. B. Nielsen. 1987. A data acquisition system for 3-dimensional measurements of natural and artificial tracers in the sea. *In: Oceanic Processes in Marine Pollution*, Vol. 4: Scientific Monitoring Strategies for Oceanic Waste Disposal. Robert E. Krieger Publishing Co., Malabar, Florida, in press.

Awagi, T., N. Imasato, and H. Kunishi. 1980. Tidal exchange through a strait: A numerical experiment using a simple model basin. *Journal of Physical Oceanography*, **10**, 1499–1508.

Harremoës, P. 1966. Prediction of pollution from planned wastewater outfalls. *Journal of Water Pollution Control Federation*, **38**, 1323–1333.

Ludwig, F. L., L. S. Gasiorek, and R. E. Ruff. 1977. Simplification of a Gaussian puff model for real-time minicomputer use. *Atmospheric Environment*, **11**, 431–436.

Maier-Reimer, E. 1975. Zum Einfluss eines mittleren Windschubes auf die Reststrome der Nordsee. *Deutsche Hydrographische Zeitschrift*, **28**, 253–262.

Sheih, C. M. 1978. A puff pollutant dispersion model with wind shear and dynamic plume rise. *Atmospheric Environment*, **12**, 1933–1938.

Van Dam, G. C. 1982. Dispersion Models. *In*: Pollutant Transfer and Transport in the Sea, Vol. 1, G. Kullenberg (Ed.). CRC Press, Boca Raton, Florida, pp. 91–160.

Chapter 5

A Modelling Approach to the Effects of Waste Disposal in the Southern Bight of the North Sea

Jean-Paul Mommaerts, Yves Adam, Philippe D'Hondt, and Thierry G. Jacques

Management Unit of the North Sea and Scheldt Estuary Mathematical Models
Ministry of Public Health and Environment
Brussels, Belgium

ABSTRACT

With a good knowledge of the instantaneous and residual velocity fields, one can develop short-term models for chemical and biological variables and also identify the zones within which such variables may be considered homogeneous. These zones then define the boundaries of the ecological box models used by the Belgian public authorities to assess alternatives in marine management. Examples of such assessment studies are presented; they deal mainly with permit applications for dumping phosphogypsum waste, acid-iron waste, hydroxyethylcellulose waste, and basic aluminate solution with alkylated compounds into the continental shelf area. The grounds on which these studies have led to acceptance, rejection, or modification of the original query are highlighted, and the advantages of this modelling approach are discussed.

5.1. INTRODUCTION

Coastal seas are characterized both by highly productive, economically valuable ecosystems and by intense human activities capable of interfering with system functions and properties. Optimizing the multiple uses of the coastal zone requires accommodating conflicting interests within a regulatory framework that ensures that disruptions remain minor, or reversible, and that marine resources are conserved. Since this requires taking account of a large number of variables and interacting processes, any public authority having to make such management decisions is faced with complex assessment problems.

The dumping of wastes at sea represents a particularly delicate issue because wastes that cannot be safely accommodated on land are usually proposed for dumping at sea. For economical reasons this activity tends to develop at the shortest distance possible from loading harbors. Substances thus disposed of may present a direct threat to sensitive coastal resources such as fish nursery grounds and productive demersal communities, and, possibly, to the nearshore ecosystem as a whole. The nature and magnitude of this threat therefore needs to be assessed carefully. An aspect that must also be considered is that dumping operations may also conflict with other legitimate uses of the sea, such as shipping and recreation.

Besides an elementary choice between environmental damage and economical loss (e.g., through stopping a particular industrial process), it should be possible to minimize the potential damage of dumping at sea by selecting an appropriate dumpsite. If impact assessments yield sufficiently objective, quantitative results, and if it can be determined with reasonable certainty how the marine ecosystem is likely to react or adjust to a dumping procedure in a particular place, then wise choices can be made. Of course, some fixed criteria may still be needed in order to decide whether a given impact is acceptable, but even simple comparative results, objectively determined, should constitute a valuable basis for management decisions.

In an environment as dynamic as the sea, adequate assessments require a full-ecosystem approach, and the quantitative, predictive tool that system modelling can provide. In this approach, numerical waste-dispersion models have been coupled to chemical and ecological models to form fate-and-effects prediction systems. Local accumulation of substances can be calculated with a simple flushing model, and concentrations can be compared with lethal and sublethal limits for important constituents of the ecosystem. An assessment that combines available information on waste properties with knowledge of ecosystem processes can thus be made.

Since 1977 these techniques have constituted an important aid in making decisions on marine management in Belgium. The experience gained makes it possible to review the ecosystem-modelling approach and examine its merits. Considerations that must be addressed are whether the tool is usable, whether comparison of the results with agreed criteria can lead to management decisions, whether decisions have proven to be acceptable, whether possible shortcomings exist, and, finally, whether improvements can be recommended.

This chapter attempts to launch such a discussion by presenting a number of case studies of actual dumping applications that were submitted to the Belgian government. From the time that the Oslo Convention was ratified in 1978, the Belgian Department of Public Health and Environment has handled eight applications, all of which underwent a preliminary screening, for permits to dump wastes at sea. Proposed dumping that contravened the rules of the Oslo Convention, that had economically viable land-based disposal or treatment solutions, or that had possible alternative production processes received no further consideration. The other applications were subjected to a toxicity evaluation and to the modelling assessment. Only one aspect of the evaluation procedure is presented here: the computer modelling of the dispersion of a waste at sea and its possible ecological effects. Modelling is not normally carried out unless the results of the preliminary screening tests favor dumping at sea, and computer simulations, therefore, constitute the last barrier raised before a permit to dump at sea can be issued.

5.2. THE MODELS USED

5.2.1. The Dispersion Model

The Southern Bight of the North Sea is a turbulent, well-mixed, shallow sea, especially so off the Belgian coast, where tidal currents are strong and storms are frequent. Ronday (1976) has developed a nonstationary, hydrodynamic, numerical model that computes the tidal variables in this area (depth-averaged current and free-surface elevation). A computer program based on this model provides the hydrodynamic information needed for the resolution of the dispersion equation, derived by Nihoul (1972). Generalizing Bowden's (1965) result, Nihoul's equation takes into account the effects of advection, shear diffusion, and biochemical interactions to model the fate of matter in a turbulent, shallow sea. The basic assumptions that have allowed the simplification of the general tridimensional equation governing the dispersion of matter are:

1. The distribution of matter is assumed to be homogeneous over the whole water column. This assumption is well justified in the turbulent and shallow portion of the North Sea off the Belgian coast.

2. The vertical variation of the horizontal component of the flow coupled with the vertical eddy diffusion generates a horizontal shear-effect diffusion that domi-

nates the diffusive process. This mathematical model is valid so long as variations of system variables are much larger on horizontal scales than on the vertical scale.

3. The horizontal current has the same direction at all depths. This allows one to parameterize the shear-effect diffusion in terms of the value of the depth-averaged velocity.

With these three points in mind, the equation that describes the evolution of a substance disposed at sea reads as follows (Nihoul, 1972):

$$\frac{\partial C}{\partial t} + \boldsymbol{u}\nabla C = H^{-1}\nabla\left[H\gamma\frac{H}{\|\boldsymbol{u}\|}\boldsymbol{u}(\boldsymbol{u}\nabla C)\right] + I_c$$
$$+ \frac{Q_c}{H(x_0,y_0)}\delta(x-x_0,y-y_0,t-\tau) \qquad 1$$

where C is the depth-averaged concentration of a substance in the water column (M L^{-3}); H is the total depth (equilibrium depth + elevation of free surface) (L); $\boldsymbol{u}$ is the horizontal depth-averaged current-velocity vector (L t^{-1}); γ is the shear-diffusion coefficient (dimensionless, assumed to be known); ∇ is the gradient operator; I_c is the chemical, biological, or physical interaction of the state variable c with the other state variables (M L^{-3} t^{-1}); Q_c is the quantity of substance added per unit area at the disposal site (M L^{-2}); δ is Dirac's function; and x_0, y_0, and τ are the geographic coordinates and time of the disposal operation, where the dimensions are given in terms of M (mass), L (length), and t (time).

Adam (1977c) has discussed the time and space scales of the validity of this equation. Because of the three basic aforementioned assumptions, the equation is valid so long as: (1) the dispersion site is sufficiently remote from the coastline (several kilometers), (2) no thermocline exists, and (3) the scale of variation of the studied concentration is less than a few tens of kilometers.

The basic input data (i.e., the tidal variables H and $\boldsymbol{u}$) are computed with the Ronday (1976) hydrodynamic, numerical model on a staggered grid covering the eastern half of the English Channel and the whole North Sea, with a mesh size of 21.7 km × 37.0 km (Fig. 5.1). In an area comprising the eastern British Channel, the Straits of Dover, and the Southern Bight, this grid is further refined to a spatial resolution of 7.7 km × 7.4

km. To allow detailed simulation of processes taking place in a 30.0-km strip along the Belgian coast, a model based on a grid with a mesh size of 1.0 km × 1.0 km is also available. Regardless of mesh size, the distributions of $\boldsymbol{u}$ and H must be interpolated to be used by the dispersion model because the special steps needed by the numerical algorithm for solving the Ronday (1976) equations apply over a scale of only a few hundred meters.

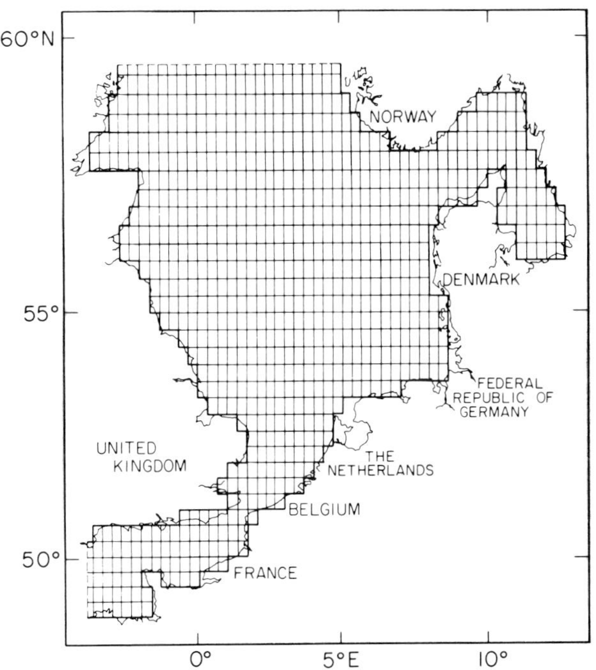

Figure 5.1. Grid of the nonstationary hydrodynamic numerical model of the North Sea.

As far as the dispersion of a passive variable is concerned, the results of the integration of the dispersion equation 1 compare well with observations made by Talbot and Talbot (1974) during rhodamine-B diffusion experiments at sea up to a few tidal periods (Adam, 1979).

To solve the dispersion equation, Adam (1977a, 1977b) has designed an algorithm that allows a superimposed computational micro-grid to surround and travel with the pollutant patch. The behavior of the patch and the displacement of the grid are computed at discrete times. In practical terms, this means that the simulation of dispersion of pollutants disposed of at sea can be numerically modelled for periods of a few days' length.

5.2.2. The Ecological Model

In addition to the dispersion model, a box model simulating the seasonal evolution of the nitrogen cycle in the Southern Bight of the North Sea provides a tool for the analysis of the events occurring in places where the dispersion simulation indicates persisting levels of pollution. This model was developed by Pichot (1980) and is based on the annual budget of nitrogen or carbon fluxes elaborated by Billen et al. (1977). It is largely validated by the *in situ* observations yielded by a permanent survey initiated in 1971 with the National Research and Development Program, Project Sea.

The model area is subdivided into 30 boxes that are interconnected by advection and diffusion flows (Fig. 5.2). If necessary, smaller boxes could be defined in order to cover specific areas. Boxes are portions of space for which state variables are assumed to be homogeneous and, thus, where their evolution may be simulated by solving a set of first-order differential equations.

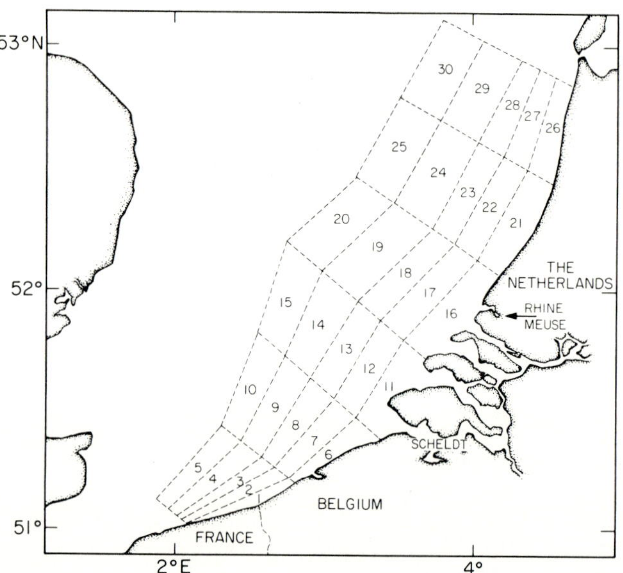

Figure 5.2.   Subdivision of the eastern Southern Bight into boxes for ecological modelling. The five longitudinal boundaries correspond to stream lines increasing seaward with water fluxes of 100, 150, 200, 217.8, and 231.1 (10 m³ s⁻¹).

The boundaries of the boxes are based on the configuration of the stream lines yielded by a mathematical model of residual circulation (Nihoul and Ronday, 1975). By definition there is no advection across longitudinal boundaries since these coincide with stream lines. Diffusion across these boundaries is expressed by a rate of mixing between two adjacent boxes, the value of which is estimated by an unidimensional analogy (Radach and Maier-Reimer, 1975). On the other hand, diffusion across the cross-boundaries proved to be negligible relative to advection.

The model involves four depth-integrated variables, all of which are expressed in units of grams of nitrogen per square meter. The variables are: (1) dissolved limiting nutrient, N, (2) phytoplankton biomass, p, (3) zooplankton biomass, Z, and (4) dissolved organic matter, OM. Figure 5.3 shows the dominant biological processes that govern these variables. A complete discussion of the functional forms adopted for relating these variables is presented by Pichot (1980), together with an extensive analysis of dynamical sensitivities of the system.

5.2.2a. The conservation equation for dissolved inorganic nitrogen

The conservation equation for dissolved inorganic nitrogen can be stated as

$$\frac{dN}{dt} = D\left(C_s - \frac{N}{H}\right) + R(OM) + EZ - GP$$

$$+ \text{ input from estuaries}$$

$$+ \text{ advection and diffusion} \qquad\qquad 2$$

where D is the diffusion coefficient for dissolved inorganic nitrogen from sediments to overlying water (L t⁻¹), C_s is the concentration of dissolved inorganic nitrogen in the interstitial water of surficial sediment (M L⁻³), N is the depth-integrated concentration of dissolved inorganic nitrogen in the water column (ML⁻²), H is the depth of the water column (L), R is the rate coefficient for remineralization of dissolved organic matter (t⁻¹), OM is the depth-integrated concentration of dissolved organic matter expressed as the equivalent amount of nitrogen (M L⁻²), E is the rate coefficient of nitrogen excretion by zooplankton (t⁻¹), Z is the depth-integrated concentration of zooplankton expressed as the equivalent amount of nitrogen (M L⁻²), G is the rate coefficient for primary production (t⁻¹), and P is the depth-integrated concentration of phytoplankton expressed as the equivalent amount of nitrogen (M L⁻²).

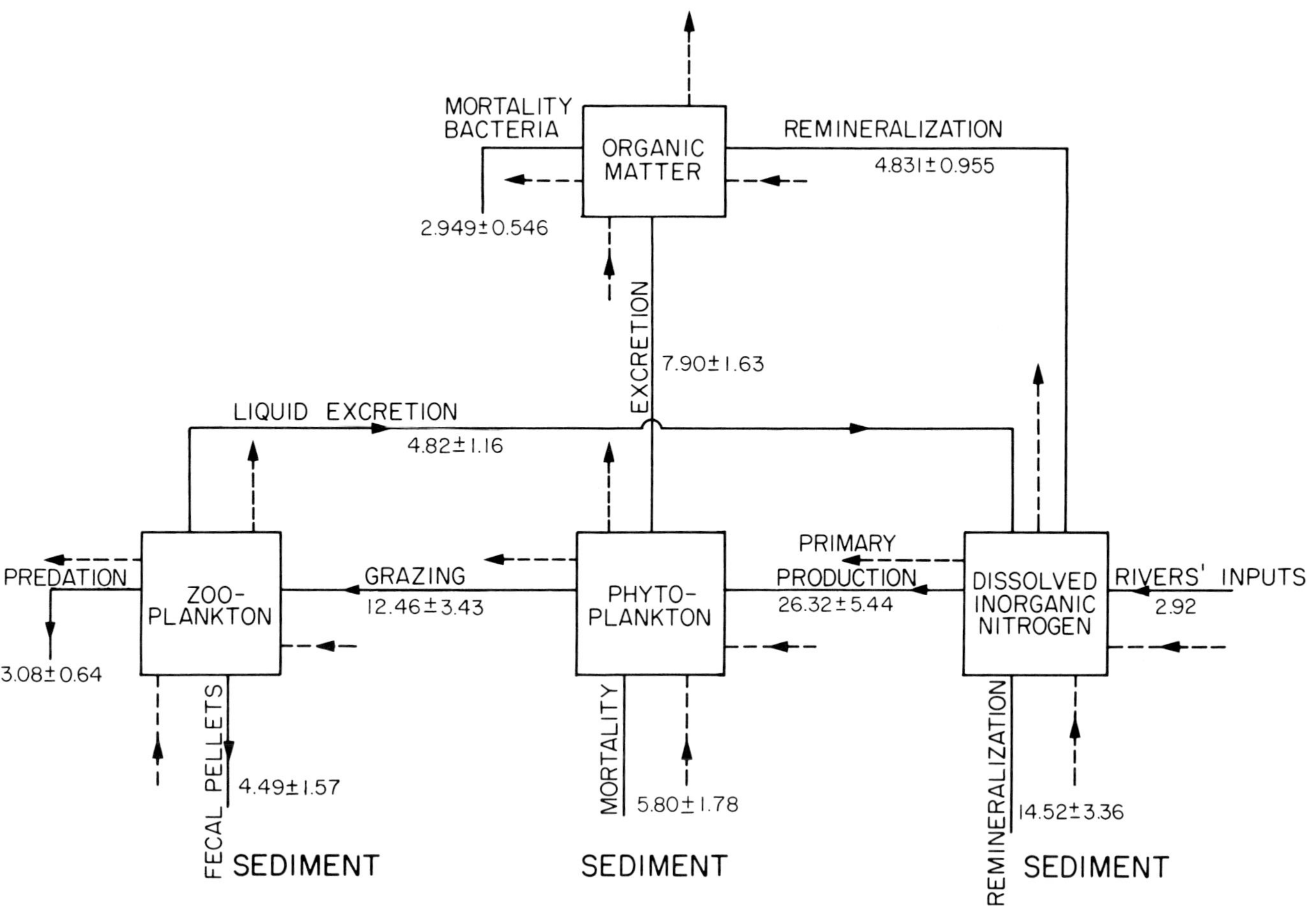

Figure 5.3. Budget of mean annual nitrogen fluxes (g m^{-2} y^{-1}) as calculated with the ecological model for the nine boxes off the Belgian coast (boxes 1–4, 6–9, and 11–14 in Fig. 5.2).

The temperature dependencies of the rate coefficients R, E, and G are all assumed to be calculable by the Arrhenius equation, so that

$$x(T) = x(T_m)A \exp\left[(T - T_m)/10\right] \qquad 3$$

where $x(T)$ represents the value of R, E, or G at temperature $T(°C)$, $x(T_m)$ represents their values at the mean temperature of seawater, and A is the Arrhenius coefficient.

The rate coefficient for primary production, G, combines the effects of temperature, available light, and concentration of dissolved inorganic nitrogen on phytoplankton production:

$$G = G_0 A \exp\left[(T - T_m)/10\right] \mathscr{L} \mathscr{N} \qquad 4$$

where G_0 is the primary production coefficient under conditions of light and nutrient saturation, $\mathscr{L}$ is the effect of light on phytoplankton production, and $\mathscr{N}$ is the effect of dissolved inorganic nitrogen on phytoplankton production.

The effect of light on phytoplankton growth, $\mathscr{L}$, is derived from the Vollenweider formula (1965) as:

$$\mathscr{L} = \frac{\lambda(1.65)}{K_e H} \arcsin \frac{y}{\left(1 + y^2\right)^{1/2}} \qquad 5$$

where λ is the fraction of a day during which there is light, H is the depth of the water column (L), K_e is the vertical light attenuation coefficient (L^{-1}), and

$$y = J/J_0(1.65) \qquad 6$$

where J is the day-averaged incident radiation in the photosynthetic range (kcal $L^{-2} t^{-1}$), and J_0 is the saturation value for incident radiation.

In equation 4 the value for K_e is determined as was done by Riley (1956) to account for nonphytoplanktonic suspended matter and self-shading, so that

$$K_e = k_{np} + k_1\left(\frac{P}{H}\right) + k_2\left(\frac{P}{H}\right)^{2/3} \qquad 7$$

where k_{np} represents the shading due to nonphytoplanktonic suspended matter, k_1 is the first self-shading coefficient, and k_2 is the second self-shading coefficient.

The effect of dissolved inorganic nitrogen on phytoplankton production is as:

$$\mathcal{N} = \frac{N}{K_s H + N} \qquad 8$$

where K_s is the half-saturation constant (M L^{-3}).

5.2.2b. The conservation equation for phytoplankton

The conservation equation for phytoplankton can be stated as

$$\frac{dP}{dt} = (1 - L)GP - WZ - SP$$

$$+ \text{ advection and diffusion} \qquad 9$$

where L represents the fraction of nitrogen uptake (by phytoplankton) that is released as dissolved organic nitrogen, W is the rate coefficient for zooplankton grazing upon phytoplankton (t^{-1}), and S is the rate coefficient for phytoplankton mortality (t^{-1}).

The temperature dependence of W is calculated by the Arrhenius equation (equation 3). Its dependence on P is calculated as

$$W = \frac{W_0 P}{I_s H + P} \qquad 10$$

where W_0 is the saturation value for grazing, and I_s is the half-saturation constant (M L^{-3}).

5.2.2c. The conservation equation for zooplankton

The conservation equation for zooplankton can be stated as

$$\frac{dZ}{dt} = WP - EZ - WFZ^2 - XZ$$

$$+ \text{ advection and diffusion} \qquad 11$$

where F is the rate coefficient for fecal pellet production ($L^2 M^{-1}$), and X is the rate coefficient for zooplankton predation by higher organisms (t^{-1}).

5.2.2d. The conservation equation for dissolved organic matter

The conservation equation for dissolved organic matter can be stated as

$$\frac{d(OM)}{dt} = LGP - R(OM) - Y(OM)$$

$$+ \text{ advection and diffusion} \qquad 12$$

where Y is the rate coefficient of marine bacterial mortality (t^{-1}).

The numerical values adopted for the various parameters in these conservation equations are listed in Table 5.1. The seasonal variations in water temperature, incident radiation, duration of daylight, and input of the parameter N from estuaries are given by the relevant meteorological and hydrographical inventories and are applied to the model as external forcing functions.

With this model, one can readily simulate toxicity effects as well as any ecological perturbation driven by such parameters as nutrient level, water transparency, or water temperature; however, the biological dynamics thus described are not sufficient to take into account more complex processes such as bioaccumulation phenomena, impacts at the population level, and metal cycling. As demonstrated in Section 5.3, the ecological model has been used for differential studies rather than for accurate predictions.

Table 5.1. Values of the Constants, Coefficients, and Other Parameters Used in the Ecological Model

Symbol	Meaning	Value or Range[a]	Units
D	Coefficient for nitrogen diffusion from sediment	0.008–0.020	$\mathrm{m\ d^{-1}}$
C_s	Concentration of nitrogen in interstitial water	3	$\mathrm{g\ m^{-3}}$ of nitrogen
H	Depth of water column	7.9–31.4	m
R	Coefficient of rate of remineralization of dissolved organic matter	0.07–0.11	$\mathrm{d^{-1}}$
E	Coefficient of rate of liquid excretion by zooplankton	0.274	$\mathrm{d^{-1}}$
G_0	Coefficient of optimal rate of primary production	2.13–3.45	$\mathrm{d^{-1}}$
A	Arrhenius coefficient (Q_{10})	2.3	none
T_m	Mean temperature of water	10–11	°C
J_0	Saturating value of incident radiation in the photosynthesis vs. light relationship	24	$\mathrm{kcal\ m^{-2}\ h^{-1}}$
k_{np}	Shading due to nonphytoplanktonic suspended matter	0.2–1.0	$\mathrm{m^{-1}}$
k_1	First self-shading coefficient	1	$\mathrm{m^2\ (g\ of\ nitrogen)^{-1}}$
k_2	Second self-shading coefficient	1.3	$\mathrm{m\ (g\ of\ nitrogen)^{-2/3}}$
k_s	Half-saturation constant of phytoplankton growth vs. nitrogen concentration relationship	0.02	$\mathrm{g\ m^{-3}}$ of nitrogen
L	Fraction of nitrogen uptake by primary production released as dissolved organic nitrogen	0.28–0.30	none
W_0	Coefficient of optimal zooplankton grazing rate	1.3–1.7	$\mathrm{d^{-1}}$
I_s	Half-saturation constant of zooplankton grazing vs. phytoplankton concentration relationship	0.04	$\mathrm{g\ m^{-3}}$ of nitrogen
F	Coefficient of production rate of fecal pellets	2.25–3.00	$\mathrm{m^2\ (g\ of\ nitrogen)^{-1}}$

Table 5.1.   Values of the Constants, Coefficients, and Other Parameters Used in the Ecological Model (*continued*)

Symbol	Meaning	Value or Range[a]	Units
X	Coefficient rate of predation on zooplankton by upper trophic levels	0.11–0.22	d^{-1}
Y	Coefficient of mortality rate of marine bacteria	0.036–0.073	d^{-1}

[a]When two values are given they indicate minimum and maximum values applied to the model.

5.3.   CASE STUDIES

Several applications for dumping permits have been studied at the Management Unit of the North Sea and Scheldt Estuary Mathematical Models in the last five years. All of the dumpsites tested with the dispersion model are presented in Fig. 5.4.

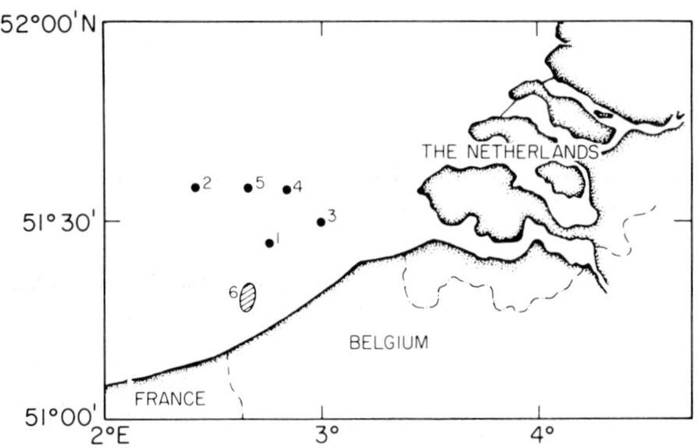

Figure 5.4.   The North Sea off the Belgian coast and the dumpsites tested with the dispersion model. Site 1 is the originally requested dumpsite for phosphogypsum waste, site 2 is the recommended dumpsite for phosphogypsum waste, site 3 is the dumpsite for acid-iron waste, site 4 is the dumpsite for waste with alkylated compounds, site 5 is the requested dumpsite for hydroxyethylcellulose waste, and site 6 is the sand- and gravel-mining area.

There is now enough material to provide examples of case studies dealing with passive or interactive dispersion and to evaluate the approach followed in Belgium. Each example demonstrates a particular feature that has challenged the expertise of the people in charge of the study. Although it is not the subject of a specific application, an example of a case study employing ecological modelling is also presented.

5.3.1.   Examples of Passive Dispersion

In such instances, the interaction term I_c (equation 1) can be neglected and concentration changes are due only to advection and diffusion.

5.3.1a.   *Phosphogypsum waste*

A firm requested a permit to dump 700 metric tons (t) of waste each day. The waste resulted from phosphoric acid production and consisted of 65–70% gypsum ($CaSO_4 \cdot 2H_2O$), 1.8% calcium phosphate, and 0.6% phosphoric acid; hence, a total of 4300 kg of phosphorus would be dumped at sea daily. The non-tidal current in the dumping area suggested by the firm [51°27'N, 2°45'E (site 1 on Fig. 5.4)] is weak.

The dispersion model was used to study the fate of the pollutant patch resulting from a single discharge. The results showed that 24 h and 50 min after the discharge (taking into account two tidal periods), the patch of pollutants would return to the vicinity of the dumpsite (Fig. 5.5).

Whenever the patch of waste tends to revolve in the same area, there is a danger for local accumulation of pollutants if that waste is dumped repeatedly. Since this was the case, a simple flushing model was used to evaluate the effects of a daily discharge into the area concerned (15 km²). In this particular case, it was calculated that only 10% of the pollutant would leave the area per day. Thus, for the area considered, the concentration of phosphorus would rise quickly to a steady-state value of 200 mg m^{-3} (half of that concentration would be reached in only 5 d). This would be a major change compared to the mean ambient concentration of 60 mg m^{-3} of phosphorus.

Situations in which new conditions are established

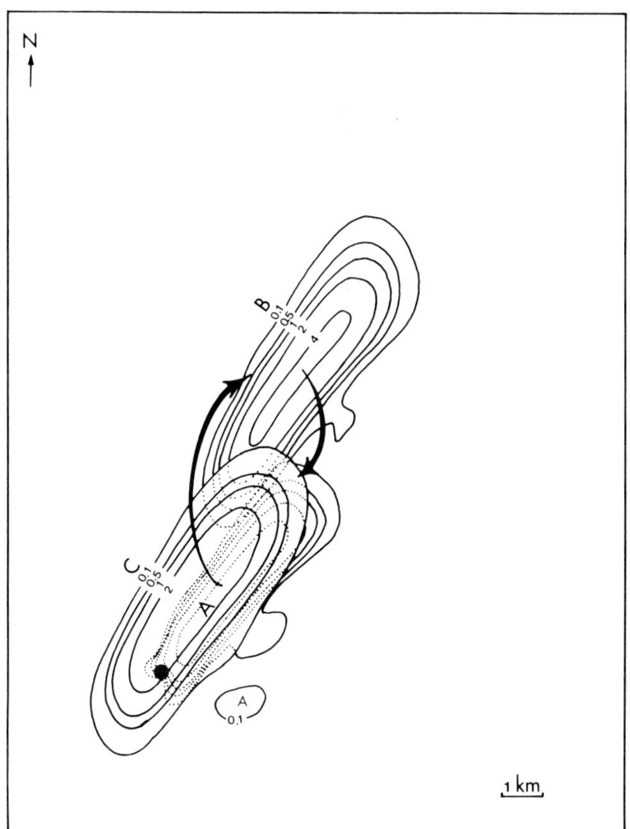

Figure 5.5. Dispersion of phosphogypsum waste disposed of at 51°27′N, 2°45′E (indicated by a black dot), as simulated with the dispersion model. The units refer to the phosphorus concentration (mg m^{-3}). Isopleths A indicate the concentration at the end of discharge; isopleths B indicate the concentration one-half of a tidal period later (maximal excursion); and isopleths C indicate the concentration two tidal periods later, when the next discharge is liable to occur.

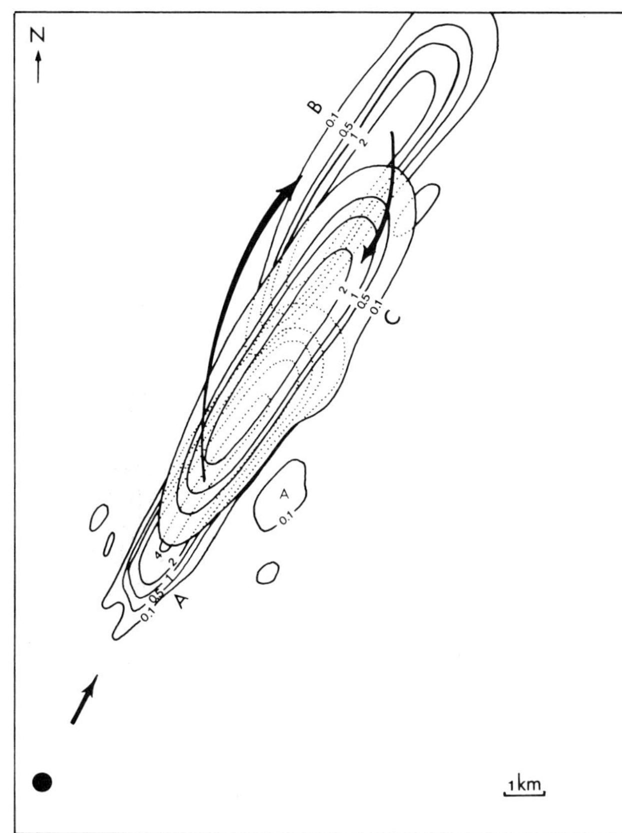

Figure 5.6. Dispersion of phosphogypsum waste disposed of at 51°35′N, 2°25′E (indicated by a black dot), as simulated with the dispersion model. The units refer to the phosphorus concentration (mg m^{-3}). Isopleths A indicate the concentration at the end of discharge; isopleths B indicate the concentration one-half of a tidal period later (maximal excursion); and isopleths C indicate the concentration two tidal periods later, when the next discharge is liable to occur.

on a more or less permanent basis are not considered desirable; therefore, an alternate dumpsite was suggested by the Management Unit of the North Sea and Scheldt Estuary Mathematical Models, on the basis of new dispersion calculations. Unlike the previous dumpsite, the new site [51°35′N, 2°25′E (site 2 on Fig. 5.4)] was characterized by a stronger non-tidal circulation. The results of the model show that 24 h after the disposal operation, the phosphorus patch was located about 12 km north of the dumpsite (Fig. 5.6); therefore, daily discharges would produce a string of well-separated patches. Furthermore, no environmentally adverse effects could be predicted from the concentrations of gypsum or phosphorus at sea after disposal occurring under the prescribed conditions.

The outcome of the application was that no permit was issued. The firm reconsidered its application, presumably in view of the extra cost involved in dumping at a greater distance from the coast than anticipated.

5.3.1b. Acid-iron waste

A firm requested a permit to dump waste from dipping metals into HCl for cleaning. The high level of FeCl$_2$ (25%), the low pH (0.28), and the presence of various metals such as chromium, nickel, and copper deserved attention. The rate of disposal at sea would average 10,000 t y^{-1}, and 7–10 dumping operations would be necessary. A dumpsite with coordinates of 51°30′N, 3°00′E (site 3 on Fig. 5.4) was first suggested by the licensing authorities.

Among several potential impacts of the iron-acid wastes, the depletion of dissolved oxygen was of particular concern. This case was unusual in that the variable was not the concentration of the substance discharged but rather a condition caused by the dumping (i.e., oxygen deficiency). Oxygen is consumed because, at a pH of 8, the $FeCl_2$ is quickly hydrolized

$$Fe^{2+} + OH^- \rightarrow Fe(OH)^+$$

and in turn, the hydroxide is oxidized to colloidal oxyhydroxide

$$Fe(OH)^+ + \tfrac{1}{2}O_2 \rightarrow FeO(OH)\downarrow$$

This process consumes 25 g of O_2 for every 100 g of $FeCl_2$ dumped at sea. Since this reaction occurs within a few minutes, it does not need to appear in the calculation of dispersion. The dispersion of the oxygen-depleted water, on the other hand, can only be simulated for as long as re-aeration from the atmosphere remains negligible with respect to dispersion (i.e., in the first hours following the discharge). The aspect of re-aeration has been ignored in the simulation.

The results of the simulation are given in Fig. 5.7. In this case, sets of isoconcentration curves are drawn every one-fourth of a tidal period, with the A set representing the situation at 3 h and 6 min after the release and the D set representing the situation 12 h and 25 min after the release. Even more clearly than in the first example, the patch of a passive element (i.e., oxygen-depleted water) revolves back to the dumpsite after one tidal period. Meanwhile, the concentration in the middle has decreased at the cost of a much wider patch extension (not shown in the figure) as a typical result of the effect of turbulent diffusion. The concentration at the periphery of the figure is roughly unchanged after one tidal cycle. Although this tendency to persist in the vicinity of the dumpsite is certainly not beneficial to the marine environment, a low discharge frequency would prevent any permanent alteration of the ecosystem in relation to this process of oxygen depletion. Moreover, even at the center of the patch a depletion of 400 mg m^{-3} of O_2 represents no more than 5% of the normally saturated oxygen concentration (about 8 mg liter^{-1} of O_2) and is therefore hardly noticeable.

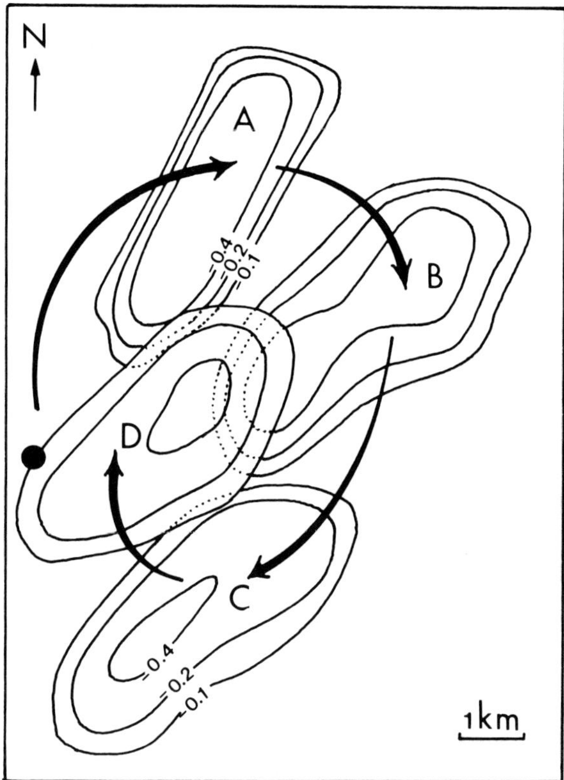

Figure 5.7. Dispersion of the oxygen-depleted zone following upon a dumping of acid-iron waste at 51°30′N, 3°00′E (indicated by a black dot), as simulated with the dispersion model. The units refer to concentration decreases (mg liter^{-1}). Isopleths A indicate the concentration one-fourth of a tidal period after discharge, isopleths B indicate the concentration decreases one-half of a tidal period after discharge, isopleths C indicate the concentration decreases three-fourths of a tidal period after discharge, and isopleths D indicate the concentration decreases one tidal period after discharge.

The application for a permit to dump an acid-iron waste was further studied in relation to the dispersion of colloidal oxyhydroxide and to a possible coprecipitation of metals. These aspects fall under the topic of interactive dispersion and are discussed in Section 5.3.2a.

5.3.2. Examples of Interactive Dispersion

In many instances the interaction term I_c (equation 1) must be taken into consideration. The interactions involved are typically short-term phenomena liable to influence directly the rate of dispersion.

5.3.2a. Acid-iron waste

The interaction term involved in the fate of oxyhydroxide due to the discharge of acid-iron waste takes into account the sinking and resuspension of particulate material. In contrast to most cases in which the interaction is chemical or biological, this example deals only with physical processes. The following equations describe the phenomena (Adam, 1977c):

$$I_1 = -\frac{1}{H} V_{sed}\left(1 - \frac{\bar{u}^2}{u_{CR}^2}\right)\bar{\rho}_1 \qquad 13a$$

$$I_2 = -\frac{1}{H^2} V_{sed}\left(1 - \frac{\bar{u}^2}{u_{CR}^2}\right)\bar{\rho}_b \qquad 13b$$

where I_1 and I_2 are the rates (kg m^{-3} s^{-1}) of sinking and of resuspension, respectively, of the bottom boundary layer; $\bar{\rho}_1$ and $\bar{\rho}_b$ are the depth-averaged concentration in the water (M L^{-3}) and the concentration over the bottom (M L^{-2}), respectively; $\bar{u}$ is the average velocity in the water column (L t^{-1}); u_{CR} is the critical value of velocity such that when $\bar{u} < u_{CR}$, sinking occurs, and when $\bar{u} \geqslant u_{CR}$, the boundary layer is eroded; and V_{sed} is the sinking rate (L t^{-1}), depending on the supposed particle size (Garrels and Mackenzie, 1971).

The term I_1 represents the rate at which suspended matter disappears from the water column through the influence of sedimentation. It depends on the ratio of the depth-averaged velocity to a critical velocity, and it decreases as depth increases. The term I_2 represents the rate at which the concentration of suspended sediment increases due to the flux of sediments eroded from the bottom layer. This rate of return of iron hydroxide to the water column is proportional to the concentration of iron hydroxide deposited onto the sediment. The variation of this rate with depth and current is largely unknown. The formulation used in this model was chosen on the basis of dimensional arguments for simplicity and for lack of a better relationship. Studies are underway to improve this part of the model.

Figure 5.8 shows the isopleths of iron concentration in water over a complete tidal cycle (sets of isocurves are drawn every one-fourth of a tidal period). As already shown for the oxygen-depleted patch, the iron-contaminated cloud returned to the area of the dumpsite. Figure 5.9 illustrates the evolution of the hydroxide

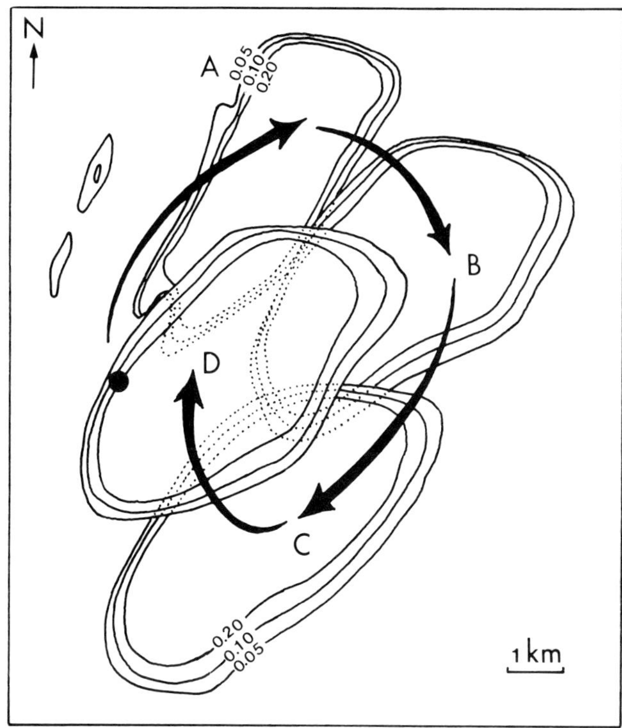

Figure 5.8. Dispersion of ferric oxyhydroxide in water after disposal of acid-iron waste at 51°30′N, 3°00′E (indicated by a black dot), as simulated with the dispersion model. The units refer to the iron concentration (mg liter^{-1}). Isopleths A indicate the concentration one-fourth of a tidal period after discharge, isopleths B indicate the concentration one-half of a tidal period after discharge, isopleths C indicate the concentration three-fourths of a tidal period after discharge, and isopleths D indicate the concentration one tidal period after discharge. Since the depth at the dumpsite is 23 m, the total amount of iron suspended in the water column is 1150, 2300, and 4600 mg m^{-2} at isopleths 0.05, 0.10, and 0.20, respectively.

in the bottom boundary layer. The cycle of sedimentation exhibits a period amounting to one-half of the tidal period. The figure shows that at one-half of a tidal period (isocurves labelled B) or one tidal period (isocurves labelled D) after dumping, high concentrations of deposited iron-rich sediments appear, whereas at intermediate times (corresponding to sets of isocurves labelled A and C in Fig. 5.8), most of the deposited sediment has been resuspended. This means that all of the newly deposited matter is resuspended twice during a tidal period. Since the deposited amount would not exceed 4% of the total hydroxide at slack water and would be fully resuspended when the current is maximum, one can conclude that this particulate matter would eventually be fully dispersed by the marine cur-

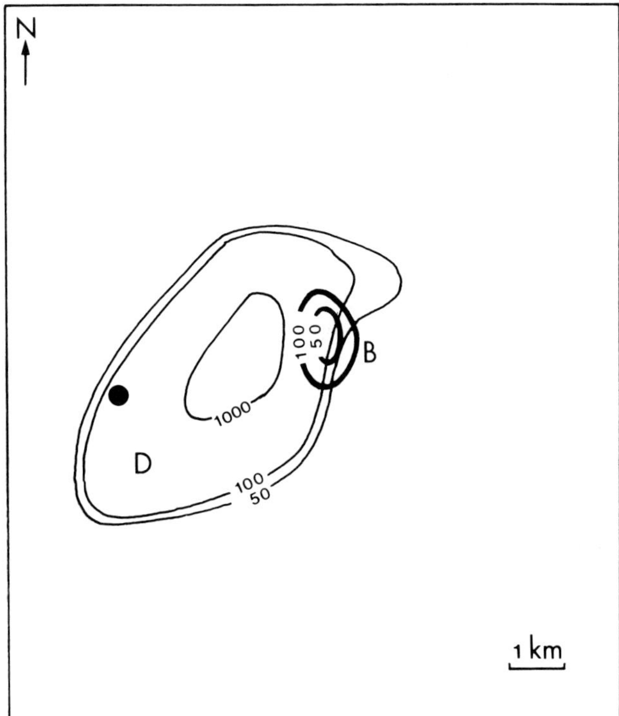

Figure 5.9. Occurrence of ferric oxyhydroxide on the bottom after disposal of acid-iron waste at 51°30′N, 3°00′E (indicated by a black dot), as simulated with the dispersion model. The units refer to the iron concentration (mg m^{-2}). Isopleths B indicate the concentration one-half of a tidal period after discharge, and isopleths D indicate the concentration one tidal period after discharge.

rents. This dispersion is dependent on the assumption made in simulating the sediment fluxes. On the other hand, the low concentrations of iron measured by Baeteman and Vyncke (1979) in the sediments of an existing nearby dumpsite where similar products, originating from the titanium dioxide industry have been discharged for more than 10 y, support the results of the interactive-dispersion model.

Where coprecipitation of metals is concerned, it was calculated that if the metals dumped together with the FeCl$_2$ were totally coprecipitated with the colloidal hydroxide, the concentrations of chromium, nickel, and copper along the 0.10 mg m^{-3} isopleth for iron would be 0.023, 0.023, and 0.017 mg m^{-3}, respectively. These concentrations are very far below the sublethal concentrations demonstrated by other studies (Table 5.2).

The outcome of the application was that although no specific detrimental effect to the marine environment could be demonstrated based on these simulations, the licensing authorities were still doubtful because of the suspicion cast on similar products from the titanium dioxide industry. Meanwhile, the firm did not pursue its application.

5.3.2b. *Waste containing alkylated compounds*

The organic fraction considered in this example represented 0.04% of an alkaline solution containing chiefly NaOH, NaCl, and sodium aluminate from the production of alkylated compounds (Table 5.3). The model simulations dealing with the salts, and especially with aluminum hydroxide formed when in contact with seawater, are not discussed in this chapter since they are basically similar to those presented in Section 5.3.1a.

Table 5.2. Toxicity of Metals to Marine Organisms Compared to the Concentrations Yielded by the Acid-Iron Waste Discharge Simulation

Metal	Concentration in Undisturbed Seawater (mg m^{-3})	Added Concentration at the 100 mg m^{-3} Isopleth for Iron (mg m^{-3})	Lethal Concentration for Several Organisms (mg m^{-3})	Sublethal Concentration for Several Organisms (mg m^{-3})
Chromium	0.2	0.023	2,200–4,300[a] 1,500–80,000[b]	12[a]
Nickel	1.8	0.023	1,500–55,000[b]	—
Copper	12.0	0.017	80–800[b]	15[c,d]

[a] From Oshida et al., 1976 [on *Neanthes arenaceodentata* (polychaetes)].
[b] From Joergensen, 1979 (on species of fish).
[c] From Berland et al., 1977 [on *Skeletonema costatum* (diatoms)].
[d] From Mandelli, 1969 (on nine species of marine phytoplankton).

Table 5.3 Composition (% and Absolute Quantities) of the Organic Fraction in Wastewater from the Production of Alkylated Compounds

Organic Compounds	Total Organic Compounds	Concentration (mg liter^{-1})
2-methyl-6-ethylanilin + 2, 6-diethylanilin	6.8	28
O-toluidin + anilin	32.2	161
2, 6-dimethylphenol	8.6	38
2, 6-dimethylanilin	21.6	95
N-butanol	19.7	106
O-cresol	11.1	50
Total	100.0	478

The application was for about 5000 t of waste to be dumped at a rate of five operations per year. At each operation, 400 kg of toxic organic compounds would be released into the marine environment. The dumpsite indicated by the licensing authorities was located at 51° 35'N, 2°50'E (site 4 on Fig. 5.4).

The interaction considered in the dispersion equation was biological since it concerned the degradation of the organic products by heterotrophic marine microorganisms. It seemed that at least 80% of the organic matter contained in the mixture would be degraded in the marine environment, based on a study by Price et al. (1974), who studied the behavior of compounds such as butanol and phenol (both present in the waste) in seawater as well as on the observations by Pitter (1976) on a large number of compounds in freshwater. The biodegradation rate varied among the compounds listed in Table 5.3 by a factor of 10; after correcting for a seawater effect the Pitter values ranged from $(4–36) \times 10^{-4}$ h^{-1}. O-cresol displayed an average behavior and was therefore chosen to simulate an average interactive dispersion. The mathematical representation of the interaction term is very simple:

$$I_c = -kC \qquad 14$$

where k is the degradation rate, and C is the concentration of the organic compound.

The result of the computer simulation is given in Fig. 5.10. Sets of isocurves are drawn for every half-tidal period. The excursion of any given patch is important, and the overlap is negligible (provided that the dumping would be repeated after at least 12 h). As in previous examples, the effect of advection is much more

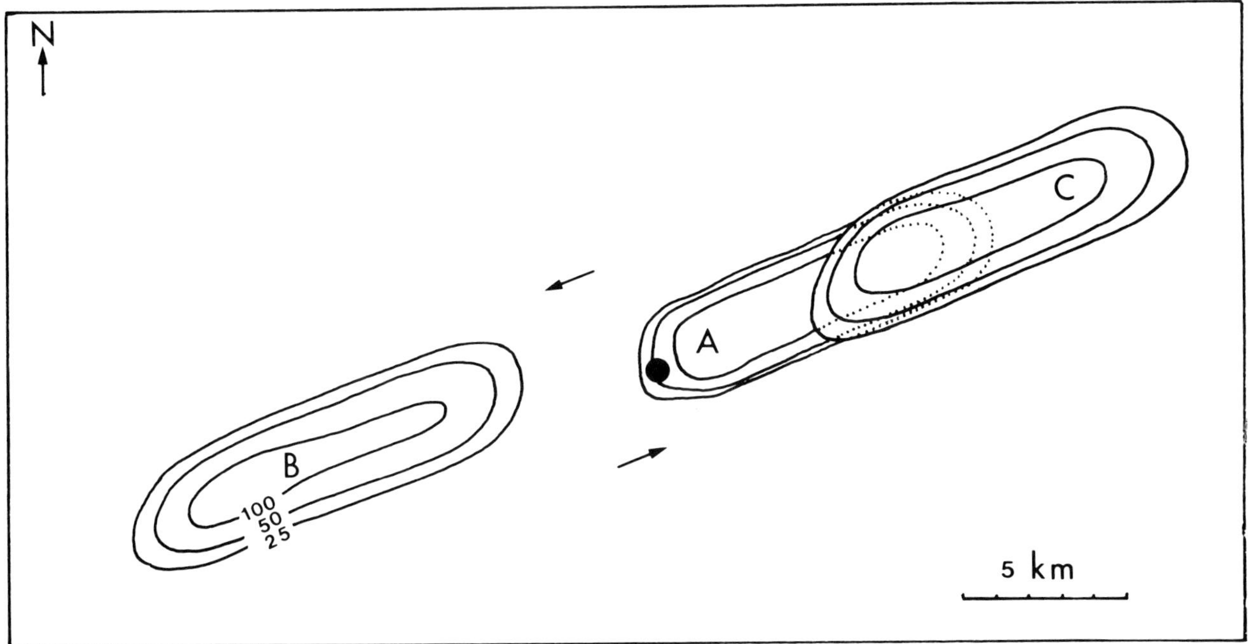

Figure 5.10. Dispersion of O-cresol after disposal of alkylated compounds at 51°35'N, 2°50'E (indicated by a black dot), as simulated with the dispersion model. The units refer to the alkylated compound concentration (mg m^{-3}). Isopleths A indicate the concentration at the end of discharge, isopleths B indicate the concentration one-half of a tidal period later, and isopleths C indicate the concentration one tidal period later.

obvious than that of diffusion, and patch spreading is a relatively slow phenomenon. In addition, the effect of biodegradation was hardly perceptible since the combined action of diffusion and biodegradation causes the concentration to decrease not more than 6.5% each hour at the center of the patch. To the extent that O-cresol is a good tracer for the other compounds listed in Table 5.3, the concentration of the organic compounds could be expected to range from 0.24 mg m^{-3} at isopleth 25 to 0.96 mg m^{-3} at isopleth 100. These concentrations are far below the lowest sublethal concentration recorded for the most toxic component of the mixture (i.e., anilin). Verschueren (1983) mentions inhibition of algal growth from a concentration of 160 mg m^{-3}. On the other hand, none of the molecules of the constituents of the mixture were likely to be accumulated by organisms, as indicated by the partition coefficients between water and octanol as calculated by the method described by Nijs and Rekker (1974).

The outcome of the application was that a permit to dump this waste was granted for a test period of one year. As with all permits, the firm was urged to find alternative land-based solutions, and it must comply with regulatory measures. In this particular case the behavior of non-organic compounds is of concern and is currently undergoing specific monitoring.

5.3.2c. *Waste from the hydroxyethylcellulose industry*

The application was for dumping a waste containing a mixture of 34% glycols and ethylcellulose and of 28% sodium acetate. About 6000 t of waste would be discharged at a rate of six operations per year. The dumping site would lie at 51°35′N, 2°40′E (site 5 on Fig. 5.4).

None of these substances occurring in the waste were highly toxic, and all were biodegradable. The biological oxygen demand of the waste at 20°C was 150 g liter^{-1}; hence, attention was focused on the effect of biodegradation of the organic waste on the oxygen content in the water. Thus, the relevant interaction in the dispersion equation was basically the same as in the case of a waste containing alkylated compounds, but, due to a reversal of priority, it was the course of oxygen depletion that was examined instead of the fate of a given organic substance.

The degradation rate constant was tentatively fixed at 20×10^{-3} h^{-1} [i.e., the rate constant of biodegrada-

tion of organic matter naturally produced in the marine environment (Billen et al., 1977)]; therefore, the depletion figures were probably overestimated. Nevertheless, the simulated oxygen deficit was very small since it amounted to 0.1% of the normal oxygen concentration in seawater.

The outcome of the application was that no dumping permit was released because it was eventually shown that an alternative land-based solution existed.

5.3.3. **Example of a Simulation with the Ecological Model**

Whenever a permanent departure from the condition normally prevailing in the marine environment is established, its effects on the functions of the ecosystem in the perturbed box of the model of the North Sea can be simulated with the ecological model. An example of this is presented in this section in an evaluation of the effects of sand and gravel extraction from the Flemish Banks off the Belgian coast.

The mining of sand and gravel from the Flemish Banks is an activity that has increased in intensity considerably in the last 10 years. The amount planned for extraction is about 5×10^{6} t y^{-1} from a rather limited area. Consequently, the Belgian authorities became worried about the possibly detrimental effects on the marine environment. The ecological effects that were particularly feared were (1) the elimination of the microbiologically active layer of the sediment, which would locally interrupt the biogeochemical cycles; and (2) a perturbation of the photosynthetic activity, due to an increase in turbidity.

A dispersion simulation showed that if the quantities of sand and gravel applied for by all firms were actually extracted, turbidity would increase by 13%, and the inorganic nitrogen flux from the sediments would decrease by 50% in a 25-km^2 area (site 6 on Fig 5.4). The ecological model was used to simulate the impact of these environmental changes. Figure 5.11 shows the normally expected time course of dissolved inorganic nitrogen, phytoplankton, and zooplankton. Figure 5.12 shows the modelled deviations from the norm, due to higher turbidity and diminished inorganic nitrogen regeneration from the bottom sediment.

During the first 20 weeks, phytoplankton growth was slowed down mostly due to increased turbidity. Hence, and notwithstanding the decreased remineralization

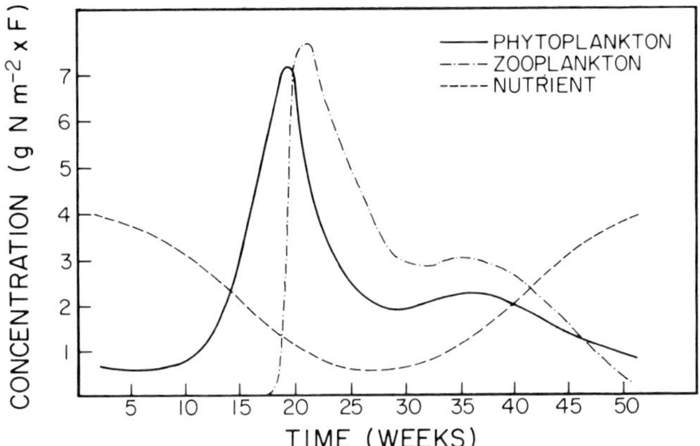

Figure 5.11. Seasonal evolution of dissolved inorganic nitrogen, phytoplankton, and zooplankton as simulated with the ecological model for a nondisturbed situation. Scaling factor F equals 1 for inorganic nitrogen, 0.1 for phytoplankton, and 0.01 for zooplankton.

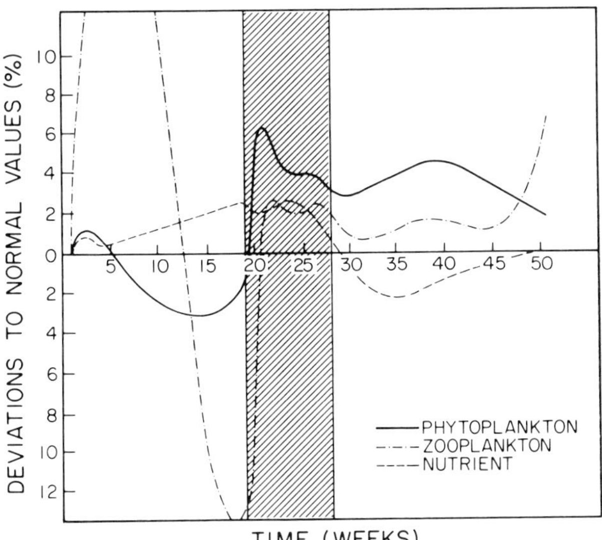

Figure 5.12. Impact of sand and gravel mining, as simulated with the ecological model. Lines refer to the deviations expressed as percent from the concentrations of phytoplankton, zooplankton, and dissolved inorganic nitrogen. The shaded area indicates a critical period.

from sediment, the nutrient concentration decreased more slowly than it did in the undisturbed situation, and zooplankton growth was delayed in the second half of this period (which corresponds to the spring bloom). The relative enhancement of phytoplankton growth in the first several weeks of the simulation and the large amplitude deviations by zooplankton to 24% over and 13% under the undisturbed situation must be consid-

ered with caution since these results involve very small stocks. In the next period of eight weeks, the decrease of phytoplankton was less pronounced than it was in the normal situation, due to lower intensity of grazing by zooplankton. In the last 24 weeks, there were less nutrients available than in the undisturbed situation, since phytoplankton were more abundant. This situation also favors zooplankton growth.

Maximal deviations from the concentrations in an undisturbed situation amounted to 2.5 and -2.3% for dissolved inorganic nitrogen (second and third period), -3.1 and 6.3% for phytoplankton (first and second period), and 24 and -13.6% for zooplankton (first period; as stated before, the large zooplankton deviations were not considered to be significant). Through its prominent influence on the availability of light for photosynthesis, turbidity appeared to be the factor most responsible for the deviations. This result was as expected. Sensitivity analysis of the ecological model showed variations of the extinction coefficient due to nonphytoplanktonic particulate matter to rank as third (after variations in the saturating parameters of the photosynthesis–light relationship and the optimal grazing rate) in influencing the output of the model.

The outcome of the study was that although the general pattern of seasonal evolution was not significantly altered by the sand and gravel extraction, deviations of 2.5–13.6% occurred during the growth period, which is an especially sensitive period of the annual cycle. Since such changes would affect at most a 25-km^2 area, it was considered unlikely that permanent damage to the entire zone would result. Consequently, the mining operations were allowed to proceed, but particular attention is devoted in monitoring programs to the biological events occurring during the period of the year during which the operations take place. Moreover, the existing decrees provide, if required, for the suspension of mining activities during the spawning season of commercially valuable fish.

5.4. DISCUSSION

Impact studies assessing the potential environmental effects of marine dumping operations conducted in Belgium during the past five years suggest the following remarks.

5.4.1. The Models and Their Results

Provided that some basic requirements could be met when carrying out the analysis (i.e., with respect to time scales and spatial scales), the modelling approach discussed in this chapter has proven useful, and management decisions could be made on the basis of the predictions it has provided. Since the dispersion model has been validated by actual experiments with rhodamine-B, it ensures a fairly good description of the short-term events following the disposal of a waste into the sea. It is therefore believed to answer adequately some of the basic questions concerning the dispersion of wastes in the vicinity of a dumpsite. More specifically, the model has been used to show whether a risk of local accumulation exists, and to compute a range of *in situ* concentrations to be compared with toxic concentrations reported in the literature.

The simulations of dispersion in waters closer to the shore and with weak residual circulation (a few centimeters per second) have consistently revealed a danger of persistence, whereas in areas further offshore and with a more intense residual circulation (up to 20 cm s^{-1}), consecutively released waste patches have shown less overlap. The isoconcentration ellipses calculated from the dispersion model for the first tidal period after dumping cover an area of about 20 km^2. Herein the concentrations are, on an average, well below the known sublethal effects for the relevant substances studied up until now. Moreover, the innermost curves (i.e., those with the more hazardous concentrations) have already started to contract during the same time (Fig. 5.7). This rules out the possibility for sublethal effects, which may possibly be associated with those higher concentrations, to expand further.

The ecological model has benefited by the findings of a multidisciplinary team of field ecologists. The multi-box coverage of the area under Belgian jurisdiction is an interesting aspect because it makes it possible to draw isopleths of concentrations of the state variables on a large spatial scale. This model has been validated by the monitoring started in 1971. As demonstrated by the sensitivity analysis and the case studies presented in this chapter, this model is, in its present state of development, driven primarily by the equation for primary production and, more specifically, by the photosynthesis–light relationship. The clearest response will therefore be obtained when relevant parameters such as light penetration, nutrient concentration, or phytoplankton physiology are significantly altered.

The available dispersion model is, at present, limited to the short-term and small-scale phenomena resulting from a discharge at sea. This situation is adequate provided that the waste degrades sufficiently rapidly. Other points needing further refinement have been identified (e.g., the hypotheses underlying sedimentation–resuspension phenomena).

Where the ecological model is concerned, it would seem from our experience that massive changes are necessary to produce demonstrable effects on the seasonal patterns of nitrogen, phytoplankton, or zooplankton. Moreover, long-term trends of ecosystem change, which are more qualitative than quantitative, remain outside the present capabilities of this model. The ecological model in its present state of development also has several of the usual shortcomings described in the literature. Stronger cohesion is needed between model elaboration and data acquisition (the problem of calibration and validation), and forced functions must be replaced as far as possible by mechanistic simulation (Kremer and Kremer, 1982); however, increasing complexity must not endanger stability (May, 1974). Precise areas of improvement have been identified as follows: (1) replacing the sediment boundary, which is acting as a source or as a sink with a set of state variables; (2) improving the modelling of heterotrophic activity in the water column (bacterial and zooplanktonic); and (3) developing the model in order to take into account the problem of persistent substances, especially with regard to the cycling of metals and other bioaccumulable substances.

5.4.2 The Environmental Management Strategy

Management decisions have been facilitated by the use both of dispersion and ecological models. In particular, the dispersion model has successfully answered two crucial questions with respect to criteria presently used by the Belgian authorities: (1) Is there a danger of local accumulation? and (2) Do the concentrations of added pollutants remain below agreed acute toxicity thresholds at all times and below the no-effect concentration, except in a reasonably small area and for a short period of time? The approach, however, has limitations. There is no agreement on criteria for defining what a ''reasonably small area'' or a ''short period of time'' must be. The same is true of the outcome of the ecological model, for which the acceptability of the deviation from the undisturbed course of evolution of the state variables must be determined. Moreover, the available dispersion model does not permit the computation of large-scale phenomena, so the problem of long-term effects in areas possibly much larger than those considered by the model remains for recurrent additions of pollutants. The criteria to be applied in the definition of such areas, referred to as receiving volumes, is still a matter of discussion.

There have been few instances in which the situation created by dumping would justify the use of a multi-compartmental and multi-box ecological model of nitrogen cycling. The computer simulations applied to this part of the North Sea can, therefore, probably only comfirm the well-known, and infamous, maxim, "the solution of pollution is dilution." In the worst cases, it has always remained possible to find a site with better dispersion characteristics and where dumping could eventually be permitted. In fact, if approximately three-fourths of the applications to dump at sea were not followed by a permit (and this proportion is even higher for the total set of applications not reviewed in this chapter), it is due more to the existence of an alternative land-based solution, or to the dissuasive effect of the longer distance to be covered at sea, than to an ascertained detrimental effect on the marine environment.

This chapter deals specifically with the well-flushed Southern Bight of the North Sea. Would there be virtually no limit to the quantities of waste that could be safely absorbed by the marine environment? The same area of the North Sea provides evidence to the contrary. For example, there are many signs of developing eutrophication (Gieskes and Kraay, 1975; Billen et al., 1977), the concentrations of metals in the biota are comparatively higher than in other areas (Oslo Commission, 1984), pollutants dumped directly into the sea must inevitably contribute to the total pollutant load. The fate of the pollutants must therefore be examined in this more global context.

5.4. CONCLUSIONS

We propose two general conclusions to the review presented in this chapter:

1. The current techniques have proven extremely useful for the preliminary assessment of direct effects of waste disposal at sea, and they will certainly be even more needed in the future.

2. Because barely measurable pollution is pervading the whole marine system, the impact assessment of long-term and large-scale effects is perhaps a more urgent challenge for the simulation modeller and the ecotoxicologist than is the assessment of the small area of seawater and the biota sacrificed with each dumping operation. Long-term dispersion models should take over from short-term models, and likewise the ecological model should be applied to the simulation of long-term effects.

Moreover, the study of sublethal effects and of bioaccumulation should in the future have priority over acute-toxicity testing, and the cycling of metals deserves as much attention as that of nutrients.

REFERENCES

Adam, Y. 1977a. Dispersion of active pollutants in a shallow sea. *Simulation*, **1**, 5–15.

Adam, Y. 1977b. Modelling the evolution of patches of pollutants in the southern North Sea. *Applied Mathematical Modelling*, **1**, 170–176.

Adam, Y. 1977c. La simulation numérique de l'évolution de polluants issus de déversements en mer du Nord. *In*: Modèles de Dispersion, Final Report of the Project Sea, 5, J. C. J. Nihoul and Y. Adam (Eds.). Ministry of Scientific Policy, Brussels, 350 pp.

Adam, Y. 1979. A higher accuracy predictor–corrector method for the simulation of sea pollution. *Applied Mathematical Modelling*, **3**, 82–88.

Baeteman, M., and W. Vyncke. 1979. Phenolic Compounds in the Water and Marine organisms off the Belgian Coast. Mededelingen van het Rijksstation voor Zeevisserij, Report 157, Rijkscentrum voor landbouwkundig Onderzoek. Ministry of Agriculture, Ghent 7 pp.

Berland, B. R., D. J. Bonin, O. J. Geurin-Ancey, V. I. Kapkov, and D. P. Arlhac. 1977. Action de métaux lourds à des doses subletales sur les caractéristiques de la croissance chez la diatomée *Skeletonema costatum. Marine Biology*, **42**, 17–30.

Billen, G., M. Bossicart, N. Daro, R. De Boever, R. De Clercq, L. De Coninck, J. Govaere, J. Hecq, C. Heip, P. Hovart, D. Janssens, C. Joiris, J. P. Mommaerts, G. Pichot, P. Polk, F. Redant, L. Thielemans, Y. Runfola, C. Van Beveren, D. Van Damme, and J. Van de Velde. 1977. Analyse van het ecosysteem van de Noordzee. *In*: Trofische Ketens en Cyclus der Nutrienten, Final Report of the Project Sea, 8, J. C. J. Nihoul and P. Polk (Eds.). Ministry of Scientific Policy, Brussels, 339 pp.

Bowden, K. F. 1965. Horizontal mixing in the sea due to a shearing current. *Journal of Fluid Mechanics*, **21**, 83–95.

Garrels, R. M., and F. T. Mackenzie. 1971. Evolution of Sedimentary Rocks. W. W. Norton and Co., New York, 397 pp.

Gieskes, W. W. C., and G. Kraay. 1975. The phytoplankton spring bloom in Dutch coastal waters of the North Sea. *Netherlands Journal of Sea Research*, **9**, 166–196.

Joergensen, S. E. (Ed.) 1979. Handbook of Environmental Data and Ecological Parameters. International Society for Ecological Modelling, Copenhagen, 1162 pp.

Kremer, J. N., and P. Kremer. 1982. A three-trophic level estuarine model: synergism of two mechanistic simulations. *Ecological Modelling*, **15**, 145–157.

Mandelli, E. F. 1969. The inhibitory effects of copper on marine phytoplankton. *Contributions in Marine Sciences*, **14**, 48–57.

May, R. M. 1974. Stability and Complexity in Model Ecosystems. Princeton University Press, Princeton, New Jersey, 265 pp.

Nihoul, J. C. J. 1972. Shear effect diffusion in open shallow seas. *Bulletin de la Société Royale des Sciences de Liège,* *41*(10), 521–526.

Nihoul, J. C. J., and F. C. Ronday. The influence of the tidal stress on the residual circulation. *Tellus,* **29**, 484–489.

Nijs, G. G., and R. F. Rekker. 1974. The concept of hydrophobic fragmental constants (f values). II. *European Journal of Medicinal Chemistry. Chimica Therapeutica,* **9**, 361–375.

Oshida, P. S., A. Mearns, D. J. Reish, and C. S. Word. 1976. The Effects of Hexavalent and Trivalent Chromium on *Neanthes arenaceodentata* (Polychaeta: Annelida). Southern California Coastal Water Research Project, El Segundo, California, 58 pp.

Oslo Commission (Ed.). 1984. Eighth Annual Report. Chameleon Press Ltd., London, 210 pp.

Paris Commission (Ed.). 1984. Sixth Annual Report. Chameleon Press Ltd., London, 115 pp.

Pichot, G. 1980. Simulation du Cycle de l'Azote à Travers l'éco-système Pélagique de la Baie Sud de la Mer du Nord. Dr. Sc. Thesis, University of Liège, Belgium, 168 pp.

Pitter, P. 1976. Determination of biological degradability of organic substances. *Water Research,* **10**, 231–236.

Price, K. S., G. T. Waggy, and R. A. Conway. 1974. Brine shrimp bioassay and seawater BOD of petrochemicals. *Journal of Water Pollution Control Federation,* **46**, 63–75.

Radach, G., and E. Maier-Reimer. 1975. The vertical structure of phytoplankton growth dynamics: a mathematical model. Sixth Liège Colloquium on Ocean Hydrodynamics, J. C. J. Nihoul (Ed.). *Mémoires de la Société Royale des Sciences de Liège,* **6**(VII), 113–146.

Riley, G. A. 1956. Oceanography of Long Island Sound 1952–1954. IX: Production and utilization of organic matter. *Bulletin of the Bingham Oceanography College,* **15**, 324–344.

Ronday, F. 1976. Modèles Hydrodynamiques de la Mer du Nord. Détermination des Circulations Transitoires et Résiduelles. Dr. Sc. Thesis, University of Liège, Belgium, 288 pp.

Talbot, J. W., and G. A. Talbot. 1974. Diffusion in shallow seas and English coastal and estuarine waters. *In*: Physical Processes Responsible for Dispersal of Pollutants in the Sea. *Rapports et Procès–Verbaux des Réunions du Conseil International pour l'Exploration de la Mer,* **167**, 43–70.

Verschueren, K. 1983. Handbook of Environmental Data on Organic Chemicals. Van Nostrand-Reinhold, New York, 1310 pp.

Vollenweider, R. A. 1965. Calculation models of photosynthesis-depth curves and some implications regarding day rate estimates in primary production measurements. *In*: Primary Productivity in Aquatic Environments, Memorie Istituto Italiano di Idrobiologia, 18(suppl.), C. R. Goldman (Ed.). University of California Press, Berkeley, pp. 425–457.

Chapter 6

Some Useful Methods of Assessing Near-Field Suspended-Sediment Dispersion

Andrey F. Lyashenko and Andrey A. Goncharov

State Committee for Hydrometeorology and Control of the Natural Environment of the U.S.S.R.
U.S.S.R. State Institute of Oceanography
Moscow, Union of the Soviet Socialist Republics

ABSTRACT

When estimating dispersion in the context of dumping dredged material, there is frequently a lack of detailed and reliable information concerning hydrodynamic conditions of the ambient water. In such cases, simplified analytical methods requiring a minimum amount of data should be used. The use of a simple (from the point of view of turbulent diffusion and water dynamics) model is justified when the usually small time and space scales of dispersion of the dumped material are taken into account.

6.1.   INTRODUCTION

The disposal of dredged material at sea increases the concentration of contaminants in the form of suspended material in the water at dumpsites. To estimate the possible effects of these suspensions on an ecosystem, it is necessary to estimate the spatial distribution and time variability of suspended-material concentrations with regard to different disposal methods under a variety of conditions. For these purposes, analytical, numerical, and statistical methods may be used. Numerical calculations based on hydrodynamic information and statistical methods are, however, rather cumbersome, and boundary conditions, diffusion coefficients, current spectra, and other necessary information is usually unavailable for most dumpsites. On the other hand, appropriate estimates can be made in a number of cases by using simple analytical methods. In such cases it does not take long to obtain the necessary source information and to do the calculations. As a result, these methods may be used in the operative practice of dumping.

6.2.   MODEL FORMULATION AND DESCRIPTION OF THE COMPUTATION METHODS

First, we shall consider an analysis of the dynamics of the suspended material by approximating the dumping process in terms of an instantaneous point source on a thin slab. Such an approximation may be applicable, for example, in the case of an instantaneous dump when the zone of release is relatively small. An expression based on a modified solution of the well-known equation for radially symmetrical diffusion is proposed for use as one of the computation methods. The basic equation is

$$\frac{\partial C}{\partial t} = \frac{1}{r} \frac{\partial}{\partial r} (\omega r) r \frac{\partial C}{\partial r} \qquad 1$$

where C is the concentration of the suspended matter, r is the distance from the center of a patch of suspended sediment, ω is the diffusion velocity, and t the time elapsed after the waste has been released. Use of this equation assumes that (1) suspended material is vertically uniform within a thin slab, and (2) horizontal turbulent diffusion is quantified by a diffusion velocity [by analogy with Joseph and Sendner (1958)].

The solution of equation 1 expressed as the depth-averaged concentration of suspended material ($\overline{C}$) and including particle loss of the seafloor may be written as

$$\overline{C}(r,t) = \frac{q}{2\pi(\omega\tau)^2}e^{-r/\omega\tau}e^{-Wt/D} \qquad 2$$

where q is the quantity of mass released per unit depth, D is the thickness of the layer containing suspended sediment; $\tau = t_0 + t$ where t_0 is the time required for the initial distribution of sediment to reach a Gaussian form (assuming a Gaussian form whereby the variance of the distribution, τ^2, equals $\omega^2 t^2$, it follows that at t_0, $\tau^2 = r_0^2/3$ where r_0 is the initial radius of the dump and so $t = r_0/\omega\sqrt{3}$); and W is the effective settling velocity, which can be found from

$$W = \left(\int_0^D \sum w_i C_i dz\right)\bigg/\left(\int_0^D \sum C_i dz\right) \qquad 3$$

where C_i is the concentration of the ith size range characterized by the settling velocity w_i.

It should be noted that equation 2 allows a more satisfactory approximation of the effects of particle-fall velocity and horizontal diffusion if particles are transported in a nearly uniform suspension with depth. Such conditions will be satisfied, for example, for a thin surface layer where active mixing by wind takes place, or for the stirring produced by a tidal stream in shallow water.

To get representative characteristics of suspended-sediment dispersion we use the following parameters:

1. The time of existence (t_e) of concentrations at or above a chosen level (C_f) is the solution of the equation

$$\frac{W}{D}t_e + 2\ln\tau_e - \ln\left(\frac{q}{2\pi\omega^2\overline{C}_f}\right) = 0 \qquad 4$$

for any fixed $\overline{C}_f$ where $\tau_e = t_0 + t_e$. Equation 4 is simply equation 2 when $r = 0$.

2. The time interval (t_m) required for any chosen isoline whose distance from the center of the "patch" is $r_{\overline{C}_f}$ to reach its maximum horizontal extent is defined by the equation

$$\frac{2W}{D}t_m + 2\ln\tau_m - \ln\left(\frac{q}{2\pi\omega^2\overline{C}_f}\right) + 2 = 0 \qquad 5$$

where $\tau_m = t_0 + t_m$. Equation 5 may be obtained from equation 2 for $\overline{C} = \overline{C}_f$ by using the natural condition

$$\frac{\partial\overline{C}_f}{\partial t} = 0 \qquad \text{at} \quad r = r_{\max}.$$

3. The maximum possible extent of isolines from the center (r_m) can be calculated from the relation

$$r_m = \ln\left(\frac{q}{2\pi\omega^2\overline{C}_f} - 2\ln\tau_m - \frac{wt_m}{D}\right)\omega\tau_m \qquad 6$$

which can be obtained from equation 2 for $t = t_m$. These parameters are insensitive to variation of the shape of the patch, so any distance r corresponds to the equivalent radius.

Continuous discharge from a pipeline is another frequently used method of dredged-material disposal. Wilson (1979) developed a model to describe this type of disposal. From a minimum of basic information, this model allows the calculation of basic characteristics of the suspended-sediment plume formed during disposal. To make the calculations, information is needed on the strength of the source of the suspended sediment, the average diameter or settling velocity, the water depth at the dumpsite, the average current velocity (u), and the average diffusion velocity. Wilson's model is based on solution of the advective-diffusion equation for a continuous discharge of material along a vertical line through the water column. The primary concern is with plumes that develop in very shallow water; as a consequence, only vertically averaged concentrations are examined. Under these conditions the solution takes the form

$$\overline{C}(x,y,t) = \frac{q}{\pi\omega^2 t}\int_0^D \frac{1}{t'^2} e^{-(x - u't'/\omega t')^2} e^{-(y/\omega t')} e^{-(\overline{\gamma})} \, dt' \qquad 7$$

In this case, $\overline{C} = 1/D\int_0^D \sum C_i \, dz$ where C_i is the particle concentration of the ith size range characterized by the settling velocity w_i, and $\overline{\gamma} = Wt/D$. Equation 7 is derived assuming that the plume is developed under the action of current parallel to the x-axis with constant velocity, u. It is important to realize that t is the time elapsed since the release of a waste, so $\overline{\gamma}$ is the ratio of the plume age to settling time. Thus, equation 7 can be solved for a given strength of release and water depth at the dumpsite after the two parameters u/W and $\overline{\gamma}$ have been specified. All calculations using equation 7 can be made quickly with the help of corresponding diagrams presented by Wilson (1979).

In some cases it is important to know how much of the discharged particle mass will reach the seafloor within a distance L of the dumpsite. For particles with relatively large fall velocities (fine sand or larger), this assessment can be easily made if the frequency distribution of fall velocities is known. For example, if it is known that f_0 is the fraction of total particles with fall velocities less than V_0 where

$$V_0 = Hu/L \qquad 8$$

with u being the depth-averaged horizontal velocity, and H the depth at distance L, then the fraction F of total particles falling to the bottom within the distance L is simply

$$F = 1 - f_0 \qquad 9$$

if the point-source approximation is used and

$$F = (1 - f_0) + \frac{1}{V_0}\int_0^{f_0} V_s \, df \qquad 10$$

if the initial distribution with depth is uniform, where f is the size fraction distribution for constituents having settling velocities $V_s \leqslant V_0$.

6.3. RESULTS OF CALCULATIONS AND DISCUSSION

The possibility of using equation 2 as well as the characteristics t_e, t_m, and r_m for an adequate description of suspended-sediment distribution was checked with data on disposals of dredged material at a number of sites (U.S. Environmental Protection Agency, 1980). Observations of the solids remaining in the water column immediately after dumping (Schubel and Carter, 1978; Bokuniewicz and Gordon, 1980) show an upper limit of 5% to be a conservatively high estimate. Therefore, in the sample calculations, a value of q was chosen to be 5% of the total amount of dredged materials dumped. The initial radius of suspended-sediment patch r_0 was assumed to equal about half the width of the dumping vessel. The vessel was 17 m wide, so r_0 was taken to be 10 m. The parameter D was selected to be 10 cm, which was the thickness of a layer accessible for visual observations. The evaluation of effective settling velocities from equation 3 gave a characteristic value of $W \sim 1.5 \times 10^{-2}$ cm s^{-1}. Estimates made for parameters t_e, t_m, and r_m with these values showed that (1) a plume with suspended-sediment concentrations ranging from 10 to 30 mg liter^{-1} should be observed in the upper layer from 30 to 60 min after discharge, with a maximum radius ranging from 40 to 60 m for $\omega = 0.5$ cm s^{-1} and $\omega = 1.0$ cm s^{-1}, respectively; and (2) a large surface patch of high turbidity can be created by the fine-sized constituents ($W \sim 1 \times 10^{-3}$ cm s^{-1}) of the dumped material. Despite their small amount, these constituents dispersing in a thin upper layer can reach concentrations of about 10 mg liter^{-1} in the top 10 cm of the upper layer over the area of approximately 50×10^3 m^2 between 3 and 7 h after the disposal operation. This is a much greater area than is calculated by using an average grain size for the suspended material.

As an example of the application of Wilson's (1979) model for a continuous discharge, data on dredged-material dumping in the shallow Dnestrovsky estuary were used (Stepanov et al., 1981). Dumping operations were carried out by a hydraulic dredge with a capacity of 0.8 m^3 s^{-1}. The dredged material consisted of silt and silt–sand. Dumping of sediment with a solid mass: water mass ratio of 1 : 6 and 1 : 7 was made at the water surface. The average depth at the dumpsite was between 1.2 and 1.5 m, and background concentrations of suspended sediment were between 15 and 35 mg liter^{-1} before the dumping operations began. During the disposal operation, the current velocity was 10 cm s^{-1}. Suspended-sediment concentrations between 3 and 5 g liter^{-1} were measured at a distance of 25 m from the pipeline head.

In general, the results of the calculations made using the diagrams of Wilson (1979) coincided with the measurements of turbidity at the surface and at the bottom

Table 6.1. Observed and Calculated Distributions of Suspended-Sediment Concentrations along the Plume Axis[a] at the Surface and along the Seafloor

Concentration (mg liter^{-1})	Distance (m)			
	Surface		Bottom	
	Observed	Calculated	Observed	Calculated
5000	25	30	—	—
750	160	140	130	150
450	200	220	250	230
250	300	290	310	330
70	360	370	400	420
55	390	390	480	450
45	410	400	—	—

[a] A dash indicates the absence of data.

(Table 6.1). Calculations were made using a value of ω/u equal to 0.1, 0.2, and 1.0, corresponding to a diffusion velocity of 1 cm s^{-1} and current velocities in the range of 1 to 10 cm s^{-1}. The average settling velocity was estimated to be about 7.5×10^{-3} cm s^{-1}, corresponding to a sediment particle diameter of about 1×10^{-3} cm. Calculations showed that in the case of a surface plume, it was appropriate to use values corresponding to the most fine-sized constituents of the dumped material instead of the mean values W; in our case it was the silt fraction with a settling velocity of 6.5×10^{-4} cm s^{-1}. The best agreement with field data was obtained for values of $\omega/u = 0.2$ (where $\omega = 1$ cm s^{-1} and $u = 5$ cm s^{-1}), and $\gamma = 0.8$ for the surface plume ($D = 10$ cm), and $\omega/u = 0.2$, and $\gamma = 1.0$ for the bottom plume ($D = 1$ m). Comparison between the measured and the calculated distributions of suspended-sediment concentrations along the plume axis for this case is shown in Table 6.1.

The use of more detailed information of hydrology and water dynamics in order to get more exact data on sediment distribution seemed to be inappropriate. The model is sensitive to the values assigned to describe the mechanical and sedimentary characteristics of dredged material, and such characteristics are frequently poorly known.

6.4. CONCLUSIONS

In order to estimate suspended-material distributions after dumping, simple analytical methods requiring minimum input data can be used. These methods allow the calculation of the suspended-sediment concentration both for the cases of overboard disposal of dredge spoil through a pipeline in shallow water and for an instantaneous disposal. The methods also allow estimates of the total content of suspended material that remains in the water and that settles to the bottom at certain distances from the dumpsite. The results of the calculations coincide well with the field data on turbidity structures at the surface and at the bottom of a body of water.

REFERENCES

Bokuniewicz, H. J., and R. B. Gordon. 1980. Deposition of dredged sediment at open water sites. *Estuarine and Coastal Marine Science*, **10**, 289–305.

Joseph, J., and H. Sendner. 1958. Uber die horisontale Diffusion in Meere. *Deutsche Hydrographische Zeitschrift*, **11**, 49–77.

Schubel, J. R., and H. H. Carter. 1978. Field Investigation of the Nature, Degree and Extent of Turbidity Generated by Open-Water Pipeline Disposal Operations. Technical Report D-78-30, U.S. Army Corps of Engineers, Waterways Experimental Station, Vicksburg, Mississippi, 115 pp.

Stepanov, V. N., P. A. Cryganovsky, and V. V. Abodovsky. 1981. Experimental investigations of diffusion of fine-size sediment in the sea. *Proceedings of the State Oceanographical Institute*, **158**, 60–65.

U.S. Environmental Protection Agency. 1980. Final Environmental Impact Statement (EIS) for Hawaii Dredged Material Disposal Sites Designation. U.S. Environmental Protection Agency Oil and Special Materials Control Division, Marine Protection Branch, Washington, D.C., 125 pp.

Wilson, R. E. 1979. A model for the estimation of the concentrations and spatial extent of suspended sediment plume. *Estuarine and Coastal Marine Science*, **9**, 65–78.

Chapter 7

Analysis of Mixing and Dispersion of Industrial Wastes Dumped into the Deep Sea

Michael Devine

National Ocean Service
U.S. National Oceanic and Atmospheric Administration
Rockville, Maryland

ABSTRACT

Wake mixing is effective in diluting liquid wastes by a factor of 5×10^3 or more. Initial oceanic dispersion is by horizontal eddy diffusion in the deep ocean, but shear dispersion becomes dominant after several hours. Large particles sink rapidly out of the surface layer, but no quantitative analysis of the process can be given yet. Medium-term processes dilute liquid wastes by a factor of 10^5 within 1–2 d. Long-term processes dilute the wastes still further, but these processes may be slow. Most particulate matter settles over an area of a few thousand square kilometers on the deep-ocean bottom within a few weeks. Because of the limited knowledge of what happens to liquid waste over the long term and of how particles move through the water column and settle onto the bottom of the sea, a conservative approach is recommended in treating the subjects of long-term dilution and settling of particulate matter.

Long-term tracking of liquid wastes, accurate modelling, quantitative measurements of particle coagulation and settling, and some bottom sampling and analysis would be necessary to refine or reduce these factors.

7.1. INTRODUCTION

The U.S. National Oceanographic and Atmospheric Administration's (NOAA) Ocean Dumping Program has carried out observations and measurements of industrial wastes dumped from a barge at three deep-ocean sites off the coasts of the United States. These field experiments have been carried out since 1976 under a wide variety of climatic conditions, and they have measured the behavior of dumped wastes from just after disposal until as much as 3 d later, although most experimental observations have lasted about 12–36 h. The wastes themselves are a complex mixture of dissolved and suspended constituents, and many of them flocculate when mixed with seawater, forming large, porous aggregates in the upper layer of the ocean. From the point of view of managing the impact of dumping, the times during which a waste is of concern range from the first few hours after dumping to the much longer period over which the waste dilutes and settles. It is relatively easy to monitor waste dilution and dispersion for the first few hours after disposal, although no one has been able to describe adequately the behavior of the particulate matter during this period, which is when acute effects of dumping can occur. The behavior of wastes over longer periods, during which the particulate matter settles through the water column and when chronic effects might be important, is less well understood.

Dumping wastes into the deep ocean at three sites [Deepwater Dumpsite-106 (DWD-106), the Western Gulf of Mexico site, and the Puerto Rico site] has been extensively discussed (Bisagni and Kester, 1981; Ichiye et al., 1981; Schwab et al., 1981), although only DWD-106 is still in use. This chapter discusses what has been learned about physical dilution and dispersion of wastes after several years of experiments and modelling, and how NOAA scientists are trying to integrate this knowledge into an approach for managing the disposal of wastes into the ocean. Although the results of individual studies vary greatly, several factors appear to be important in all cases. The U.S. Ocean Dumping Program does not require a complete, analytical description of the dilution and dispersion processes. Instead, it is necessary to know when the concentrations of a waste are reduced to levels that can be considered innocuous. The program must also define carefully terms such as *dilution* and *dispersion* because in some waste mixtures almost the entire concentration of some contaminants may be in particulate form (Presley et al., 1981). These results must be correlated with biological and chemical information to determine if present dumping practices are acceptable, what the safety margin is, and how much more dumping activity could be accommodated. Although the volume of dumped wastes has declined in recent years, these questions are particularly important in the United States, where oceanic dumping is once again being considered for many types of wastes (Swanson and Devine, 1982).

In this chapter, discharged wastes are considered as a two-component system, liquid and suspended solids, and four time scales are considered: mixing in the wake of the dumping vessel, initial dispersion (up to a few hours after disposal), medium-term processes (from a few hours up to about 2 d after disposal), and long-term processes (> 2 d). Only industrial wastes that are primarily liquid are discussed; suspensions such as coal waste constitute a separate problem.

Dispersion of the liquid component of a waste consists of active dispersion from a barge, followed by oceanic dispersion over small to medium scales. The liquid quickly becomes neutrally buoyant and therefore tends to remain in the upper water column (Csanady, 1981). The particulate component of the waste also undergoes active dispersion in the wake of the barge, but particulate matter may also be generated by precipitation (Gibbs, 1982a). Particulate matter may undergo coagulation in the turbulent wake of the barge, making the particulate fraction and size distribution of the waste–seawater mixture quite different than would be

expected from direct dilution of the waste. After initial mixing, the particulate matter continues to coagulate due to oceanic shear and mixing, but at a slower rate (Hunt, 1982a). Initially, the suspended solids mix laterally with the liquid component, but eventually they sink out of the upper water column at a rate controlled by oceanic conditions, particle size, concentration, and density (Hunt, 1982b). The rate at which the concentration of particles in the upper water column is decreased by sinking and the relative concentrations of liquid and particulate wastes are among the most important questions relating to waste dispersion.

7.2. FOUR TIME SCALES CONSIDERED IN WASTE DISPERSION

7.2.1. Wake Mixing

When wastes are dumped from a moving barge, the rate of initial dilution is controlled mainly by the speed and size of the barge and by the rate and manner of dumping. The wastes are usually diluted so rapidly that the density of the waste is not important, and neutral buoyancy is quickly attained (Csanady, 1981). An exception to this may occur when the wastes contain high amounts of particulate matter and act to some extent as a suspension rather than as a dispersing liquid. Observations of sewage sludge (Orr and Baxter, 1983) indicated that particles sink rapidly to depths of 60–100 m. Positive or negative buoyancy of the waste may also become important at high discharge rates.

At DWD-106, in the Gulf of Mexico, and off Puerto Rico, mean dilutions of $(5-10) \times 10^3$ were rapidly reached. Figure 7.1 compares calculated values for initial dilution (D_0) as functions of mixing area (A) and inverse spreading area (U_B/R, where U_B is barge speed and R is the dumping rate) for each of these sites by using the relation (Csanady, 1981)

$$D_0 = A U_B / R \qquad\qquad 1$$

Because dumping practices and initial dilutions changed considerably with time, four examples are shown for DWD-106; the more recent dumps have the highest dilutions (Devine et al., 1981). Actual dilutions are often quite patchy for several minutes after dumping (Ichiye and Carnes, 1981). Dilution varies directly with mixing area, which can be approximated by 2.5 times the barge width multiplied by the mixing depth (H), about 10 m (Csanady, 1981). Dilution also varies

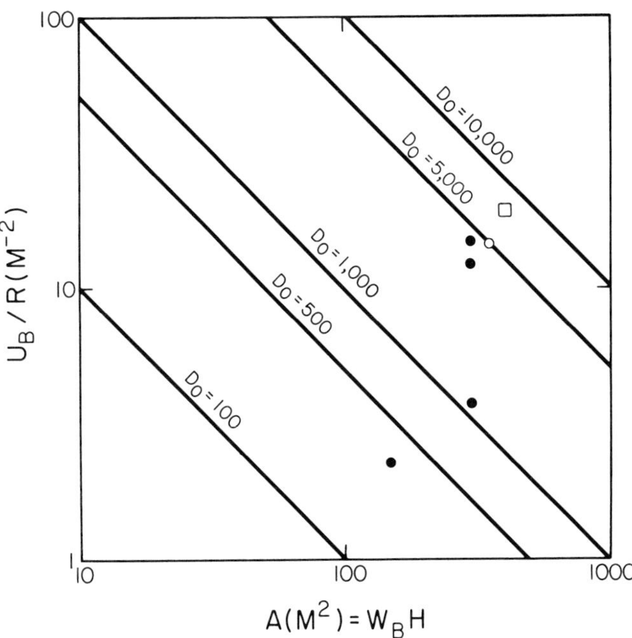

Figure 7.1. Calculated initial dilution (D_0) of dumped industrial waste as a function of mixing area (A) and the inverse of spreading area (U_B/R). A is wake width (W_B) times mixing depth (H), both given in meters; barge speed (U_B) is given in m s^{-1}; and the dumping rate is given in m^3 s^{-1}. Closed circles denote values obtained at DWD-106, the open circle denotes those for Puerto Rico site, and an open square denotes that for the Gulf of Mexico site.

directly with the inverse spreading area, which is the barge speed (U_B, expressed as m s^{-1}) divided by the dumping rate (R, expressed as m^3 s^{-1}). For moderate barge speeds ($U_B \geq 2.5$ m s^{-1}) and dumping rates ($R \leq 0.2$ m^3 s^{-1}), quite high initial dilutions are achieved. For the two cases of markedly lower initial dilution at DWD-106 (Fig. 7.1), slower barge speeds and relatively rapid dumping contribute to these low dilutions (Kohn and Rowe, 1981). The lowest dilution shown in Fig. 7.1 was in the case of a slightly buoyant waste, for which the observed mixing depth was only 5 m.

Most dumped wastes contain several percent solids, sometimes as a result of precipitation (Kester et al., 1981), although the particle concentration and size distribution have not been quantitatively measured in the ocean. Coagulation of particles can generate flocs with particle sizes much greater in the wake of the barge than in the barge itself (Hunt, 1982a). Inhomogeneities due to settling during transport while in the barge can also

be important. The Stokes settling rate of uncoagulated particles with a density of 2.6 g cm^{-3} averaging 5 μm in diameter in the ocean is about 2.5×10^{-3} cm s^{-1}, or just over 2 m d^{-1}, although under actual oceanic conditions, settling velocities of such small particles may be several times higher (Chase, 1979; Hawley, 1982). Coagulation into 50-μm particles, such as has been observed experimentally (Gibbs, 1982a), would produce settling rates of 0.25 cm s^{-1}, and particles would remain in the upper 10 m of the ocean for only about an hour. The mechanism that dominates coagulation of particles or floc in this 5- to 50-μm size range is turbulent shear coagulation (Saffman and Turner, 1956), the process by which turbulent velocity gradients displace sinking particles and result in collision probabilities that are much higher than those in a passive system. Large particles can be produced by coagulation of flocs (Gibbs, 1982a), but these coagulated flocs are much more porous and buoyant than uncoagulated solid particles. As a result, a 1000-μm floc will settle only as fast as a 30-μm Stokes particle (Gibbs, 1982b). Coagulation by differential settling, in which faster falling particles overtake and adhere to smaller, more slowly settling particles, is the dominant coagulation mechanism for solid particles over 100 μm in diameter (Friedlander, 1977) and may also be important for porous, buoyant particles in this size range.

Although the zone of initial mixing is complex in terms of turbulence structure, salinity, particle concentration, and coagulation processes, it is necessary to have some quantitative understanding of this region in order to evaluate its effect on particle coagulation and breakup. Although estimates can be given for dilution of the liquid component in the wake of a barge, no estimate beyond an assumption of being uniformly distributed with the liquid phase can be given for suspended particulate matter; furthermore, where flocculation occurs, it is difficult to give even approximate concentrations. Values reported for the suspended-solid content of wastes dumped at DWD-106 range from 0.2 to 7.6% (Murray, 1977); however, it is possible that there are localized regions of high particle concentration and that these regions would have a much more rapid coagulation rate than would be expected from mean concentrations measured in a large volume of fluid (Hunt, 1982b). Most of the particle aggregation may occur in this zone of initial mixing, where the turbulence of the fluid is much higher than is normally encountered in oceanic waters. Low ambient turbulence and low particle concentration after initial dilution may greatly decrease the importance of coagulation at later times.

7.2.2. Initial Oceanic Dispersion

After dispersion by the wake of the barge, wastes dumped in a line will be dispersed initially by small-scale horizontal eddies and later primarily by horizontal shear dispersion (Taylor, 1954; Bowden, 1965). Horizontal shear dispersion is the process by which a vertical shear in the horizontal velocity interacts with vertical turbulence to spread a waste horizontally and mix it vertically, resulting in enhanced horizontal dispersion. The initial stage of passive oceanic dispersion, during which conventional small-scale, horizontal turbulent diffusion dominates will usually last from a few minutes to several hours after disposal. An initial period is required for wake turbulence to reduce initially large concentration gradients; oceanic conditions will then determine how long turbulent diffusion dominates.

The passive-dispersion stage of waste dilution has been extensively observed in the past few years under a wide variety of oceanographic and meteorological conditions. Dilution has been calculated from concentrations of dyes and natural tracers such as iron (Kester et al., 1981; Kohn and Rowe, 1981). Ships, aircraft, and satellites have also been used to observe and study waste dispersion (Ichiye and Carnes, 1981; Ohlhorst, 1981). The presence of particulate matter has been detected by acoustic backscattering (Proni and Hansen, 1981) and transmissometry (Schwab et al., 1981); however, neither instrument system of detection gives a quantitative estimate of particle concentration.

It can be considered that waste disperses passively after being dumped in the form of a uniformly mixed slab of width W and depth H. Parameters derived from point-source diffusion concepts (Csanady, 1973) can then be employed to elucidate dispersion. By definition

$$\sigma_{xk} = \sqrt{2k_x t} \qquad\qquad 2$$

where σ_{xk} is the standard deviation of a Gaussian plume (in cm), t is time (in s), and k_x is the coefficient of horizontal eddy diffusion (in cm^2 s^{-1}). However, the initial distribution is not Gaussian, and, as the width of the uniformly mixed slab doubles, the maximum concentration of waste decreases only slightly. For a k_x of 500 cm^2 s^{-1}, a rather high diffusivity (Table 7.1), maximum concentrations decrease by only about 20% in the first 25 min. By the time the width has tripled, the assumption of a Gaussian plume is reasonable. The plume width, which defines the plume limit by approximately 5% of maximum-concentration line, is taken as

Table 7.1. Horizontal Diffusion Coefficient (k_x) for the Early Stage of Dispersion, Time of Transition to Dominance by Shear Dispersion (t_ω), and Diffusion Velocity (ω) for Six Selected Cases

Location	k_x (cm^2 s^{-1})	t_ω (s)	ω (cm s^{-1})
DWD-106[a]	225	1.1×10^4	0.20
DWD-106[b]	275	2.5×10^4	0.14
DWD-106[c]	—	—	0.15
DWD-106[d]	—	—	0.54
Gulf of Mexico[e]	225	—	—
Puerto Rico[f]	325	1.1×10^4	0.24

[a] Average of two cases. From Kohn and Rowe, 1981.
[b] From Csanady, 1981.
[c] Average of two satellite observations 25–40 h after dumping began. From Ohlhorst, 1981.
[d] Average of four satellite observations 10.5–15.5 h after dumping began. From Ohlhorst, 1981.
[e] Early dispersion only. From Proni et al., 1981.
[f] Average of four observations. From Ichiye (unpublished report to NOAA).

$$W = 5\sigma_{xk} \qquad\qquad 3$$

It would take 8 h for eddy diffusion with a k_x of 500 cm s^{-1} to change the width of a 100-m-wide plume by 50 m because the rate at which eddy diffusion controls the growth of the plume width is inversely proportional to W (equations 2 and 3). By contrast, a vertical velocity shear of 1 cm s^{-1} over the upper 10 m of the ocean (0.001 s^{-1}) would, in about 1.4 h, spread particles horizontally at the surface from those at a depth of 10 m by a distance of 50 m. At some point in time, therefore, shear diffusion should be greater than eddy diffusion.

There are several time constants that are important in the initial stage of plume spreading. The time for horizontally dispersing material to diffuse over a given depth (t_h) is important for the development of shear dispersion. For example, for vertical diffusion that would spread material over a 10-m layer of water, assuming a Gaussian form in which the vertical diffusion coefficient k_z is related to the variance by

$$\sigma_z = \sqrt{2k_z t} \qquad\qquad 4$$

requires t_h to equal $2 \times 10^4/k_z$ seconds. Except for conditions of very little vertical diffusion, this will occur in well under an hour (often less than 10 min) and will not be a limiting factor in the development of shear dispersion.

The processes of small-scale horizontal dispersion and shear dispersion occur simultaneously. If the plume has spread out to a few tens of meters or more, then standard deviations for both processes can be defined as

$$\sigma_{xk} = \sqrt{2k_x t} \qquad \text{5a}$$

$$\sigma_x = \omega t \qquad \text{5b}$$

where ω is the diffusion velocity. The physical significance of ω is discussed below. For small times and a likely k_x, the rate of initial spreading will be dominated by σ_{xk}.

The widths of spreading as determined by each process alone will be equal when

$$t = 2k_x/\omega^2 \qquad \text{6}$$

At this time, shear diffusion becomes the dominant factor determining the width of the plume. For eddy diffusion coefficients of more than 100 cm^2 s^{-1} and diffusion velocities less than 1 cm s^{-1}, this can occur anywhere from a few minutes to many hours after dumping. For most conditions eddy diffusion will be dominant for several hours (Table 7.1).

During this period of initial mixing, particulate matter will be dispersed passively with the fluid and will also sink at a rate controlled by size, density, and oceanic conditions. In fact, concentrations of particles on the presumed density microstructure as well as on the thermocline have been observed acoustically (Orr and Baxter, 1983), but no quantitative analysis can be given. The shear-coagulation model by Hunt (1980) predicts that areas of low shear and turbulence, such as occur just above the thermocline, should be regions in which small particles have less tendency to coagulate, and therefore remain slowly sinking. Again, no quantitative analysis can as yet be given. Shear coagulation is of some concern because, although particle sinking in general should make concentrations of particles in contaminants decrease more rapidly in the water column

than do concentrations of pure fluids, such a concentration of particles into a small vertical region could produce locally and temporarily higher concentrations of particles than would be deduced from fluid-dispersion considerations.

7.2.3. Medium-Term Dispersion

Medium-term dispersion occurs during the period when shear dispersion begins to dominate until the time that present methods allow no further measurements (<24 h in most cases). Minimum dilution measured either directly from dye or tracer concentrations or from plume spread observed from aircraft or satellites can be taken as a measure of dispersion. If the plume has at least tripled in width, the minimum-dilution factor D can be converted to width as defined by the 5% concentration line by

$$W/W_0 = 2D/D_0 \qquad \text{7}$$

where W_0 and D_0 are initial width and dilution factors, respectively (Csanady, 1973).

Measurements of minimum dilution are assumed to be for a random cross-section of the plume; direct photographs of plume width can then be related to measured dilution by taking a mean width. Tables 7.1 and 7.2 show eddy diffusion coefficients, diffusion velocities, plume width, and dilution for a number of cases at the Puerto Rico site and at DWD-106. The single observation from the Gulf of Mexico site is only of early eddy diffusion, but it fits with the other observations. The values deduced for the horizontal eddy diffusivity, the dominant process for about 3–7 h, are quite small (from 225 to 325 cm^2 s^{-1}). The values for the diffusion velocity are about 0.14–0.54 cm s^{-1}, giving deduced spreads of 590–2300 m after 24 h. In most cases, actual observations were for less than 24 h, and the value of the diffusion velocity was used to extrapolate to a width at 24 h. Since plume spreading is associated with the different rates of horizontal movement of the top and the bottom of the plume, diffusion velocity might be expected to be related directly to the vertical shear in horizontal velocity. However, if plume spread were considered simply as given by the vertical gradient in velocity, then for a plume 10 m deep, equations 3 and 5b give

$$V_0 - V_{10} = \Delta W/\Delta t = 5\omega \qquad \text{8}$$

Table 7.2. Plume Widths (m) after 1 (W_1), 4 (W_4), and 24 (W_{24}) h

Location	W_1	W_4	W_{24}^a	Dilution Factor[b] after 24 h
DWD-106[c]	65	105	890[a]	10,000
DWD-106[d]	80	130	590[a]	45,000
DWD-106[e]	—	110	650	50,000[i]
DWD-106[f]	—	380	2,300[a]	170,000[i]
Gulf of Mexico[g]	85	—	—	—
Puerto Rico[h]	75	180	1,100	80,000

[a] Calculated from equation 5b for cases where observations did not extend to 24 h.

[b] Calculated from equation 7.

[c] Average of two cases. From Kohn and Rowe, 1981.

[d] From Csanady, 1981.

[e] Average of two satellite observations 25–40 h after dumping began. From Ohlhorst, 1981.

[f] Average of four satellite observations 10.5–15.5 h after dumping began. From Ohlhorst, 1981.

[g] Early dispersion only. From Proni et al., 1981.

[h] Average of four observations. From Ichiye (unpublished report to NOAA).

[i] Assuming $D_0 = 4500$ and $W_0 = 30$ m.

where V_0 is the velocity of the plume at the surface of the ocean, and V_{10} is the velocity of the plume at 10-m depth. In this case, much higher diffusion velocities would be expected in the upper water column than those deduced from observations of plume spreading. With a wind speed greater than 5 m s^{-1}, for example, as is usually the case in the Puerto Rico site (Brower, 1974) $\omega \geqslant 1$ cm s^{-1} would be expected if diffusion velocity were directly related to mean shear, since high wind speeds are associated with high velocity shears in the upper water column (Neumann and Pierson, 1966). The diffusion velocity parameter appears, therefore, to be related to a more complex oceanic process than direct velocity shear.

With normal dumping procedures, dilutions of $(1–5) \times 10^4$ or more are readily obtainable within 24 h. The lowest dilution factor, $1{:}10^4$ after 24 h (Table 7.2), is not anomalous as far as spreading is concerned. Passive oceanic dispersion was still within the range deduced for the other cases; the anomaly was entirely caused by dumping procedures.

After shear diffusion spreads the waste uniformly over the entire mixing depth, an asymptotic stage of diffusion, in which the plume once again grows as $t^{1/2}$, is expected; it may be regarded as representing an

equilibrium between shear effects and vertical mixing (Csanady, 1981). Short-term observations do not give any indication of when such a stage is reached. Under moderate wind conditions, it would not be reached for at least 2 d (Csanady, 1981).

Evidence concerning the behavior of particles during medium-term dispersion is incomplete and inconsistent. Experiments using high concentrations and high stirring rates have shown rapid coagulation (Gibbs, 1982a), implying that 90% or more of the particulate matter would sink out of the plume within 24 h (Gibbs, 1982a). This sinking would produce effective dilution factors of particles in the range of 10^5–10^6. If the waste were composed of 1% solids by volume, concentrations ranging from $1{:}10^7$–$1{:}10^8$ would be produced in the upper water column. Qualitative observations of particulate concentrations have, however, been made 27–74 h after waste dumping (Kester et al., 1981; Orr and Baxter, 1983), indicating that oceanic microstructure may retain some particulate matter within the plume for extended periods of time.

7.2.4. Long-Term Processes

The long-term movement and dilution of the liquid component of dumped wastes depends on large-scale oceanic processes that are not yet fully understood. Dispersive processes in the first 24–48 h after dumping may produce dilution factors of 5×10^4–10^5 but will not necessarily transport the waste away from the dumpsite through advection. Mean-current analyses at DWD-106 and the Puerto Rico site predict that the wastes should be carried away at a rate of at least 10 km d^{-1}, a rate that has sometimes been observed (Bisagni, 1981); however, under many other oceanographic situations, drogues, which should simulate the movement of the mass center, have not been advected away from the dumpsite in any predictable manner (Williams, 1981). Gulf-stream eddies and intrusions of shelf water can alter motions at DWD-106 for several days to weeks, and mesoscale eddy or wave motions at the Puerto Rico site can result in a very small net water movement, even over periods of weeks or months.

Dispersive processes over large space and time scales, which should lead to further dilution regardless of advection, have not been observed for dumped wastes. At some point, plume growth reaches an asymptotic stage, after which dilution is controlled by an apparent eddy process but with a diffusion coefficient (K) of 10^5 cm^2 s^{-1} or larger (Csanady, 1969). Once this asymptotic phase is reached dilution increases with $t^{1/2}$. The time of transition from shear diffusion to the asymptotic stage can be calculated as

$$t = 2K_A/\omega^2 \qquad\qquad 9$$

where K_A is the eddy diffusion coefficient characterizing the asymptotic stage. If K_A is chosen to be 10^5 cm^2 s^{-1}, then the transition will occur between 8 and 120 d after dumping for the 0.14- to 0.54-cm s^{-1} range of diffusion velocities shown in Table 7.1. With those velocities, dilution will reach a factor of 10^6 within 6 to 22 d and occur during the shear-diffusion stage. Dilution to 10^7 would occur in the asymptotic stage and require about 400 d. This time estimate is certainly too conservative, because episodic storm events and large-scale oceanic processes will accelerate dilution beyond the rates implied by the diffusion velocities in Table 7.1 and by constant K_A of 10^5 cm^2 s^{-1}. A dilution factor of 10^7 has been selected as acceptable, however, for wastes displaying acute toxicity in laboratory tests at a dilution of 10^5 (Csanady et al., 1979). Any attempt to establish time scales for dilutions to $1:10^6$ or $1:10^7$ would involve extensive and costly field studies, as well as a realistic model of mesoscale oceanic processes, including dispersion generated by storms.

Empirical data is also not available either for sinking rates or for spreading rates of the particulate component of wastes. To maximize the hypothetical impact on the seafloor, it can be assumed that all of the particulate matter will settle within a few weeks, that it will not be dispersed beyond the dumpsite area, which will be taken as 10^3 km^2, and that there is no mixing into the sediment. The concentrations of a contaminant in the sediment (C_s) is then the concentration in the particulate component of the waste (C_w), times the ratio of particulate waste flux to total annual particulate flux (F). If Q is the amount of particulate matter dumped per year and measured in thousands of tons, then

$$C_s = C_w \frac{Q}{Q + 10^4 F} \qquad\qquad 10$$

7.3. CONCLUSIONS

The questions of whether dumping wastes into the deep ocean is acceptable at present levels of activity and whether different types of wastes and greatly increased amounts could be disposed of without ill effects cannot be answered by physical studies alone, but information on the physical consequences of an increase in dumping provides a necessary input for analysis of effects both at the surface and on the bottom of the ocean. As far as the surface is concerned, the increase need only be carried out in such a way that individual dumps are not superimposed and so that the overall diluting capacity of the site is not exceeded. Biological and chemical studies carried out for lower levels of dumping activity would then remain valid. The same level of studies that would be needed to assess how present dumping activities affect the seafloor would provide projections for an increase. In practice, because any accumulation on the seafloor would be very slow, the simplest and most important monitoring requirement would be benthic sampling near the dumpsite.

In summary, the physical aspects of dumping industrial wastes into the deep sea appear to be manageable. Present dumping and future scenarios can be analyzed sufficiently to provide adequate margins of safety for wastes in the upper water column. When considering the problem of the accumulation of wastes on the seafloor, a subject on which there is the least direct knowledge, the dispersion of waste in the water column and mixing into the sediment provide an implicit margin of safety. In the absence of field information, very conservative assumptions about accumulation rates can be made, provided that the loading of particulate matter is known.

REFERENCES

Bisagni, J. J. 1981. Lagrangian Measures of Near-Surface Waters at the 106-Mile Dumpsite. Technical Memo OMPA-11, National Oceanic and Atmospheric Administration, Office of Marine Pollution Assessment, Boulder, Colorado, 23 pp.

Bisagni, J. J., and D. R. Kester. 1981. Physical variability at an east coast United States offshore dumpsite. *In*: Ocean Dumping of Industrial Wastes, B. H. Ketchum, D. A. Kester, and P. K. Park (Eds.). Plenum Press, New York, pp. 89–107.

Bowden, K. F. 1965. Horizontal mixing in the sea due to a shearing current. *Journal of Fluid Mechanics*, **21**, 83–95.

Brower, W. A. 1974. Environmental Guide for the Mona Passage Area. National Oceanic and Atmospheric Administration/Environmental Data Service National Climatic Center, Asheville, North Carolina, 145 pp.

Chase, R. R. P. 1979. Settling behavior of natural aquatic particles. *Limnology and Oceanography*, **24**, 417–426.

Csanady, G. T. 1969. Diffusion in an Ekman layer. *Journal of Atmospheric Science*, **26**, 414–426.

Csanady, G. T. 1973. Turbulent Diffusion in the Environment. D. Reidel Publishing Co., Boston, 248 pp.

Csanady, G. T. 1981. An analysis of dumpsite diffusion experiments. *In*: Ocean Dumping of Industrial Wastes, B. H. Ketchum, D. A. Kester, and P. K. Park (Eds.). Plenum Press, New York. pp. 109–129.

Csanady, G., G. Flierl, D. Karl, D. Kester, T. P. O'Connor, P. Ortner, and W. Philpot. 1979. Deepwater Dumpsite 106. *In*: Assimilative Capacity of U.S. Coastal Waters for Pollutants, E. D. Goldberg (Ed.). U.S. National Oceanic and Atmospheric Administration, Boulder, Colorado, pp. 123–147.

Devine, M., E. R. Meyer, and T. O'Connor. 1981. NOS assessment, technical summary. *In*: Assessment Report on the Effects of Waste Dumping in the 106-Mile Ocean Waste Disposal Site. Dumpsite Evaluation Report 81-1, National Oceanic and Atmospheric Administration, Boulder, Colorado, pp. 11–24.

Friedlander, S. K. 1977. Smoke, Dust, and Haze. Wiley-Interscience, New York, 317 pp.

Gibbs, R. J. 1982a. Effect of pipetting on mineral flocs. *Environmental Science and Technology*, **16**(2), 119.

Gibbs, R. J. 1982b. Particle dynamics of sewage sludge dumping in the ocean. *In*: Marine Pollution Papers, Oceans '82. Marine Technology Society/The Institute of Electrical and Electronics Engineers, Washington, D.C., pp. 1058–1062.

Hawley, N. 1982. Settling velocity distribution of natural aggregates. *Journal of Geophysical Research*, **87**(C12), 9489–9498.

Hunt, J. R. 1980. Prediction of oceanic particle size distributions from coagulation and sedimentation mechanisms. *In*: Particulates in Water, M. C. Kavannaugh and J. O. Leckie (Eds.). *Advances in Chemistry*, **189**, 243–257.

Hunt, J. R. 1982a. Self-similar particle size distributions during coagulation: theory and experimental verification. *Journal of Fluid Mechanics*, **122**, 169–185.

Hunt, J. R. 1982b. Particle dynamics in seawater: Implications for predicting the fate of discharged particles. *Environmental Science and Technology*, **16**(6), 303–309.

Ichiye, T., and M. Carnes. 1981. Application of aerial photography to the study of small-scale upper ocean phenomena. *Pageoph*, **119**, 293–308.

Ichiye, T., M. Inoue, and M. Carnes. 1981. Horizontal diffusion in ocean dumping experiments. *In*: Ocean Dumping of Industrial Wastes, B. H. Ketchum, D. A. Kester, and P. K. Park (Eds.). Plenum Press, New York, pp. 131–159.

Kester, D. R., R. C. Hittinger, and P. Mukherji. 1981. Transition and heavy metals associated with acid-iron waste disposal at Deep Water Dumpsite 106. *In*: Ocean Dumping of Industrial Wastes, B. H. Ketchum, D. A. Kester, and P. K. Park (Eds.). Plenum Press, New York, pp. 215–232.

Kohn, B., and G. T. Rowe. 1981. Dispersion of two liquid industrial wastes dumped at 106-mile site. *In*: Assessment Report on the Effects of Waste Dumping in 106-Mile Ocean Waste Disposal Site. Dumpsite Evaluation Report 81-1, National Oceanic and Atmospheric Administration, Boulder, Colorado, pp. 133–156.

Murray, T. E. 1977. Summary of waste inputs to 106-Mile Site during 1976 and 1977. *In*: Assessment Report on the Effects of Waste Dumping in 106-Mile Ocean Waste Disposal Site. Dumpsite Evaluation Report 81-1, National Oceanic and Atmospheric Administration, Boulder, Colorado, pp. 33–38.

Neumann, G., and W. J. Pierson, Jr. 1966. Principles of Physical Oceanography. Prentice Hall, Englewood Cliffs, New Jersey, 545 pp.

Ohlhorst, C. W. 1981. The use of landsat to monitor deep water dumpsite 106. *Environmental Monitoring and Assessment*, **1**, 75–81.

Orr, M. H., and L. Baxter II. 1983. Dispersion of particles after disposal of industrial and sewage wastes. *In*: Wastes in the Ocean, Vol. 1: Industrial and Sewage Wastes in the Ocean, I. W. Duedall, B. H. Ketchum, P. K. Park, and D. W. Kester (Eds.). Wiley-Interscience, New York, pp. 117–137.

Presley, B. J., J. H. Trefrey, and J. S. Schofield. 1981. Trace metal geochemistry of industrial waste materials shell biosludge. *In*: Western Gulf of Mexico Dumping Site Assessment Report. Dumpsite Evaluation Report 81-2, National Oceanic and Atmospheric Administration, Boulder, Colorado, pp. 119–150.

Proni, J. R., and D. V. Hansen. 1981. Dispersion of particulates in the ocean studied acoustically: the importance of gradient surfaces in the ocean. *In*: Ocean Dumping of Industrial Wastes, B. H. Ketchum, D. L. Kester, and P. K. Park (Eds.). Plenum Press, New York, pp. 161–174.

Proni, J. R., J. J. Tsai, D. S. Walter, F. C. Newman, and M. Devine. 1981. Gulf of Mexico ocean dumping experiment: results of acoustic studies. *In*: Western Gulf of Mexico Dumping Site Assessment Report. Dumpsite Evaluation Report 81-2, National Oceanic and Atmospheric Administration, Boulder, Colorado, pp. 185–203.

Saffman, P., and J. Turner. 1956. On the collision of drops in turbulent clouds. *Journal of Fluid Mechanics*, **9**, 16–30.

Schwab, C. R., T. C. Sauer, Jr., G. F. Hubbard, H. Abdel-Reheim, and J. M. Brooks. 1981. Chemical and biological aspects of ocean dumping at the Puerto Rico dumpsite. *In*: Ocean Dumping of Industrial Wastes, B. H. Ketchum, D. A. Kester, and P. K. Park (Eds.). Plenum Press, New York, pp. 247–274.

Swanson, R. L., and M. Devine. 1982. Ocean dumping policy. *Environment*, **24**(5), pp. 14–25.

Taylor, G. I. 1954. The dispersion of matter in turbulent flow through a pipe. *Proceedings of the Royal Society of London*, **A223**, 446–467.

Williams, W. G. 1981. Meridional displacements of surface water north of Puerto Rico. *Transactions of the American Geophysical Union, EOS*, **62**(17), 305.

Chapter 8

A Review of Problems in Modelling the Hazards of Wastes Dumped into the Deep Sea

George T. Needler

Bedford Institute of Oceanography
Dartmouth, Nova Scotia, Canada

Dr. Needler's present address: Institute of Oceanographic Sciences, Brook Road, Wormley, Godalming, Surrey GU8 5UB, United Kingdom

ABSTRACT

Assessment of the potential hazards associated with waste disposal into the sea includes modelling the distribution of the contamination that will result from that disposal. This chapter discusses the need to define the assessment problem well enough so that a suitable model can be chosen. The differences among models lie in their temporal and spatial scales of application. With increasing scale it becomes necessary to parameterize transfer processes and, correspondingly, to lose resolution on smaller scales. The availability of models as a function of scale is reviewed, and recommendations are made for choosing among them. In particular, the incorporation into models of chemical and biological processes that influence contaminant distributions is exemplified.

8.1. INTRODUCTION

Considerable attention has been given to the estimation of the hazards of dumping wastes into the deep sea, particularly those wastes containing radioactivity. Usually the transfer rate of a contaminant from a dumpsite to humans has been estimated by using relatively simple models. Because simple models omit mechanisms that can be perceived as important, this has often led to criticism from those who wish to protect the environment. On the other hand, these same simple models have been criticized as being overly conservative by those who see the ocean as the safest area in which to dispose of certain wastes. It is difficult to provide a realistic assessment of how conservative any model truly is.

In 1980, because of their need to review the regulations controlling the dumping of radioactive wastes in the deep sea, the International Atomic Energy Agency requested the United Nations IMCO/FAO/UNES-

CO/WMO/IAEA/UN/UNEP Joint Group of Experts on the Scientific Aspects of Marine Pollution (GESAMP) to provide advice on how best to model the transport of contaminants from deep-sea dumpsites. GESAMP agreed to this in the context of all contaminants and gave a working group terms of reference that included (1) reviewing existing knowledge of pathways by which substances might be transferred from a deep-sea dumpsite to humans, (2) recommending methods for calculating concentrations of contaminants throughout an oceanic basin, (3) assessing the reliability of the models recommended for the calculation of such concentrations, and (4) recommending research that might be needed to answer important unresolved issues. This chapter discusses some of the problems facing experts attempting to deal realistically with quantifying the hazards of dumping wastes into the deep sea, and it provides some ideas as to what may be done now. The latter part of this chapter is based directly on the work of the GESAMP working group.

8.2. SOME GENERAL PROBLEMS FACING MODELLERS

8.2.1. The Acceptability of Models

Modellers of environmental problems often face criticism of their efforts to quantify processes important in hazard assessment, and the fear is often expressed that only quantifiable aspects of environmental detriment have been taken into account in any particular situation. That is to say, there is a distrust in quantitative estimates and the perhaps justified concern that, if they are detailed in nature, they will give an unjustifiable impression of being reliable when, in fact, important considerations are being ignored. One can particularly see this distrust in considerations of the possibility of sublethal damage to marine ecosystems. Such concerns pose a dilemma for those who would model the marine environment. It is pointless to spend a lot of time creating and testing models if they will not be applied in a useful fashion. Models must therefore often be constructed in ways that will answer the questions of the interested parties, both pro- and anti-dumping, and not necessarily in the way that seems to be the most effective practical or scientific approach. This is a difficulty that apparently cannot be avoided and must be kept in mind by those who are involved in such endeavors.

8.2.2. The Need to Define a Question

Another problem facing modellers of contaminant transport is that there is often no single well-posed problem that must be addressed. Because the ocean is both complicated and, in many aspects, badly understood, this causes difficulty because models must be designed to answer relatively well-specified questions if their accuracy is to be maximized. Although no model is capable of addressing all questions, some models can address some questions rather well. For example, the requirements for models to predict the concentration very near a localized dumpsite are very different from those for models that can give the concentration in the far field (e.g., on the continental shelf) after the contaminant has been transported thousands of kilometers decades to centuries after it has been introduced into the marine environment. Similarly, there is usually a great difference between models that are suitable for estimating transfers along critical paths (where one is interested in estimating the maximum likely transfer of contaminants to unique groups of people) and models suitable for collective-dose assessments. Since collective-dose assessments are used for comparing one disposal option to another, the models used for this purpose must give the best estimate of transfer rates (usually to large numbers of people) in order to make an effective comparison.

The nature of the waste release is also important in selecting a model. In general, it is easier to deal with a continuous release than with one that varies with time, especially when the variation is on the order of decades. This arises in part because one can have some confidence that models that reproduce steady distributions of naturally occurring substances will also deal reasonably well with the steady release of a long-lived contaminant, even if the processes controlling the distribution are not well described by the model. This is especially true for substances that cycle enough times through the ocean before removal so that the nature of their source is relatively unimportant. The adaption of such models to a situation that varies with time may be completely unjustified if important processes are omitted or badly parameterized.

8.2.3. The Availability of Models

Although many oceanic processes are poorly understood, many models that have some credibility for analyzing marine pollution already exist; thus, the question often arises as to whether an existing model is adequate for a particular disposal scenario. For example, one could use simple point-source diffusion models to describe the concentration of a waste near its source. One could increase the complexity by including the far-field effects arising from the finite nature of the ocean or localized scavenging by falling particulate matter or deep-sea sediments. Simple plume models can also describe situations in which advection is important near the dumpsite. All of the aforementioned models are simple but have the advantage of having analytical solutions that can be examined to test the validity of the model and its dependence on its parameters. More complicated models of the near field could include time-dependent effects, which can be dealt with by numerical methods of varying resolution. Stochastic models can include more explicitly the effects of fluctuations, and high-resolution eddy-resolving models may be used to describe the effects of quasi-geostrophic turbulence at a dumpsite.

In the far field (e.g., the whole ocean), the simplest model is perhaps that of a well-mixed box. This model has some validity for extremely long-lived contaminants that do not interact strongly with particulate matter or with sediments (i.e., that are distributed relatively homogeneously through the ocean). One-dimensional models have been widely used to describe naturally occurring substances and can also be used to describe the transfer of relatively long-lived contaminants from dumpsites or even the average transfer of short-lived elements. There are various box models as well as medium-resolution finite-difference numerical models for ocean basins or the global ocean. Higher-resolution models exist for smaller regions and may be of the eddy-resolving type that provide explicit quasi-geostrophic fluctuating currents.

8.2.4. Parameterization

Given such a range in model complexity, one must determine what is appropriate to use for a given task. Clearly, one is not in a position to provide a single model that covers all of the scales from those of salt fingers at a size of 1 mm to those of the large oceanic currents over a distance of 10^3 km. Thus, one is always forced to parameterize some processes for which the resolution is inconsequential to the problem at hand. This is done, for example, through the use of the classical eddy-diffusion coefficients, which are intended to include a whole range of processes that may either

mix a substance vertically across density surfaces or horizontally across an oceanic basin. The basic question is how well the transfers of a contaminant by smaller-scale processes may be represented by their parameterization. A similar question is whether more complicated models, which usually have more detailed parameterizations, are more reliable than are simple ones for a given problem. In dealing with this issue, the GESAMP working group took the approach that one should use the simplest model that is able to address the particular question with realistic accuracy; it also recommended that, in many cases, more is to be learned by varying the parameterizations in simple models than is to be obtained by using more complex models (GESAMP, 1983). Such an approach usually allows one to understand the limitations of a given model and to test the importance of more complicated transport mechanisms by using the simple model as a basis of comparison or as a framework in which to analyze the effects of more complex processes. The working group decided, however, that more complex models are useful research tools for evaluating certain processes.

Much remains unknown about oceanic transfer mechanisms, and the question is often posed as to whether one should wait to use the ocean as a disposal site until the oceanic system is better understood. Although this is an attractive option, it is overly simplistic. A better question is whether one understands the oceans well enough to be assured that the hazard to human populations will be minimal or to know how the hazard compares with that from other disposal options. There is some amount of material that can be confidently disposed of in the deep sea without being detrimental to humans or to the marine ecosystems. This amount may be large or small. As one seeks to define the limits ever more precisely, the probability of error increases. Perhaps the most difficult task that faces modellers involved in this issue is to put reliability estimates on their models for transferring contaminants. The GESAMP working group found this very difficult to do, even with the models they recommended. They suggested that in most cases the question of reliability was best addressed by using simple models for which limiting transports could be obtained as a function of changes in the parameters (GESAMP, 1983).

The GESAMP working group examined particular oceanic processes with the aim of determining whether certain ones are important either in general or in particular situations. Interactions of contaminants with marine sediments or with particulate matter in the water column, as well as the possibility of transfer by biological processes, were considered. Some of these aspects are discussed in Sections 8.3.1 and 8.3.2; a more detailed description may be found in a report by GESAMP (1983).

8.3. INTERACTIONS WITH SEDIMENTS, PARTICULATE MATTER, AND BIOLOGICAL PROCESSES

8.3.1. The Use of Deposition Velocities

Of basic importance for substances that are reactive with marine sediments is the question of the rate of removal of the substances from the water to the sediments with which they are in contact. When one is dealing with a steady-state problem and when the water and sedimentary phases of the substances in question are in chemical equilibrium, it is useful in the case of a large number of removal processes to parameterize the removal rate in terms of a deposition velocity. In particular, for all those removal processes for which the removal rate is proportional to the concentration of the substance in water (C), one may define the deposition velocity (V_d) by the equation

$$\text{Surface Flux} = - V_d C \qquad 1$$

The calculation of V_d as a function of the decay rate (λ) of a contaminant and its reactivity as defined by a distribution coefficient (K_D), the dimensionless equilibrium ratio of the concentration of a contaminant in the solid phase to that in the dissolved phase, is not difficult for a wide variety of sedimentary models. An example is given in Fig. 8.1; this sedimentary model includes a near-surface sedimentary layer (of depth h_b), which is stirred by bioturbation at a rate defined by a diffusivity (K_b) ranging from 10^{-12} to 10^{-13} m^2 s^{-1}. Under this layer lies a region where the substance may be diffused through the pore water at a rate described by a diffusivity K_{pw} of about 10^{-10} m^2 s^{-1}, while being adsorbed onto the sediments yielding net diffusivity $K_{eff} = K_{pw}/K_D$ of about 10^{-14} m^2 s^{-1} for a K_D of 10^4. At the water–sediment interface, new sediments are burying the old ones at the rate w_s of, for example, 1 cm per thousand years. Figure 8.2 shows V_d for each of these three removal processes acting independently of each other. The net accumulation of sediment on the surface buries the contaminant in the sediment at the rate of $w_s K_D$ independently of the half-life of the material. Bioturbation keeps mixing the substance into the sediment, where it decays. In order to keep the concentration of the waste in the surface sediment in equilibrium with that in the water, the contaminant must be added to the sediment at the rate $\lambda K_D h_b$ times the concentration so that in this case $V_d = \lambda K_D h_b$. If λ is large enough, material is not diffused across the bioturbated layer, and the deposition velocity is modified as shown in Fig. 8.2. Similarly, diffusion in the underlying pore water yields a deposition velocity of $(K_D K_{pw} \lambda)^{\frac{1}{2}}$, which again must be modified as shown for a large value for λ because the substance will not decay

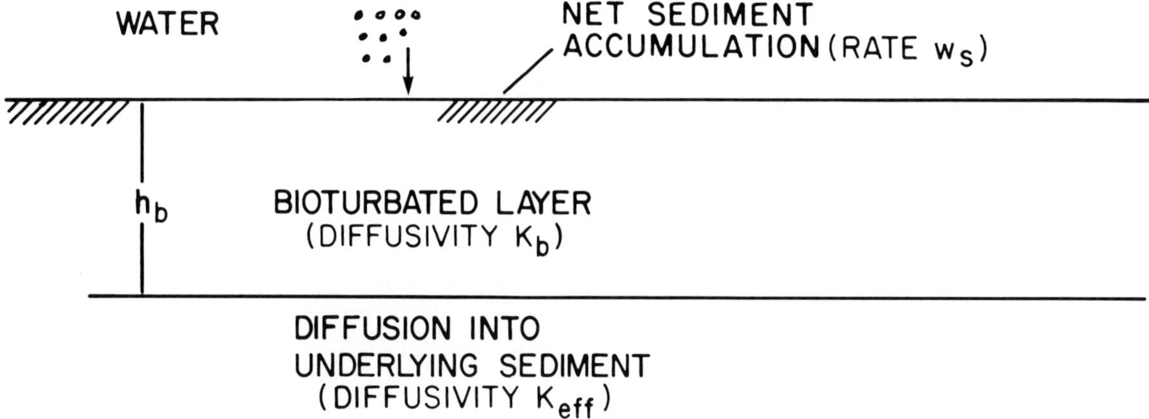

Figure 8.1. Schematic diagram of a sedimentary model with deep sediment overlaid with a layer disturbed by biological activity and with accumulating sediment.

before crossing the bioturbated layer. It may be seen that, for the chosen values, burial by accumulating sediments is the most important mechanism for long-lived substances, whereas mixing in the bioturbated layer is the most important for short-lived substances.

Although the details of these deposition velocities are dependent on the sedimentary model and parameters that have been chosen for this illustration, the comparisons form a basis for understanding which processes are important enough to be included in a model for a contaminant with the given K_D and λ. Using all processes simultaneously would yield a value of V_d slightly greater than that given for the dominant process of each λ.

It is worth noting that the mechanism of burial as described is independent of the nature of the transport of a contaminant to the bottom via falling particles. In general, incoming particles partially dissolve relatively quickly and release a fraction of the substances they carry to the water. Removal by burial depends only on that fraction of the particulate matter that contributes to the net sediment-accumulation rate.

8.3.2. Effects of Strong Local Scavenging

The question has often been raised as to whether reactive contaminants can be transported far from a dumpsite or if they are scavenged locally within or near the dumpsite by the sediment. This question may be investigated by considering a simple diffusive model (allowing for radioactive decay) for the transfer of a contaminant from a finite-sized dumpsite and by taking removal at a boundary into account through the use of a deposition velocity (Fig. 8.3). The appropriate steady-state equation for the interior of the ocean is

$$K_V \frac{\partial C}{\partial z^2} + K_H \nabla_H^2 C - \lambda C = 0 \qquad 2$$

where

$$-K_V \frac{\partial C}{\partial z} = Qf(r) - V_d C \qquad 3$$

at the water–sediment interface, and wherein K_V and

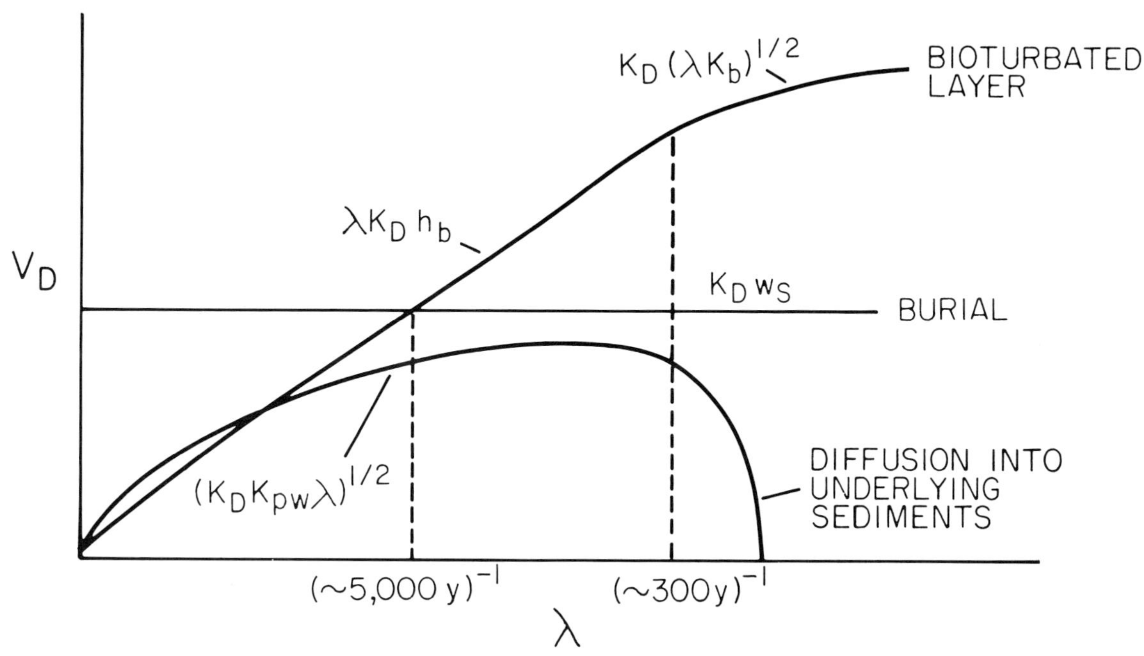

Figure 8.2. Deposition velocity as a function of radioactive decay rate for various processes taken separately. K_D has a fixed value.

Figure 8.3. Schematic diagram of the sediment–water interface near a source, with removal to the sediments described by a deposition velocity.

K_H are the vertical and horizontal diffusivities, respectively; $f(r)$ describes a distributed source of unit integral strength; Q is the source strength; z is the vertical coordinate (positive upward), and ∇_H^2 is the horizontal Laplacian.

Solution of this problem, which is described in detail by GESAMP (1983), is shown schematically in Fig. 8.4, in which the nature of the solution is shown as a function of scavenging strength as defined by the non-dimensional parameter $V_d H / K_V$ and of decay rate as defined by the nondimensional parameter $(\lambda H^2 / K_V)^{1/2}$, where H is the depth of the ocean. As one might expect, strong scavenging or strong decay restricts the contaminant to a region close to the source. On the other hand, for smaller values of these parameters, the contaminant will be widespread and exist either in the water or in the sediment. Quantitatively similar results

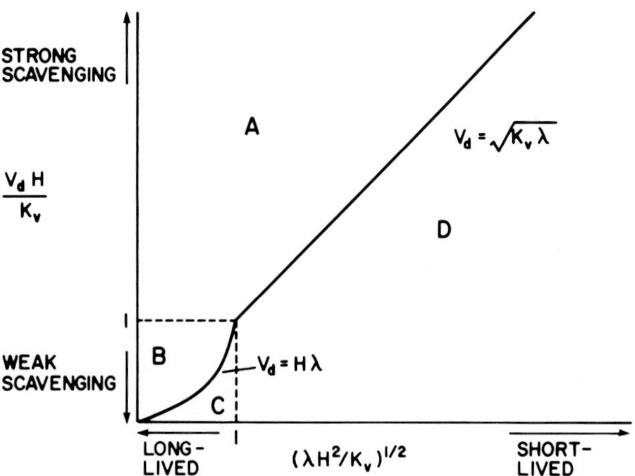

Figure 8.4. The boundaries between regions of different characteristic solutions of the diffusion-scavenging model as a function of nondimensionalized scavenging and radioactive decay. In regions A and D the contaminant is confined near the source by scavenging and decay, respectively. In regions B and C the contaminant is widespread and exists predominantly in the sediments or the water column, respectively.

are obtained if one takes into account the effects of scavenging by particles falling through the water column. Although the boundaries between the regions in Fig. 8.4 are not distinct (i.e., more than one mechanism is important near a boundary), the results are indicative of the understanding that can be gained using a relatively simple model about the distribution of a contaminant. Such results are not likely to be greatly changed in a qualitative sense when advection is taken into account.

8.3.3. Vertical Scavenging Models

The role of removal into sediments and scavenging by particles may also be effectively dealt with by one-dimensional models in certain circumstances. Such models can be appropriate when a substance has a long enough half-life or low enough reactivity that one expects it to be well mixed horizontally before it decays or is substantially removed to the sediments by scavenging. Even if this should not be the case, the average penetration of the substance into the upper ocean may often be adequately represented by a one-dimensional model. Care must always be taken, however, in one-dimensional models to ensure that the fluxes of the contaminant are properly described and that the model does not concern itself solely with the local balances in the water column (e.g., with profile fitting). In particular, it is necessary to take into account the fact that the upward flux of certain substances through the upwelling of deep water over most of the oceanic basin may be at least partially balanced by the sinking of equal amounts of water carrying substantial amounts of a substance to the bottom in polar regions.

A schematic diagram of a model that takes return flows into account is shown in Fig. 8.5. In the interior of the ocean, upwelling and diffusion are included, and sinking particulate matter scavenges the dissolved phase of a substance through first-order reversible exchange mechanisms. In the surface layer, which is assumed to be well mixed, other particles referred to in this chapter as biological particles take up the contaminant and redistribute it at depth as they dissolve at a prescribed rate on their way to the bottom. In the bottom boundary layer the same processes are included, and account is taken of the injection of water, and of the contaminant it carries, from the region of sinking at the poles. Various sedimentary models may be used, but the results that follow are for a sedimentary layer including burial and bioturbation along with first-order reversible exchanges between the dissolved and sedimentary

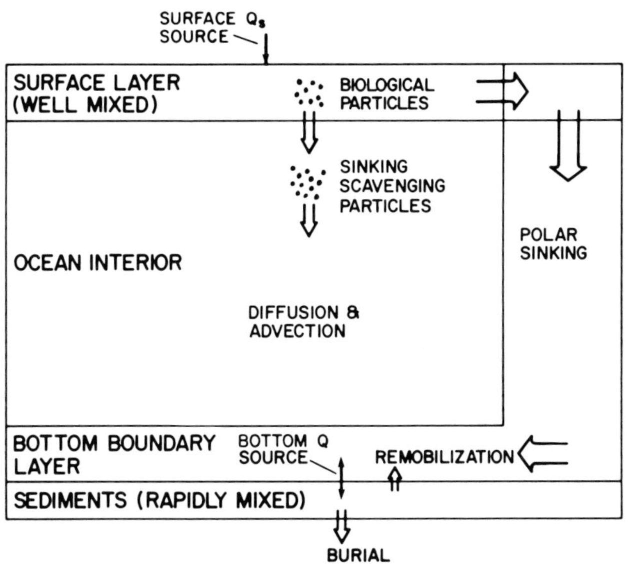

Figure 8.5. Schematic diagram of a model of an ocean with upwelling, diffusion, particle scavenging, removal to the sediments, and sinking at the poles.

phases. The upper pore water is assumed to have the same concentration as the overlying oceanic water. The details of this steady-state model and its application may be found in a report by GESAMP (1983).

The model has been used to describe the distributions of naturally occurring elements such as Th^{230}, Ra^{226}, Mn, Cu, Ni, and Cd, and consistent results can be obtained without excessive juggling of parameters. Indeed, a single set of parameters can describe the basic features of the vertical distribution of all of the elements that do not involve large transfers by the biological particles. For such elements, a limited number of types of biological particles (i.e., several choices of dissolution times) are needed to describe adequately the major features. Using the parameters determined in this way, the model has been used to predict the steady-state distribution of a contaminant released from a source of a strength of 1 unit $m^{-2}d^{-1}$ at the seabed. The results are shown in Fig. 8.6 for several depths as a function of λ and K_D; contour intervals are in units per m^3. As one might expect, long-lived and nonreactive (low K_D) substances have relatively high concentrations at all depths.

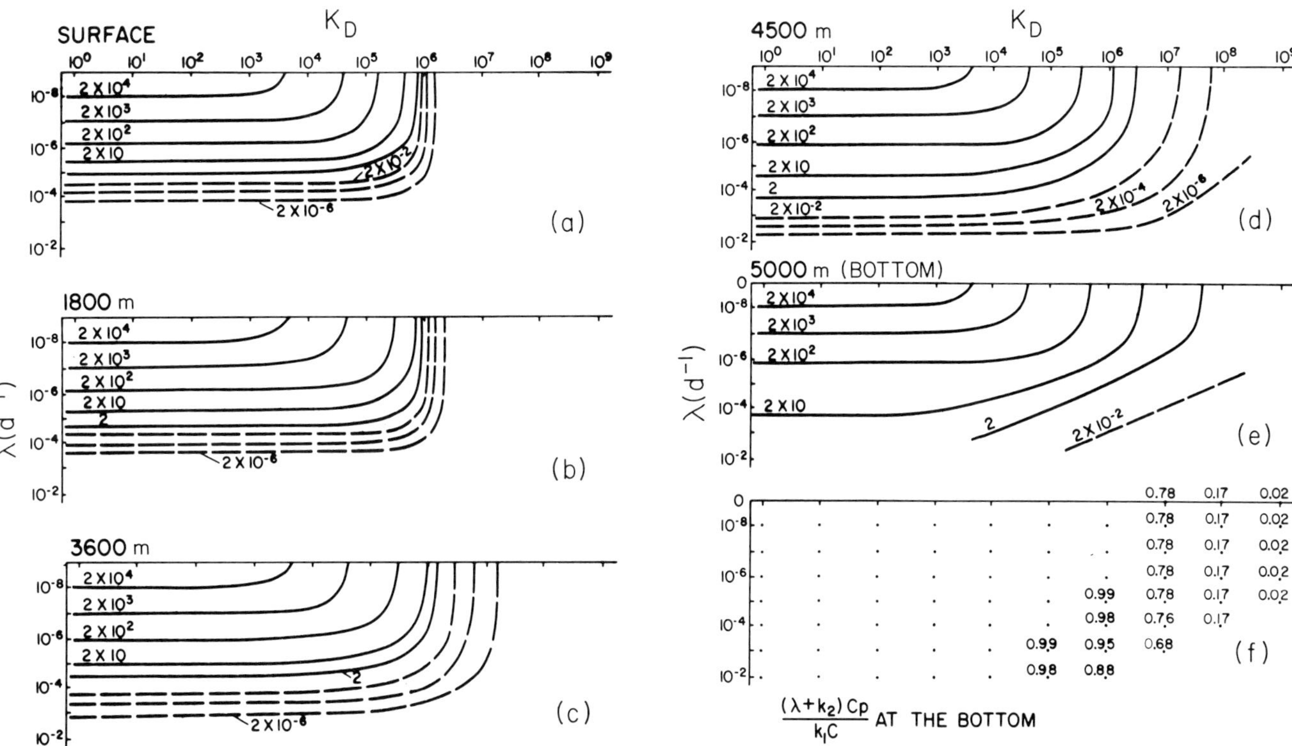

Figure 8.6. Results of the model illustrated in Fig. 8.5. Panels (a) to (e) show contours of equal concentration at various depths as a function of λ and K_D, and (f) illustrates the degree to which the water and particle phases are in equilibrium.

More reactive and short-lived elements are less concentrated in the water column because of increased decay or removal to the sediments and are confined increasingly to the region near the bottom. One can also note the very strong effect that occurs in this model for a K_D of about 10^6. In this case, for the particular parameters chosen, upwelling of the contaminant is in balance with its downward flux on scavenging particles. For smaller values of K_D, the particles are not effective in keeping the contaminant close to the bottom, whereas for larger values of K_D, only small amounts of the contaminant reach the surface.

The adequacy of the usual assumption of chemical equilibrium of particulate and dissolved phases of the contaminant is illustrated in Fig. 8.6f. Shown are the values for which the ratio $(\lambda + k_2)C_p/k_1C$ at the seafloor is other than unity (where C and C_p are the dissolved and particle-phase concentrations, and k_1 and k_2 are first-order adsorption and desorption rate constants). This ratio equals 1 for equilibrium. One can see that it is only for large values of K_D that the model predicts that the particles will not have time to come into equilibrium. This is a consequence of large concentration gradients in the water phase near the bottom. Although this provides some indications of the parameter range needed for the detailed equilibrium models, it should be pointed out that the removal mechanism for long-lived reactive contaminants is primarily by burial. Thus, the concentration at the bottom is almost equal to that necessary in the steady state to balance the input from the source; it will be much the same for equilibrium and nonequilibrium models. Changes in the water-column inventory as given by equilibrium and nonequilibrium models for long-lived reactive contaminants therefore only arise from differences in the shape of the vertical profiles. For the model discussed in this section, the inventories differ only by a factor of about two over the ranges of K_D and λ shown, even though the profile shapes are rather different.

The model can also illustrate the effect that burial of pore water can have on removing even nonreactive contaminants from the water column. This is illustrated in Fig. 8.7, in which the concentration at the seafloor is given as a function of K_D for conservative elements ($\lambda = 0$). For the parameters chosen, the concentration for nonreactive contaminants near the seafloor is three or four orders of magnitude greater than that for moderately reactive contaminants (i.e., for one with a K_D of 10^4). One should note, however, that given the same source strength there is nevertheless a finite steady-state concentration so that this removal mechanism may be

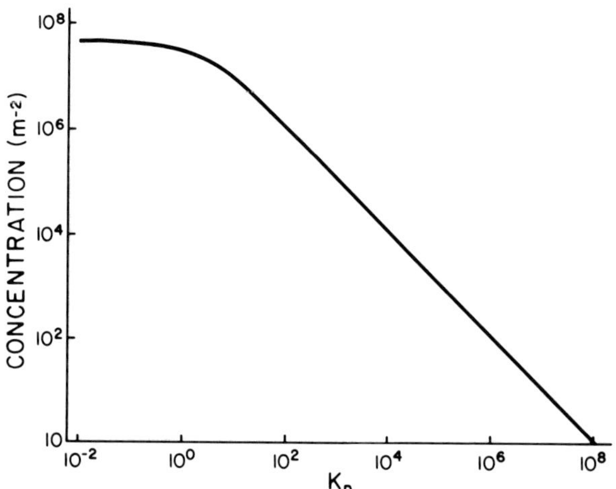

Figure 8.7. Concentration at the seabed for a stable contaminant ($\lambda = 0$) as a function of K_D for the model illustrated in Fig. 8.5.

an important factor in estimating the hazard of very long-lived nonreactive constituents of disposed wastes. This can help to place in context the real hazard of those that are minor constituents in wastes.

8.3.4. Biological Pathways

In general, it may be shown that due to the very small portion of the water column that is occupied by biological matter, biological mechanisms averaged over large space and time scales are small compared with physical transfer mechanisms or those associated with scavenging particulate matter. The question arises, however, whether this may not be the case in local regions or over short times, especially if certain critical pathways are involved. For example, it has been suggested by a number of authors [e.g., Angel (1983)] that buoyant eggs might carry a substantial amount of a contaminant from the seafloor to the surface. In this situation, a contaminant is released from a dumpsite and diluted to some extent by physical processes over a limited area (Fig. 8.8). Fish arrive at the dumpsite, their eggs equilibrate with the local concentration of the contaminant, and, upon spawning, the eggs float to the surface, where they might result in a surface "hot spot" of a given contaminant.

Since deep-sea productivity can be estimated with some accuracy, it is not difficult to put an upper bound on this biological transfer mechanism. For conservative elements ($\lambda = 0$), the concentration resulting from this

DOUBLE "HOT" SPOT

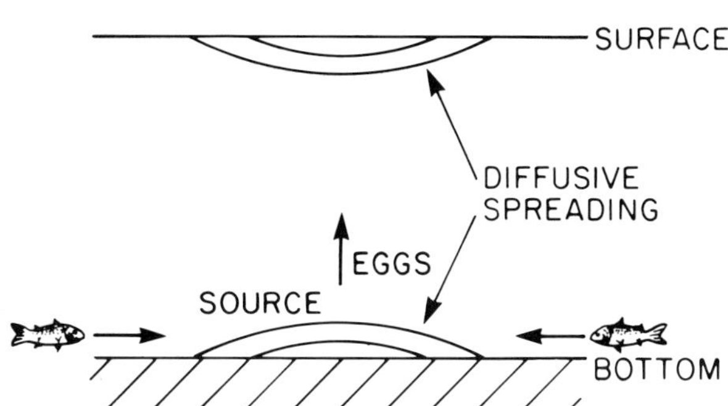

Figure 8.8. Schematic diagram of the possible transfer of a contaminant to the surface by buoyant eggs.

contaminant-transfer mechanism averaged over an oceanic basin will be negligible when compared to that resulting from physical processes (GESAMP, 1983). It is also negligible in the area over the dumpsite unless the concentration factor for the contaminant in the eggs is approximately 10^4 or 10^5. For short-lived contaminants, the relatively fast transport of eggs to the surface would result in a concentration at the surface greater than those arising from physical processes; this can also occur if the eggs are released over a short period.

One important question is whether such a transport process is important in the absolute rather than in the relative sense. Estimates of the limits of this particular transport process can be provided. It is more difficult, however, to determine whether the contaminant carried by the eggs can be taken into a food chain, thereby creating an important critical pathway, before the eggs are widely dispersed throughout the surface of the ocean. It should be noted, however, that any such pathway is unlikely to be more important than one directly involving the deep-sea fish in the area of a dumpsite, even though the low productivity of the deep sea would limit the sustainable yield of fish from a dumpsite. In this context the GESAMP working group estimated that a sustainable pathway to a limited critical group will on the average require about 100 km^2 of seafloor (GESAMP, 1983).

Many other biological mechanisms have been proposed as being important in assessing the hazard of deep-sea dumping. Most are amenable to limiting estimates of maximum transfer rates into food chains involving humans. Although it is possible that modellers will miss critical pathways of this type, it seems unlikely

that, if one takes a sensible and general enough approach to this problem, substantial effects will be ignored. As in the case of modelling physical and geochemical mechanisms, quantitative limits can and should be placed on biological transport mechanisms as they are identified. It is only in this way that these mechanisms can be put into proper context.

Of perhaps greater importance is the question as to whether assessments should include potential critical pathways that concern fishing or other activities in deepwater, especially adjacent to or within dumpsites. Such pathways, often termed *hypothetical*, are often more limiting than are pathways that are based on activities at the surface of the ocean, especially for contaminants with half-lives of less than several decades. Modellers and those involved in hazard assessments are often given little guidance on how to treat this situation, which is often more of a question as to how one wishes to manage the environment (or protect it so that it may always be available for unforeseen uses) than a question of the probability of a particular activity. This problem is more of a factor in critical-path assessments than in those concerning collective dose, since for most elements the latter will still be dominated by pathways involving activities at the surface of the ocean.

8.4. MODEL SELECTION AND ACCURACY

As is apparent from the discussion in Section 8.3.4, the choice of a model for a particular contaminant released in a particular way must be made with care. The GESAMP working group took the approach that it is best in most circumstances to use the simplest model that can provide both the needed results and remain consistent with one's level of understanding of the processes involved. Often more than one model will be needed to predict the input to different food webs or parts of the environment.

Although it would be useful to give a detailed prescription for calculating oceanic concentrations, the GESAMP working group found this difficult to do and liable to misinterpretation. They therefore listed a number of observations based on detailed analyses of parts of the overall problem. They recommended that, keeping these observations in mind, simple models should be used that include at least the processes of physical horizontal and vertical transport by advection and/or diffusion, and the interaction of reactive contaminants with falling particulate matter and the sediments. Some models that could meet these requirements were given but many others can be devised.

On the subject of reliability, a very important aspect of any hazard assessment, the GESAMP working group noted that a model's ability to predict accurately the concentrations arising from the release of a contaminant depends primarily on whether the model provides a realistic representation of all the important processes affecting the contaminant's dispersal. This is usually difficult to determine, especially quantitatively. They thus recommended that, as far as possible, models should be evaluated over the full range of possible variability of their parameters and that the range of predicted concentrations should always be presented as part of the output. The sensitivity of the prediction to the model's parameters thus obtained provides valuable information about the model's performance, since for those cases in which a "best estimate" of the concentration field is needed, it indicates the range of possible uncertainty about this estimate, and where estimates of the maximum likely concentration field are needed, it enables the determination of the maximum realistic concentration field within the model's assumptions and indicates its difference from the "best estimate." It also indicates the parameters to which the model's results are most sensitive and thus, one might hope, to an evaluation as to whether this is due to the model's realistic parameterization of significant processes or to a lack of knowledge of those processes.

Models will always have to be examined for the possibility that important processes are being ignored. This can often be done effectively on a case-by-case basis without resorting to the use of a much more complicated model. Models must always be checked against existing data describing the distributions of natural and synthetic substances. Although this may not in itself prove that multiparameter models are valid, it

certainly must be of concern if agreement is not obtained at an appropriate level of accuracy. It is, after all, these distributions that provide most of the information that is available about oceanic transport mechanisms and their parameterization.

8.5. SUMMARY

Many difficulties, not all of which are scientific, face those who seek to provide answers to the problems of deep-sea dumping. Given well-posed problems, however, it is usually possible to provide estimates of transport rates or oceanic concentrations of a contaminant released into the deep sea that are accurate enough for the task at hand. No simple prescriptions, however, exist for providing such estimates or for putting bounds on their accuracy. Thus, it is essential that oceanographic expertise be applied at all stages of the assessment of potentially hazardous deep-sea dumping. The GESAMP working group has provided some steps in this direction. It is to be hoped that the oceanographic community will carefully examine their work and modify or expand it as necessary. Only through such participation will it be possible to identify and solve the real problems inherent in modelling releases of contaminants.

REFERENCES

Angel, M. V. 1983. Are there any potentially important routes whereby radionuclides can be transferred by biological processes from the seabed towards the surface? *In*: Ecological Aspects of Radionuclide Release. Special Publication Series of the British Ecological Society No. 3, Blackwell Science Publishers, pp. 161–176.

United Nations IMCO/FAO/UNESCO/WMO/IAEA/UN/UNEP Joint Group of Experts on the Scientific Aspects of Marine Pollution (GESAMP). 1983. An Oceanographic Model for the Dispersion of Wastes Disposed of in the Deep Sea. GESAMP Reports and Studies No. 19, International Atomic Energy Agency, Vienna, 182 pp.

Chapter 9

Development of a Compartment Model of the Ocean

Shelly F. Mobbs and Marion D. Hill

National Radiological Protection Board
Chilton Didcot
Oxfordshire, United Kingdom

Paul A. Gurbutt and John G. Shepherd

Fisheries Laboratory
Ministry of Agriculture, Fisheries and Food
Lowestoft, Suffolk, United Kingdom

Peter D. Killworth

Department of Applied Mathematics and Theoretical Physics
University of Cambridge
Cambridge, Cambridgeshire, United Kingdom

Peter Killworth's present address: Robert Hooke Institute, Department of Atmospheric Physics, Oxford, Oxfordshire, United Kingdom

ABSTRACT

A compartment model of the Atlantic Ocean is being developed for use in assessments of the impact of the disposal of radioactive waste into the ocean. In the model, the ocean is divided into compartments by using isopycnal coordinates in the vertical direction and bottom topography in the horizontal. The average salinity and temperature of the water in each compartment were derived from existing data. Mixing and advection are represented by positive volume-exchange rates between the compartments, and values for these exchanges are being determined from the salinity and temperature distributions by using inverse methods. The two methods tried so far are based on matrix inversion and quadratic programming. The former produced a flow pattern with some negative volume-exchange rates; this pattern is consistent with the salinity and temperature distributions. The latter produced positive volume-exchange rates, but the flow pattern is not entirely consistent with the salinity and temperature distributions. The solutions were very sensitive to the input data, and these data are therefore being refined before further calculations of exchange rates are made.

9.1. INTRODUCTION

The radiological impacts of disposal of many types of radioactive wastes extend over long time periods. For this reason, the only practical means of assessing these impacts, and hence providing an input to decisions on the acceptability of disposal methods, is by using predictive mathematical models. The form that modelling studies should take, and the type of results required, depend largely on the criteria used to judge the acceptability of disposal methods from a radiological-protection point of view. The criteria presently in use internationally and in most countries are those contained in the system of dose limitation recommended by the International Commission on Radiological Protection (ICRP) (1977). In order to demonstrate that a disposal method complies with the ICRP system of dose limitation, it is necessary to calculate both potential maximum annual doses to individuals and also collective doses to populations integrated over appropriate time periods. The maximum annual individual doses are to be compared with dose limits set by the ICRP, while the collective doses are used mainly in comparisons between waste management and disposal options.

It is important to realize that, since dose rates to an individual and to a population are defined in terms of the exposure over a period of one or more years, it is not necessary to resolve all of the fine-scale processes that will bring the radioactivity from the disposed waste back to humans, provided that these processes can be adequately parameterized in coarser-scale models. Simple models that maximize the potential dose at each stage are often adequate for individual-dose calculations. Even though more realistic models are needed for

estimation of collective dose, because of their use in comparison with other disposal options, coarse time- and space-scale models can be used because the doses are integrated over substantial time periods (typically thousands of years). While the adequacy of some parameterizations is still an important research question, progress can still be made and doses estimated with the required accuracy for the appropriate management decisions.

In the disposal of radioactive waste onto the deep-ocean seabed, the overall system to be modelled in order to calculate doses to humans can be divided into a number of subunits (Fig. 9.1). A model of the release of the activity from the wastes is needed to estimate the source term, describing the input of radionuclides into the water column. Near the bottom, activity may react with the sediments and disperse into the interior of the ocean. Once the radionuclides are in the interior, they can be dispersed by physical oceanographic processes (advection and diffusion), and the rates of dispersion can be modified by sediment scavenging and biologically mediated processes. The concentration field thus produced provides the necessary input data for calculations of the uptake of activity by biota and the doses to humans via marine food chains and other pathways (e.g., external irradiation from beach sediments). This chapter is concerned with one of the most important steps in producing an overall system model: the con-

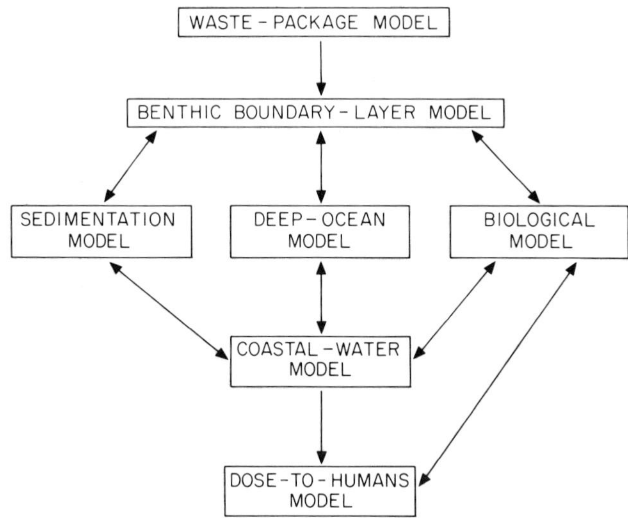

Figure 9.1. Models for assessing the radiological impact of disposal on the ocean bed.

struction of a model to predict the physical dispersion of radioactivity in the interior of the ocean by advection and diffusion.

Before describing progress in developing a new physical dispersion model, it is worth outlining the models that have been used in past assessments. Webb and Morley (1973) devised a model to examine the dispersion of radioactivity in the deep ocean in the vicinity of the present Nuclear Energy Agency (NEA) dumpsite in the northeastern Atlantic Ocean. This model was criticized because it implicitly assumed that the ocean is infinite and static. Shepherd (1976), using a recirculating-channel ocean, overcame these criticisms by introducing horizontal advection and three-dimensional diffusion into his model. The International Atomic Energy Agency (IAEA) used a one-dimensional diffusion model as part of the oceanographic basis of the IAEA definition of and recommendations on radioactive wastes unsuitable for dumping at sea (IAEA, 1978). This one-dimensional model gave similar results (concentrations within a factor of a few) to Shepherd's model of a recirculating ocean.

The aforementioned models are useful in generic assessments designed to predict maximum individual doses. There is, however, a need to develop more realistic and more site-specific ocean models for several purposes, including assessing the continued suitability of the present dumpsite for radioactive wastes in the northeastern Atlantic, reviewing international recommendations on dumping at sea, and carrying out assessments of the type required to compare dumping at sea with land-based disposal methods. Several methods are possible for producing improved models, but all require the generation of an internally consistent oceanic circulation pattern with which to perform the tracer calculations. One model with realistic coasts and ocean bottom topography, and which can also be tailored to a specific site, is Bryan's (1969) general circulation model. Three countries (the United States, the United Kingdom, and the Federal Republic of Germany) are already using this model for tracer-dispersion studies in relation to radioactive-waste disposal (Anderson, 1982; Robinson, 1984) and an intercomparison exercise is currently underway (Gurbutt, 1983).

Shepherd (1980) has used a two-dimensional meridional ocean section model to interpret the data from the Geochemical Oceans Sections Study (GEOSECS) experiment (Bainbridge, 1972). As mixing and motion in the ocean is thought to take place preferentially along isopycnal (constant density) surfaces, this model uses density coordinates instead of level surfaces in the vertical direction. The circulation pattern and eddy diffusivities in the model are adjusted by trial and error until the observed temperature and salinity distributions can be generated by the model. Flow fields can also be generated from the observed temperature and salinity distributions in more objective ways by using simplifying assumptions about the general balance of forces in the interior of the ocean. The calculation of geostrophic velocity has been used for many years in oceanography but has always suffered from the problem of not knowing what depth-independent (barotropic) component of velocity needs to be added to the baroclinic results obtained (the problem of the "level of no motion"). Methods of calculating the barotropic component have been recently derived (Stommel and Schott, 1977; Killworth, 1980; Needler, 1982), and methods have been developed using generalized inverse techniques (Wunsch, 1978) for calculating the flow across several hydrographic sections. For the purpose of radiological assessment, the use of inverse techniques is particularly attractive because it ensures that the model produced will give predictions comparable to observed temperature and salinity profiles in the ocean. Such profiles give a good indication of the pattern of dispersion of tracers, which is the objective of radiological modelling.

One other important consideration in developing a physical-dispersion model to be used in radiological assessments is the ease of interfacing with models of the other parts of the overall system. Compartment, or box, models and other coarse-grid, finite-difference models have the advantages of reasonable structural definition and also have the ability to represent short-circuit processes, as well as being relatively simple to interface with models of geochemical and biologically mediated processes. A compartment model that appears to be potentially suitable for use in assessments was devised by Worthington (1976); however, Section 9.2 of this chapter describes exploratory calculations using this model and explains why it would be difficult to apply it in assessments.

The remainder of this chapter describes the progress being made toward developing a new compartment model of the Atlantic Ocean, based on density coordinates (Shepherd, 1980) and using inverse techniques. The mathematical formulation of the problem is presented together with the methods used to derive transfer coefficients between compartments (singular value decomposition and quadratic programming), along with some of the results. The difficulties encountered in developing the new model are discussed.

9.2. THE WORTHINGTON MODEL OF THE NORTH ATLANTIC OCEAN

As noted in Section 9.1, Worthington's (1976) compartment model of the North Atlantic Ocean appears to be potentially suitable for radiological-assessment purposes, although it has never been used in this way. In the model the North Atlantic Ocean is divided into 6 areas, and each area is vertically divided into 4 or 5 layers (on the basis of temperature ranges), giving a total of 28 compartments (Fig. 9.2). Worthington used measured water flow to build a purely advective model of the transport of water between the compartments, rounding up to the nearest Sverdrup [1 Sverdrup (SV) $= 10^6$ m^3 s^{-1}]. Where there were no flow measurements, he inferred the water flows by considering the conservation of water within each compartment. The model was created as a possible description of water movement in the North Atlantic; Worthington did not test its predictive capability, nor did he attempt to validate the models by using data on the distribution of nonradioactive tracers, fallout, or naturally occurring radionuclides.

In order to determine whether the Worthington model is appropriate for use in assessments, it is necessary to find out how well it represents the general features of the North Atlantic Ocean. Predicting the salinity of the various ocean sections was chosen for this purpose. The model was set up on the computer by using Worthington's values for the water flows; however, since Worthington gave values for flows into only 21 of the ocean compartments, the other 7 were effectively excluded from the model. In order to calculate salinities, saline inputs from the Norwegian and Mediterranean Seas and from the Arctic and South Atlantic oceans were introduced into the model. These inputs were calculated from water flow rates given by Worthington (1976) and estimates of the corresponding salinity of the water from data provided by Wright and Worthington (1970) and Sverdrup et al. (1942). The model was then used to predict the salinity of each compartment of the model, and the average salinity in each compartment was calculated from measured salinity profiles in the ocean given by Worthington (1976) and Wright and Worthington (1970) and compared with the value obtained from the model. The results of the comparison are given in Table 9.1. The predicted salinities lie within the range of measured salinities (34.9–36.5°/oo), indicating that the saline inputs were approximately correct; however, the predicted salinities for individual compartments do not agree well with the measured values [errors greater than about 10% of the difference between maximum and minimum salinities

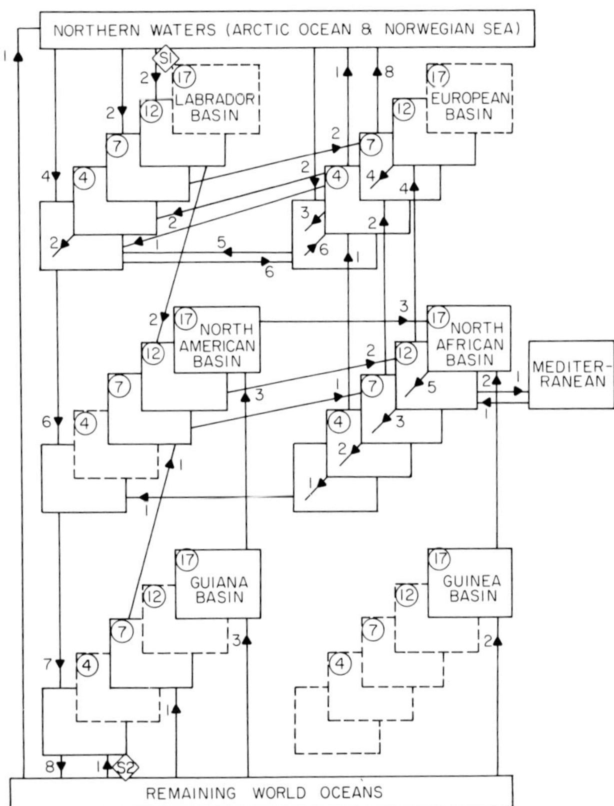

Figure 9.2. Three-dimensional representation of the Worthington Box Model. The circled numbers are temperature (°C), the values along the arrows represent flow rates (10^6 m^3 s^{-1}), and S1 and S2 in diamonds are two flows modified as diffusive fluxes (see text).

(i.e., 0.2°/oo) are considered unacceptable]. More seriously, the model consistently overestimates the salinity in the warmest layers. This suggests that the model does not adequately describe the vertical distribution of salt in the North Atlantic.

Since the original, purely advective Worthington model is unsatisfactory, parameters within the model were varied in order to ascertain whether any improvement in predicting the salt distribution could be obtained and to investigate the sensitivity of the results to changes in the flow rates (Table 9.1). A diffusive exchange between adjacent compartments was introduced to incorporate the seven null compartments into the model and to provide an additional exchange mechanism in the vertical direction. Although this tended to smooth out the distribution of salt in the ocean by reducing the difference in salinity between adjacent compartments, it did not produce a substantial im-

Table 9.1. Worthington Model: comparison of Calculated Salinities (‰) with Measured Values[a]

Area	Temperature (°C)	Measured Value	Standard Model	With Diffusion	Flows Increased by 0.05 SV	With Evaporation and Precipitation	S1[b] Increased by 10% with Diffusion	S2[b] Decreased by 10% with Diffusion
European basin	<4	− 0.05	0.08	0.19	0.14	0.01	0.40	− 0.03
	4–7	0.06	0.16	0.32	0.11	0.03	0.65	0.16
	4–12	0.45	0.60	0.52	0.14	0.00	1.09	0.46
	12–17	0.58	0.92	0.67	0.29	− 0.05	1.30	0.63
North African basin	<4	− 0.05	0.64	0.20	0.49	0.16	0.41	− 0.08
	4–7	0.13	0.64	0.42	0.42	0.16	0.80	0.29
	7–12	0.50	0.64	0.59	0.14	0.16	1.06	0.53
	12–17	0.82	0.92	0.78	0.08	0.28	1.23	0.75
	>17	1.63	1.00	0.76	− 0.64	0.39	1.14	0.73
Guinea basin	<4	—	—	0.19	—	—	0.41	− 0.13
	4–7	—	—	0.42	—	—	0.80	0.29
	7–12	—	—	0.57	—	—	0.97	0.51
	12–17	—	—	0.71	—	—	1.05	0.68
	>17	0.77	0.70	0.74	− 0.01	− 0.73	1.00	0.72
Labrador basin	<4	− 0.07	0.05	0.18	0.14	0.01	0.38	− 0.04
	4–7	− 0.01	0.16	0.34	0.16	0.03	0.57	0.21
	7–12	0.26	− 0.08	0.37	− 0.20	− 0.08	1.01	0.32
	12–17	0.82	− 0.08	0.35	− 0.86	− 1.09	1.70	0.32
North American basin	<4	− 0.07	0.14	0.19	0.21	0.02	0.40	− 0.08
	4–7	—	—	0.40	—	—	0.80	0.28
	7–12	0.24	0.20	0.51	− 0.16	− 0.20	1.04	0.45
	12–17	0.94	− 0.08	0.58	− 0.82	− 1.09	1.26	0.54
	>17	1.46	1.20	0.74	− 0.29	1.08	1.17	0.71
Guinea basin	<4	− 0.07	0.11	0.19	0.19	0.01	0.40	− 0.17
	4–7	—	—	0.41	—	—	0.80	0.29
	7–12	− 0.10	− 0.20	0.50	− 0.01	− 0.20	0.87	0.45
	12–17	—	—	0.71	—	—	1.05	0.68
	>17	1.07	1.20	0.87	0.13	0.64	1.10	0.85

[a] The table gives x (‰) where x = salinity − 35. Dashes indicate the absence of data.
[b] The water flows modified to give saline inputs S1 and S2 are shown in Fig. 9.2.

provement in the results. Increasing the flow rates by 0.5 SV caused small fluctuations in the predicted salinities. Altering the salt input and output parameters, either by including evaporation and precipitation or by varying the salinity of the exchanges with the other oceans, changed the average salinity of the ocean and also increased the salinity gradient between adjacent compartments. None of these variations, however, produced closer agreement between the predictions of the distribution of salt in the North Atlantic and measured values.

It would be possible to improve the results by introducing additional vertical current flows between some of the compartments; however, all of the other flows would then have to be altered to ensure that water is conserved within each compartment, and thus the whole model would have to be revised. Similarly, obtaining a more accurate prediction of the salinity distribution by including a greater number of exchanges with the South Atlantic, while possible, would result in a completely different model. It was therefore concluded that it would be better to develop a new model than to adapt the Worthington model.

9.3. THE NEW COMPARTMENT MODEL OF THE ATLANTIC OCEAN

9.3.1. Definition and Structure of the Model

The compartment model to be discussed in this chapter covers the area of the Atlantic Ocean from 50°S to

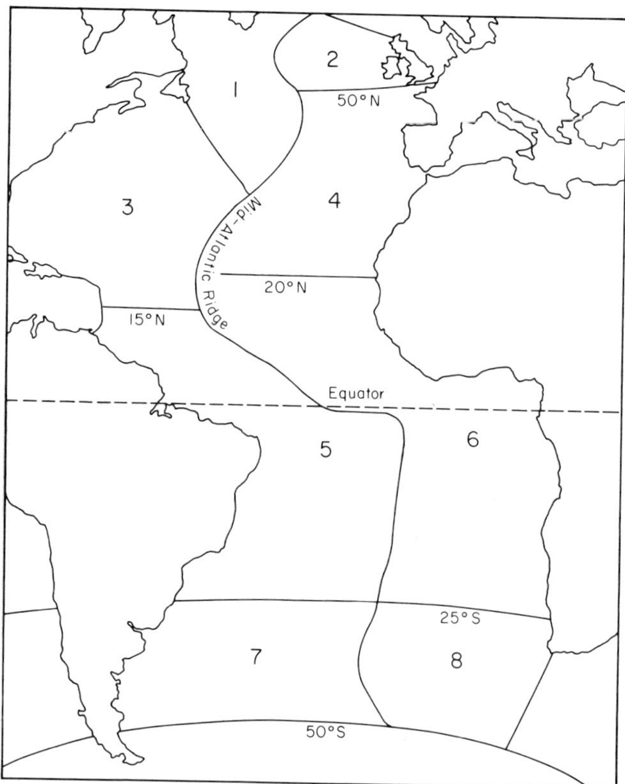

Figure 9.3.   The new ocean model: division into areas.

65°N and uses isopycnal (density) coordinates in the vertical direction, since mixing and movement in the ocean are believed to occur primarily along isopycnal surfaces. Exchanges with other oceans are also included since radionuclides entering the Atlantic will eventually disperse into all of the oceans. The exchanges are with the Arctic Ocean and Norwegian Sea to the north, the Mediterranean Sea to the east, and the Pacific and Indian oceans, via the Antarctic circumpolar current, to the south.

The first stage in developing the model was to define its basic structure. The Atlantic Ocean was divided into eight areas (Fig. 9.3), with the central dividing line corresponding to the Mid-Atlantic Ridge, and the areas extending to the edge of the continental shelf. Areas 1 and 3 are based on areas chosen by Worthington (1976) on the basis of bottom topography, and area 2 was chosen to enable the patterns of current flow in the north to be adequately represented. The southern boundary of area 4 (20°N) corresponds approximately to the position of the Cape Verde Islands. Areas 7 and 8 cover the remainder of the South Atlantic Ocean and

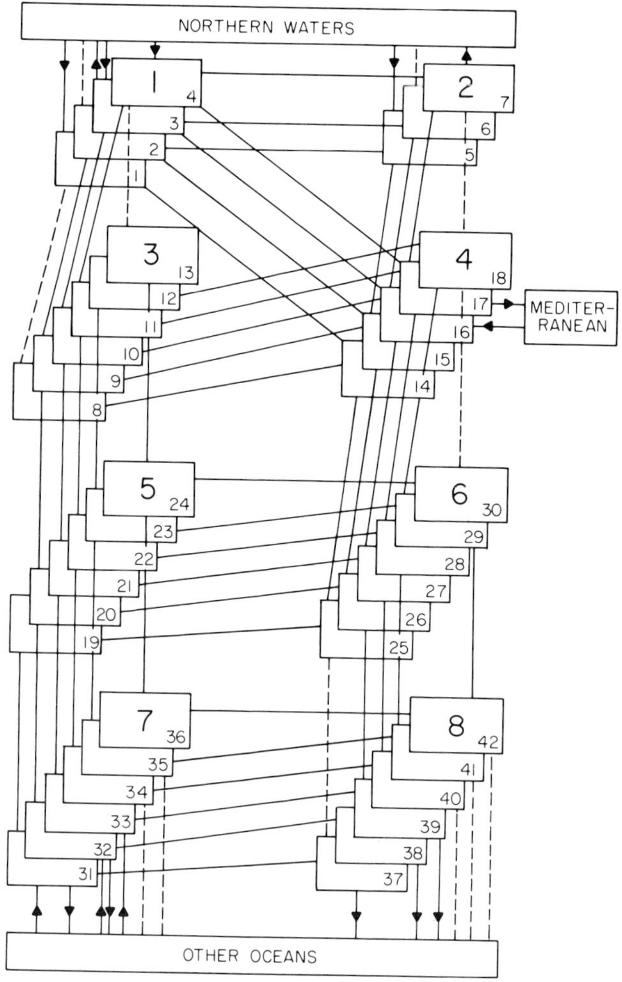

Figure 9.4.   The new ocean model. Each compartment exchanges vertically with its adjacent compartments. Horizontal exchanges occur between compartments connected by solid lines. Arrows designate source and sink exchanges. The broken lines indicate horizontal exchanges hypothesized to occur in one formulation of the SVD approach (see text).

extend to 50°S, which is approximately the latitude of the Falkland Islands in the southwest. The present NEA dumpsite is in area 4.

It is possible to identify well-defined water masses at specific depths throughout the Atlantic Ocean, and therefore the eight ocean areas were divided into layers so that each compartment would represent a separate water mass. The following water types were identified: (1) Antarctic bottom water (only in areas 7 and 8), (2) North Atlantic deep water, (3) North Atlantic central water, (4) 18°C water, and (5) surface water. The boundaries of the compartments were obtained by tracing the neutral (density) surfaces through the ocean by using the method described by Ivers (1975) and the data from the GEOSECS survey (Bainbridge, 1972). Surveys from the International Geophysical Year (Fuglister, 1960) and unpublished data from field work by the U.K. Ministry of Agriculture, Fisheries and Food were also used. The initial choice of 6 levels in the vertical direction led to 42 compartments in the model (Fig. 9.4). Each compartment was assumed to exchange with all adjacent compartments unless topography suggested otherwise, giving rise to a total of 84 exchanges (Fig. 9.4).

The second stage was to define the modelling approach. It was decided to attempt to determine the exchanges between the compartments from the existing temperature and salinity distributions in the ocean rather than from measurements of flow rates. This approach was taken for two reasons. First, measurements of mean flow are difficult, time-consuming, and the data base is inadequate. Second, as the exploratory calculations with the Worthington model show, use of measured current flow does not ensure compatibility between model predictions and the observed temperature and salinity distributions. The chosen modelling approach is termed the *inverse method* and has been used by Wunsch (1978), Thompson and Veronis (1980), Stommel and Veronis (1981), and Wunsch and Grant (1982).

9.3.2. Data Base

The data used to define the structure of the model were also used to derive the mean values for temperature and salinity in each compartment. Initially, this was done by averaging the values along the two surfaces defining the compartment and then taking the average for the compartment as the average of these two values. The calculated values are presented in Table 9.2. An alternative method of using bilinear interpolation between successive profiles was also considered, and work is still in progress on this revised data set.

The data for the exchanges with the other oceans were compiled from several sources (Sverdrup et al., 1942; Wright and Worthington, 1970; Worthington, 1976; and Georgi, 1981). Estimates of rainfall, evaporation, and heat flux for the surface of the ocean were obtained from Baumgartner and Reichel (1975), Bunker (1976), and Dorman and Bourke (1981). Since the ocean model is based on assumed steady-state tracer

Table 9.2. Salinity and Temperature Data for the Ocean[a]

Compartment Number	Salinity (‰)	Temperature (°C)
1	34.897 ± 0.055	1.883 ± 0.176
2	34.902 ± 0.072	2.549 ± 0.394
3	34.941 ± 0.228	4.958 ± 1.460
4	34.798 ± 0.388	8.132 ± 2.292
5	34.941 ± 0.008	2.856 ± 0.171
6	35.2 ± 0.011	6.969 ± 0.050
7	35.479 ± 0.043	10.690 ± 0.373
8	34.86 ± 0.016	1.623 ± 0.105
9	34.926 ± 0.051	2.566 ± 0.092
10	35.185 ± 0.051	6.933 ± 0.208
11	35.731 ± 0.062	12.854 ± 0.234
12	36.298 ± 0.060	16.642 ± 0.191
13	36.510 ± 0.096	17.946 ± 0.298
14	34.923 ± 0.008	2.275 ± 0.052
15	35.019 ± 0.066	3.420 ± 0.332
16	35.283 ± 0.083	7.381 ± 0.391
17	35.617 ± 0.074	11.661 ± 0.375
18	35.907 ± 0.254	13.907 ± 1.508
19	34.768 ± 0.057	0.736 ± 0.220
20	34.878 ± 0.053	2.034 ± 0.314
21	34.781 ± 0.052	4.936 ± 0.600
22	34.929 ± 0.147	9.755 ± 0.986
23	35.57 ± 0.197	14.981 ± 1.113
24	35.887 ± 0.282	17.171 ± 1.464
25	34.872 ± 0.033	2.015 ± 0.188
26	34.902 ± 0.033	2.870 ± 0.118
27	34.756 ± 0.063	4.926 ± 0.320
28	34.735 ± 0.085	8.002 ± 0.391
29	35.409 ± 0.202	16.065 ± 0.593
30	35.938 ± 0.386	22.498 ± 1.074
31	34.691 ± 0.007	0.351 ± 0.065
32	34.761 ± 0.034	1.561 ± 0.185
33	34.767 ± 0.147	3.746 ± 0.391
34	34.804 ± 0.264	8.144 ± 0.908
35	35.249 ± 0.402	14.213 ± 1.411
36	35.616 ± 0.670	17.099 ± 2.268
37	34.7 ± 0.013	0.409 ± 0.163
38	34.73 ± 0.026	1.380 ± 0.163
39	34.80 ± 0.112	2.320 ± 0.823
40	34.469 ± 0.256	3.209 ± 1.486
41	34.235 ± 0.328	5.333 ± 1.761
42	35.018 ± 0.464	14.505 ± 2.491

[a]Compartments are shown in Fig. 9.4.

concentrations, it is important that the sources and sinks of each tracer in the ocean are balanced. Unfortunately, the references for these data did not produce a balanced set of sources and sinks, and it was therefore necessary to adjust them slightly to achieve a balance. Two approaches were taken. In the first, the heat, salt, and water balances of the Atlantic Ocean were calcu-

Table 9.3. Source Terms for the Ocean Model

| | A. Exchanges with Other Oceans | | | | | |
| | Set 1 | | | Set 2 | | |
Compartment Number	Transfer Rate (SV)	Salinity (‰)	Temperature (°C)	Transfer Rate (SV)	Salinity (‰)	Temperature (°C)
1	4.0	34.86	3.4	2.0	34.95	2.8
2	—	—	—	2.0	34.95	3.4
3	2.1	34.3	2.0	2.1	—	—
3	− 1.1	34.94	4.96	− 2.0	34.94	4.96
4	2	34.3	2.0	− 0.5	34.80	8.13
5	2	35.0	3.4	0.3	34.98	4.5
6	—	—	—	2.0	34.98	4.5
7	− 8.4	35.48	10.69	− 9.0	35.48	10.69
16	1.0	36.4	11.0	4.0	35.08	10.5
17	− 1.0	35.62	11.66	− 1.0	35.62	11.66
31	7.0	34.4	0.0	4.7	34.71	0.2
31	− 4.0	34.69	0.35	− 2.0	34.69	0.35
32	1.0	34.38	1.0	—	—	—
32	− 0.4	34.76	1.56	− 8.0	34.76	1.56
33	2.0	34.0	4.0	1.5	34.4	2.5
34	—	—	—	2.8	34.2	4.0
35	—	—	—	2.0	34.3	5.5
37	− 2.8	34.7	0.41	2.0	34.67	0.3
38	− 2.0	34.73	1.38	2.0	34.69	0.8
39	− 1.1	34.8	2.32	− 1.0	34.8	2.32
40	—	—	—	− 1.0	34.47	3.2
41	—	—	—	3.0	34.2	4.0
42	—	—	—	− 0.7	34.24	5.33
42	—	—	—	− 2.8	35.02	14.5

| | B. Exchanges with the Atmosphere | | | |
| | Set 1 | | Set 2 | |
Compartment Number	Net Water Input (SV)	Net Heat Input (GW)	Net Water Input (SV)	Net Heat Input (GW)
4	3.93 E-2	− 5.8748 E5	3.93 E-2	− 1.0 E5
7	1.34 E-2	− 3.2553 E5	1.34 E-2	− 5.0 E4
13	− 8.0 E-2	2.5260 E4	− 8.0 E-2	− 8.0 E4
18	− 4.86 E-2	4.1040 E4	− 4.86 E-2	− 9.0 E4
24	0.0	5.5598 E5	0.0	3.52 E5
30	− 1.522 E-1	9.1290 E5	− 1.522 E-1	4.0 E5
36	− 2.17 E-2	− 2.0445 E5	− 2.17 E-2	− 7.4 E4
42	− 5.02 E-2	− 1.5722 E5	− 5.02 E-2	− 4.0 E4

lated with minimal change to the initial estimates of the sources and sinks and assuming a net water exchange with the atmosphere of 0.3 SV. In the second, a similar procedure was followed, but additional exchanges with the other oceans were included. The two sets of source terms are presented in Table 9.3.

9.3.3. Mathematical Basis

The model was set up so that each exchange is represented by a pair of transfer coefficients, one in each direction. These transfer coefficients, which by definition are assumed to be positive, can be interpreted in terms of advection and mixing; to a first approximation, by using the upwind (also termed "upstream" or "backward") differencing scheme, the mixing coefficient is the lesser of the two values and the advection is the difference.

Expressing the transfer coefficients in terms of column transfer coefficients W_{ij} (expressed in SV) we obtain from conservation of mass:

$$-Q(W)_i = \sum_j W_{ji}\rho_j - \sum_j W_{ij}\rho_i \qquad i \neq j \qquad 1$$

where $Q(W)_i$ is the input of water into compartment i [in megatons (10^{12} g) of water per second], and ρ_i is the density of water in compartment i in tons per m^3.

Assuming that salt is transported by the movement of water, conservation of salt gives

$$-Q(S)_i = \sum_j W_{ji}S_j\rho_j - \sum_j W_{ij}S_i\rho_i \qquad i \neq j \qquad 2$$

where $Q(S)_i$ is the salt input into compartment i [in kilotons (10^9 g) per second] and S_i is the mean salinity (in parts per thousand) in compartment i. Similarly, conservation of heat gives

$$-Q(H)_i = \sum_j W_{ji}C_pT_j\rho_j - \sum_j W_{ij}C_pT_i\rho_i \qquad i \neq j \qquad 3$$

where $Q(H)_i$ is the heat flux into compartment i (in 10^9 W), T_i is the mean temperature of water in compartment i (in °C), and C_p is the specific heat capacity of water.

Writing these three equations for each compartment gives a system of 126 equations in 168 volume-transfer coefficients (the unknowns). This is written in matrix form as

$$P \cdot W = R \qquad 4$$

where P is the property matrix (containing terms in ρ, $S\rho$, and $C_pT\rho$), W is the vector of volume-transfer coefficients, and R is the vector of sources and sinks. Initially it was assumed that ρ and C_p are constant throughout the ocean. This not only simplifies the property matrix but also implies that the volume, not the mass, of water is conserved. In order to improve the accuracy of the calculation, the salinity in each compartment was expressed as the sum of 35 ‰ and the residual, or salinity difference. The salt-conservation equations were then rewritten in terms of salinity difference and were substituted into P and R.

The system of equations obtained by the method described above is underdetermined in some places and overdetermined in others; therefore, it has an infinite number of solutions, none of which may be acceptable. These solutions can be expressed as the sum of a particular solution (usually the least-squares solution) that is determined by the sources and sinks, and a series of null-space solutions that are independent of the sources and sinks. The least-squares solution was considered to be the most useful starting point since it represents the flow pattern corresponding to the minimum energy in the ocean.

Two methods of obtaining the solution were considered. The first, based on work by Wunsch (1978), uses a matrix technique called singular value decomposition (SVD) to reduce the matrix of the set of conservation equations to a more manageable form. From this the least-squares solution and the null-space solutions are easily derived. The method [subroutine EBIOA described by Hopper (1981)] was based on a report by Golub and Reinsch (1970).

Since there is no method by which the SVD can be constrained to produce a purely positive least-squares solution, it is necessary to search for a positive solution by adding different combinations of the null-space solutions to the least-squares solution. If this is done in a least-squares sense (i.e., by using quadratic programming) (Wunsch and Minster, 1982) to minimize W (i.e., $\sum_{ij} W_{ij}^2$), subject to $W + \sum_k \alpha_k V_k > 0$, where V_k is the null-space solution vector, and W is the least-squares solution vector, then the final solution is the minimum-energy, purely positive solution to water flow in the ocean.

The second method of solution uses quadratic programming to search for a positive solution that produces the best fit to the conservation equation. The three formulations used were as follows:

1. Minimize W subject to

$$-Q(S)_i = \sum_j W_{ji} S_j \rho_j - \sum_j W_{ij} S_i \rho_i \qquad 5a$$

$$-Q(W)_i = \sum_j W_{ji} \rho_j - \sum_j W_{ij} \rho_i \qquad 5b$$

$$-Q(H)_i = \sum_j W_{ji} C_p T_{ji} \rho_j - \sum_j W_{ij} C_p T_i \rho_i \qquad 5c$$

for all i, $(i \neq j)$ and subject to $W_{ij} \geqslant 0$ for all i and j $(i \neq j)$. This formulation minimizes the energy of the occan and finds the flow pattern that exactly fits all the input data.

2. Minimize

$$\sum_i \left[\left(\sum_j W_{ji} S_j \rho_j - \sum_j W_{ij} S_i \rho_i + Q(S)_i \right)^2 \right.$$

$$+ \left(\sum_j W_{ji} \rho_j - \sum_j W_{ij} \rho_i + Q(W)_i \right)^2$$

$$\left. + \left(\sum_j W_{ji} \rho_j T_j C_p - \sum_j W_{ij} \rho_i T_i C_p + Q(H)_i \right)^2 \right] \qquad 6$$

subject to $W_{ij} = 0$ for all i and j $(i \neq j)$. This formulation finds the flow pattern that gives the best approximation to the input data.

3. Minimize

$$\sum_i \left[\left(\sum_j W_{ji} S_j \rho_j - \sum_j W_{ij} S_i \rho_i + Q(S)_i \right)^2 \right.$$

$$\left. + \left(\sum_j W_{ji} \rho_j T_j C_p - \sum_j W_{ij} \rho_i T_i C_p + Q(H)_i \right)^2 \right] \qquad 7$$

subject to $-Q(W_i) = \sum_j W_{ji} \rho_j - \sum_j W_{ij} \rho_i$ for all i $(i \neq j)$ and $W_{ij} \geqslant 0$ for all i and j $(i \neq j)$. This formulation conserves water exactly and finds the flow pattern that gives the best approximation to the salt and heat data.

The solutions were checked against the original data in two ways. First, the fit to the sources and sinks was determined by multiplying the original matrix of the conservation equations and the solution. This gives a new estimate of the sources and sinks. The second check was made on the predicted values of volume, salinity, and temperature in the ocean. The conservation equations were reformulated to give

$$F \cdot X = R \qquad 8$$

where F is the matrix of volume transfer coefficients, $F_{ii} = -\sum_j W_{ij}$ and $F_{ij} = W_{ji}$; X is the vector of salinities, volumes, and temperatures and R is the vector of sources. This equation was solved by using a variation of Gaussian elimination on F [subroutine MA21A described by Hopper (1981)] to give values for X. These were then compared with the original values.

Initially each of the conservation equations was given equal weight by scaling them so that all of the coefficients of the matrix P were on the order of unity. This also improves the accuracy of the calculation since the matrix is now well scaled.

9.4. RESULTS

The results presented here are preliminary and work on this model is continuing. The early calculations were done using the first set of sources and sinks (Table 9.3).

9.4.1. The SVD Approach

The SVD was used to produce a least-squares solution that fitted the original salinity and temperature distributions to within 10^{-2} (i.e., 0.2 ‰ and 0.1°C) and the first set of sources and sinks to within 10^{-6} (i.e., 10^{-6} SV, 2×10^{-5} ‰, and 1×10^{-5}°C). The solution contained both positive and negative volume-transfer coefficients (Figs. 9.5 and 9.6). This solution indicates that the energy is highest in the western basin, at the equator, and in the areas of deep-water formation; this agrees well with the expected spatial distribution of energy in the ocean.

The null-space solutions were confirmed as true null-space solutions by adding them all to the least-squares solution. The predicted values of sources, sinks, salinity,

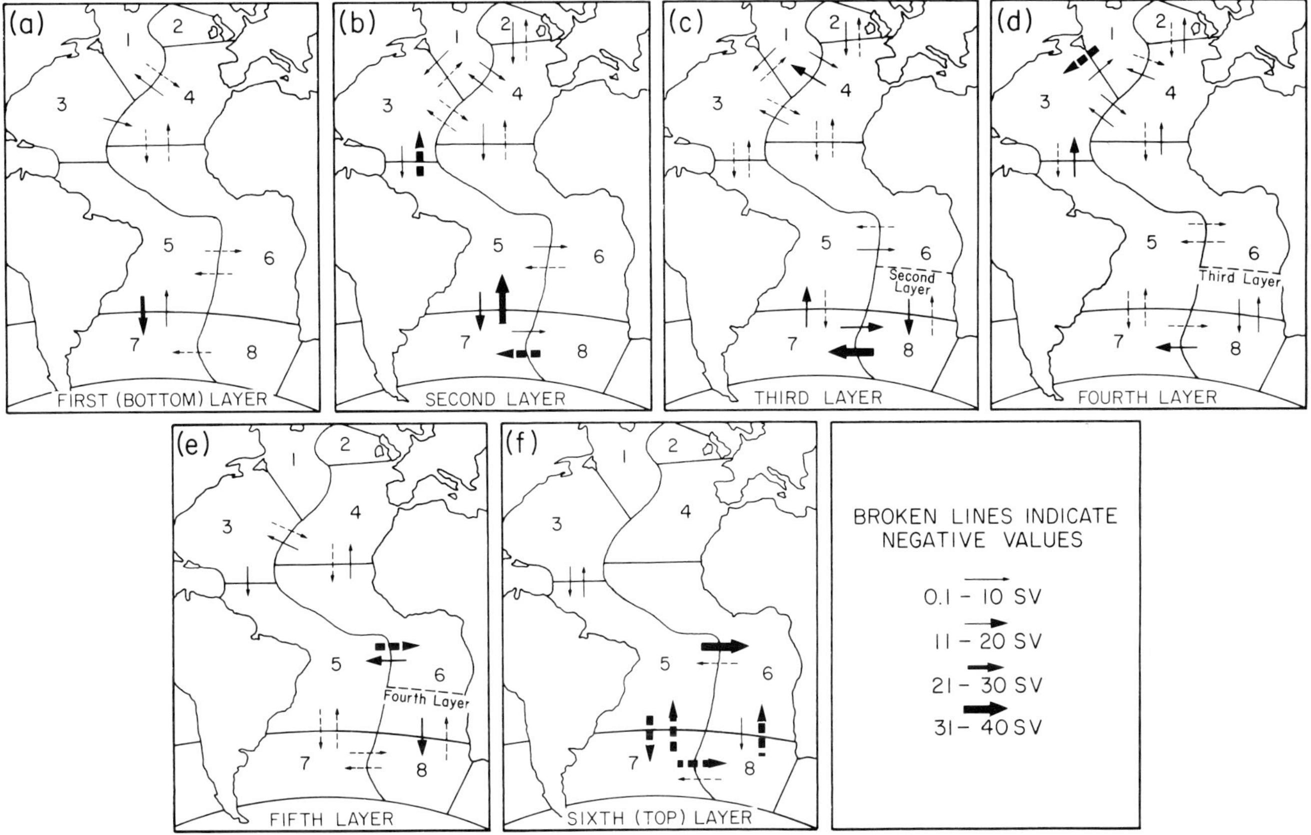

Figure 9.5. Calculated horizontal volume-transfer rates in the ocean: (a) first (bottom) layer, (b) second layer, (c) third layer, (d) fourth layer, (e) fifth layer, and (f) sixth (top) layer. Solid lines indicate positive values, and broken lines indicate negative values. Calculations based on the first set of source terms (Table 9.3) and the SVD approach (see text).

volume, and temperature were unchanged; however, no combination of null-space solutions could be found that would produce a positive solution. This was considered to be due to the structure of the model since, although it was undetermined on the whole, there were parts of the model in which the volume-transfer coefficients were determined. This occurred where one compartment exchanged with only three or less unknowns. This determined part of the matrix resulted in these volume transfer coefficients having no projection onto the null space, and therefore no positive value if the least-squares value was negative. Five very small ($<10^{-12}$) projections onto the null space were found and five additional exchanges (10 volume-transfer coefficients) were therefore included to free these areas. These five exchanges are shown as broken lines in Fig. 9.4.

The SVD was used on this enlarged system of 126 equations with 178 unknowns and produced a least-

squares solution that fitted all of the original data with an error of 10^{-10} (i.e., 10^{-11} SV, 10^{-9} ‰, and 10^{-9} °C). This indicated that the additional freedom had allowed the accuracy of the calculation to be greatly improved; however, a positive solution still could not be found. Using the improved accuracy obtained above as an important clue, it was decided to increase the freedom of the system even further in the hopes that doing so would allow a positive solution. Two approaches were taken.

In the first approach, since least-squares solution is dependent on the source terms, the extra freedom was allotted to the source terms. This is acceptable because the source terms were derived from inexact estimates and, therefore, small variations could be allowed. The freedom was added by introducing a slack variable (δ) into each of the conservation equations for compart-

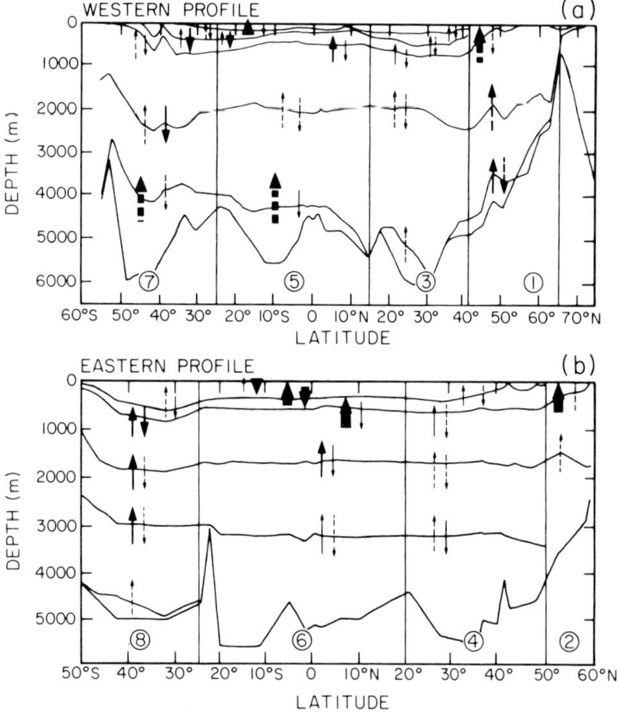

Figure 9.6. Vertical volume-transfer rates in the ocean. Calculated using the first set of source terms (Table 9.3) and the SVD approach (see text). Circled numbers refer to area numbers. Solid lines indicate positive values, and broken lines indicate negative values.

ments with sources and sinks, as in the following example:

$$ -Q(W)_i = \sum_j W_{ji}\rho_j - \sum_j W_{ij}\rho_i + \delta_1 \qquad 9 $$

Since 20 compartments have source terms, there are 60 slack variables. By allocating different weights to the slack variables, different accuracies could be assigned to the source terms. All slack variables were given the same weighting, and weights of 10^{-8} (i.e., very close to the given values), 1.0, and 10.0 were tried. The weight of 10^{-8} gave rise to the same least-squares solution but to slightly different null-space solutions. The weights of 1.0 and 10.0 changed both the least-squares solution and the null-space solutions as expected and gave differences of 10^{-4} between the original and predicted source vector **R**. The derived values of the slack variables are given in Tables 9.4 and 9.5; however, no positive solution could be found for any of these particular weighting values.

In the second approach, freedom was added to the structure of the model by introducing 20 more exchanges. These exchanges connected compartments 13 and 18 (the unconnected surface boxes in the North Atlantic Ocean), compartments 30 and 42 (the unconnected surface boxes in the South Atlantic Ocean), areas 3 and 6 (six exchanges across the Mid-Atlantic Ridge in the North Atlantic Ocean), areas 5 and 8 (six exchanges diagonally across the Mid-Atlantic Ridge in the South Atlantic), and areas 6 and 7 (six more exchanges diagonally across the Mid-Atlantic Ridge in the South Atlantic Ocean). The SVD produced a least-squares solution that gave predicted source terms to within 10^{-10} of the original source terms; however, a positive solution still was not found.

9.4.2. Quadratic Programming Approach

9.4.2a. First formulation

The quadratic programming approach was first used on the original system (with 168 volume-transfer coefficients), and initial values of unity were given to all of the volume-transfer coefficients. This formulation of the problem was found to be inefficient in terms of computer time. A feasible point routine (Hopper, 1981) was tried to give the quadratic programming routine a better set of initial values, but this was not any more efficient. It was concluded that there was some inconsistency in the constraint equations and that this formulation would not be productive. Since this formulation is the same as the one used by the SVD, this conclusion is not surprising.

9.4.2b. Second formulation

This formulation allows considerably more freedom in the solution. It was used on the original system, and values of the objective function (the expression to be minimized) were output at each iteration. These values indicated that this formulation would be costly to pursue and may give residuals that are too large to be interpreted as revised source terms.

9.4.2c. Third formulation

This formulation has the attraction that it provides for conservation of water within the interior of the ocean. It was used on the original system but was found to be very costly. Therefore, a different approach was taken that reduced the number of variables and constraints,

Table 9.4.   Source Terms Deduced from Slack Variables Chosen by SVD

A. Weighting of Slack Variables = 1.0			
Compartment Number	Transfer Rate (SV)	Salinity (‰)	Temperature (°C)
1	6.03	34.99	1.09
3	1.66	34.49	5.24
3	− 1.1	34.94	4.96
4	0.39	31.47	222.3
5	1.86	34.90	0.0
7	− 10.05	34.98	3.30
13	− 1.27	35.03	10.0
16	− 0.16	26.76	− 72.44
17	− 2.21	34.72	− 5.11
18	− 1.20	35.96	10.0
24	− 1.70	35.01	11.0
30	− 0.89	35.18	15.0
31	11.24	34.99	0.44
31	− 4.0	34.69	0.35
32	3.11	34.83	2.08
32	− 0.4	34.76	1.56
33	6.0	34.92	5.51
36	− 1.99	35.11	9.0
37	− 1.34	35.26	0.24
38	− 0.89	35.31	0.74
39	− 1.87	35.24	3.56
40	− 1.07	35.15	− 0.78
42	− 1.73	35.17	5.0

B. Weighting of Slack Variables = 10.0			
Compartment Number	Transfer Rate (SV)	Salinity (‰)	Temperature (°C)
1	4.55	34.96	2.7
3	3.22	34.78	1.27
3	− 1.1	34.94	4.96
4	2.26	34.67	6.68
5	2.05	34.99	3.02
7	− 7.28	34.93	10.80
13	− 0.9	33.85	10.0
16	0.31	39.54	36.74
17	− 0.99	34.36	11.34
18	− 0.67	34.98	10.0
24	− 0.63	35.01	11.0
30	− 1.55	34.97	15.0
31	9.2	34.93	0.03
31	− 4.0	34.69	0.35
32	1.7	34.63	0.53
32	− 0.4	34.76	1.56
33	3.58	34.72	2.23
36	− 1.00	35.00	9.00
37	− 3.12	35.09	0.36
38	− 2.35	35.11	1.11
39	− 1.21	35.16	2.07
40	− 0.23	34.99	− 0.43
42	− 1.3	35.01	5.0

Table 9.5. Atmospheric Exchanges Deduced from Slack Variables

A. Weighting of Slack Variables = 1.0		
Compartment Number	Transfer Rate (SV)	Heat Input (GW)
4	3.93 E-2	− 5.87 E5
7	1.34 E-2	− 3.25 E5
13	1.5 E-1	8.46 E4
18	2.14 E-2	− 1.59 E4
24	− 1.0 E-2	2.73 E5
30	1.28 E-1	6.20 E5
36	− 2.17 E-2	− 4.84 E4
42	− 5.0 E-2	− 5.81 E4

B. Weighting of Slack Variables = 10.0		
4	3.93 E-2	− 5.87 E5
7	1.34 E-2	− 3.26 E5
13	1.10 E-1	6.63 E4
18	− 4.86 E-2	6.60 E4
24	0.0	5.21 E5
30	− 1.52 E-1	9.44 E5
36	− 2.17 E-2	− 1.51 E5
42	− 5.0 E-2	− 1.18 E5

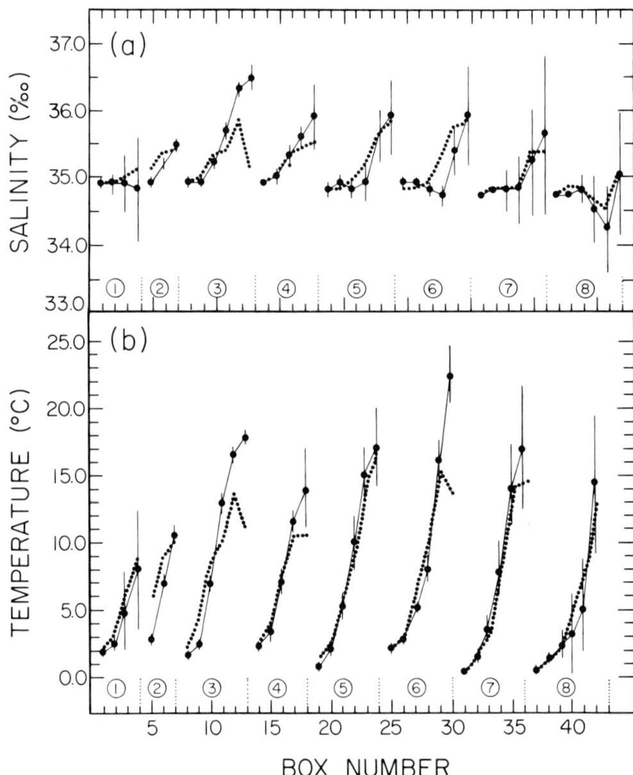

Figure 9.7. Comparison of original values (solid lines) $\pm 2\sigma$ and those calculated (dotted lines) for salinity (a) and temperature (b), by using the second set of source terms (Table 9.3) and the third formulation of the quadratic programming approach (see text). Numbers in circles refer to area numbers.

thereby improving the efficiency of the calculation. The SVD was used to obtain the least-squares solution and the null-space solutions to the volume-conservation equations only. Quadratic programming was then used to find the coefficients of the null-space vectors that optimized the fit to the temperature and salinity source terms and gave positive volume-transfer coefficients. Conservation of water was automatically achieved by confining the search to the null space of the volume-conservation equations. This was used on the system with 178 volume-transfer coefficients. The accuracy of the calculation was such that for the first set of source terms the routine returned with a few small (10^{-2}) negative volume-transfer coefficients. An offset of 0.05 was therefore applied to the positive solution constraints (i.e., subject to $W + \sum_k \alpha_k V_k = 0.05$). The resulting flow pattern has minimum volume-transfer coefficients of 5 SV and maximum values of 9 SV.

With the second set of source terms it proved unnecessary to apply the offset. The solution and the fit to the original data are shown in Figs. 9.7 through 9.9. Figure 9.7 compares the predicted and original values of temperature and salinity in the ocean by overlaying the corresponding profiles of temperature and salinity with depth for each ocean area. Many of the salinity and temperature results lie within the range of the original values, but there are discrepancies. In Figs. 9.8 and 9.9,

it can be seen that the solution reflects some of the general features of the circulation. For example, in most layers the influence of the Antarctic circumpolar current is in the expected direction (Fig. 9.8), but problems with the box structure and the sources and sinks in Area 8 give spurious deep-water results. The southward flow of the North Atlantic Deep Water and the northward movement of the Antarctic Intermediate Water are also reproduced (Fig. 9.9).

9.5. DISCUSSION

Although a solution that contains only positive volume-transfer coefficients and that conserves water has been produced for the ocean model, the values of temperature and salinity that it predicts for the interior of the ocean are not as close to the original data as desired. Since the system is underdetermined, it should be possi-

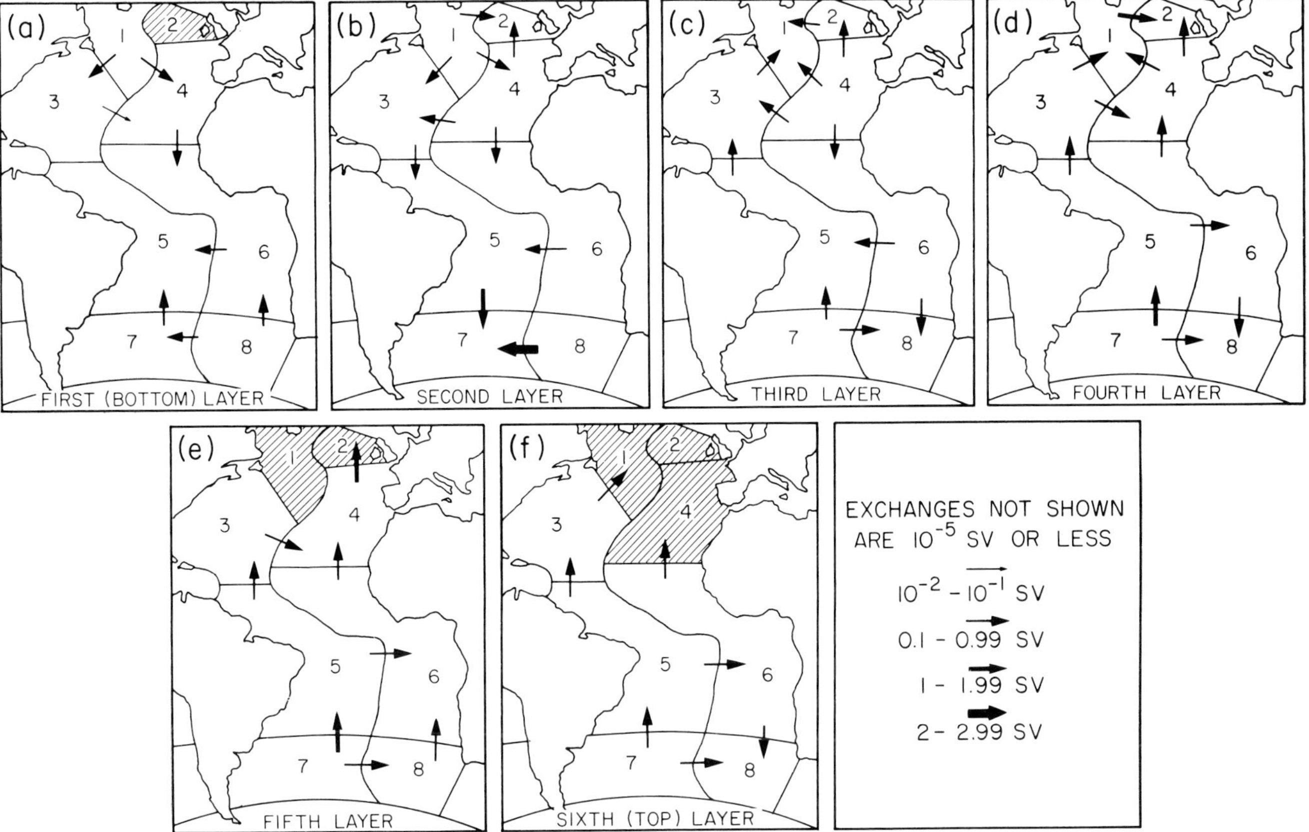

Figure 9.8. Calculated horizontal volume-transfer rates in the ocean: (a) first (bottom) layer, (b) second layer, (c) third layer, (d) fourth layer, (e) fifth layer, and (f) sixth (top) layer. Calculation based on the second set of source terms (Table 9.3) and the third formulation of the quadratic programming approach (see text).

ble to reproduce exactly the data used to set up the model. This has been done with the solution containing both negative and positive volume-transfer coefficients.

The difficulty in obtaining a positive solution by requiring an exact fit to the data is probably due to the existence of temperature and salinity gradients within the model that can only be sustained by a negative volume-transfer coefficient. If these gradients can be removed by allowing the model increased freedom to choose its source terms and by revising the original temperature and salinity distributions, then both a positive solution and a good fit to the data may be possible. For example, significant improvements may be obtained by revising the exchanges with the other oceans for Area 8. The SVD formulation does not easily allow the system flexibility in its sources and optimization techniques will be given more emphasis.

9.6. CONCLUSIONS

A compartment model for the Atlantic Ocean has been defined and attempts have been made to use the observed temperature and salinity distributions in the ocean to derive a flow pattern. There are some inconsistencies in the data, some of which may occur if the values are not representative of the area spanning the width of half of the ocean. The assumptions made in setting up the model discussed in this chapter will be reassessed, and the same techniques reapplied.

In conclusion, this experience with inverse techniques on large systems suggests that the apparently limitless freedom of an underdetermined system is illusory and that if this method is to be applied successfully, great care is needed in formulating the problem.

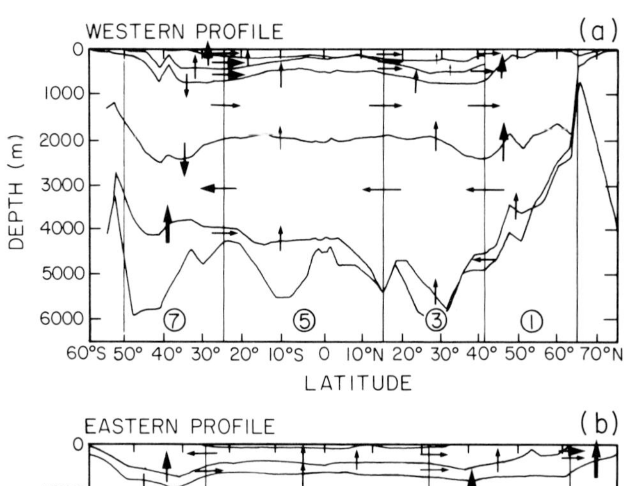

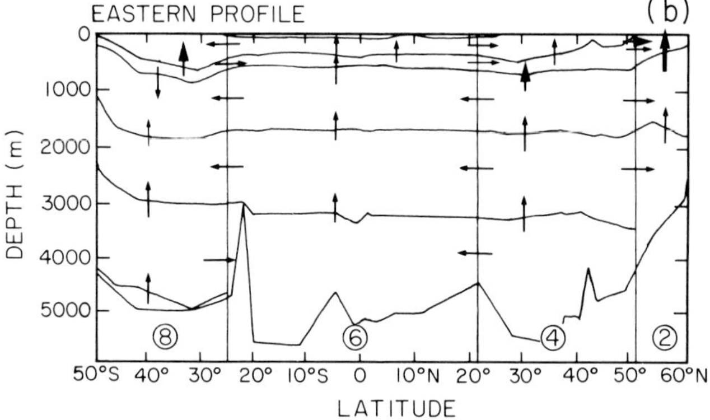

Figure 9.9. Calculated vertical volume-transfer rates in the ocean. Calculation based on second set of source terms (Table 9.3) and the third formulation of the quadratic programming approach (see text). Broken lines indicate negative values. The arrows indicate (in terms of increasing width) 5.0–5.4, 5.5–5.9, 6.0–6.9, 7.0–7.9, 8.0–8.9, and 9.0–9.9 SV, respectively. The circled numbers refer to area numbers.

Since this chapter was originally written, the model has been enlarged to include the entire world's oceans. A more detailed resolution of the water masses in the Atlantic Ocean and subsequent recalculation of the average temperature and salinity in each (new) box removed the difficulties encountered in describing the exchanges between Area 8 and the other oceans. Unconstrained optimization techniques have been used to obtain positive solutions, but these solutions are very dependent on the initial values used in the optimization. Needler (personal communication) pointed out that the term *inverse technique* describing this modelling approach usually implies that dynamical constraints (such as geostrophic balance) have been included in the minimization function. This is not the case here, and recent work has shown that the simple use of just tracer constraints in an underdetermined system, with which

this chapter deals, can lend to an unrealistic model of circulations. Consequently, this model is currently being run with exchanges selected in a conventional manner from flows inferred from observation and augmented by constant lateral and diapycnal mixing rates. Ways of improving and verifying the model are being considered.

REFERENCES

Anderson, D. R. (Ed.). 1982. Seventh International NEA Seabed Working Group (March 15–19, 1982), LaJolla, California. Sandia Report SAND 82-0460, Sandia National Laboratories, Albuquerque, New Mexico, 221 pp.

Bainbridge, A. E. 1972. GEOSECS Atlantic Expedition, Vol. 2: Sections and Profiles. Stock No. 038-000-00435-2, U.S. Government Printing Office, Washington, D.C., 198 pp.

Baumgartner, A., and E. Reichel. 1975. The World Water Balance. R. Oldenburg, Munich, 179 pp.

Bryan, K. 1969. A Numerical Model for the study of the circulation of the World Ocean. *Journal of Computing Physics*, **4**, 347–376.

Bunker, A. F. 1976. Computations of surface energy flux and annual air–sea interaction cycles of the North Atlantic Ocean. *Monthly Weather Review*, **104**, 1122–1140.

Dorman, C. E., and R. H. Bourke. 1981. Precipitation over the Atlantic Ocean 30°S to 70°N. *Monthly Weather Review*, **109**, 554–563.

Fuglister, F. C. 1960. Atlantic Ocean Atlas of Temperature and Salinity Profiles and Data from the International Geophysical Year of 1957–1958. Woods Hole Oceanographic Institution Atlas Series, Woods Hole, Massachusetts, 209 pp.

Georgi, D. T. 1981. Circulation of bottom waters in the southwestern South Atlantic. *Deep-Sea Research*, **28**(A), 959–979.

Golub, G. H., and C. Reinsch. 1970. Singular value decomposition and least squares solutions. *Numerical Mathematics*, **14**, 403–420.

Gurbutt, P. A. 1983. International Model Intercomparison Exercise. OCMAG 3 (The Newsletter of the NERC Ocean Modelling Group) February 1983, D. Webb and K. Richards (Eds.). Institute of Oceanographic Services, Wormley, U.K., unpublished manuscript.

Hopper, M. J. 1981. Harwell Subroutine Library: A Catalogue of Subroutines. Atomic Energy Research Establishment Report AERE-R19185, Harwell, Oxon, United Kingdom, 71 pp.

International Atomic Energy Agency. 1978. The Oceanographic Basis of the IAEA Revised Definition and Recommendations Concerning High-Level Radioactive Waste Unsuitable for Dumping at Sea. IAEA-210, International Atomic Energy Agency, Vienna, 65 pp.

International Commission on Radiological Protection. 1977. Recommendations of the International Commission on Radiological Protection. ICRP Publication 26. *Annals of the International Commission of Radiological Protection*, **1**(3), 1–55.

Ivers, D. W. 1975. The Deep Circulation of the Northern North

Atlantic with Especial Reference to the Labrador Sea. Ph.D. Thesis, University of California at San Diego, 179 pp.

Killworth, P. D. 1980. On the determination of absolute velocities and density gradients in the ocean from a single hydrographic section. *Deep-Sea Research*, **27**, 901–929.

Needler, G. T. 1982. On determining the velocity from the density field-including a closed form. Unpublished manuscript in *Ocean Modelling*, Vol. 46, available on request from A. E. Gill (Ed.). Robert Hooke Institute, Department of Atmospheric Physics, Oxford, U.K., unpaginated.

Robinson, A. R. (Ed.). 1984. Interim Meeting of the Physical Oceanography Task Group of the International NEA Sea Bed Working Group Held in Cambridge, U.K., 1982. Sandia Report SAND 83-1808, Sandia National Laboratories, Albuquerque, New Mexico, 187 pp.

Shepherd, J. G. 1976. A Simple Model for the Dispersion of Radioactive Wastes Dumped on the Deep-Sea Bed. Fisheries Research Report No. 29, Ministry of Agriculture, Fisheries and Food, Lowestoft, U.K., 29 pp.

Shepherd, J. G. 1980. A two dimensional model of the meridional transport of tracers in the deep ocean. Unpublished manuscript in *Ocean Modelling*, Vol. 31, available on request from A. E. Gill (Ed.). Robert Hooke Institute, Department of Atmospheric Physics, Oxford, U.K., unpaginated.

Stommel, H., and F. Schott. 1977. The beta spiral and the determination of the absolute velocity field from hydrographic station data. *Deep-Sea Research*, **24**, 325–329.

Stommel, H., and G. Veronis. 1981. Variational inverse method for study of ocean circulation. *Deep-Sea Research*, **28**(A), 1147–1160.

Sverdrup, H. F., M. W. Johnson, and R. H. Fleming. 1942. The Oceans. Prentice-Hall, New York, 1060 pp.

Thompson, R. O. R. Y., and G. Veronis. 1980. Transport calculations in the Tasman and Coral Seas. *Deep-Sea Research*, **27**(A), 303–323.

Webb, G. A. M., and F. Morley. 1973. A Model for the Evaluation of the Deep Ocean Disposal of Radioactive Waste. National Radiological Protection Board Report NRPB-R14, Her Majesty's Stationery Office, London, 26 pp.

Worthington, L. V. 1976. On the North Atlantic Circulation. Johns Hopkins University Press, Baltimore, Maryland, 110 pp.

Wright, W. R., and L. V. Worthington. 1970. The Water Masses of the North Atlantic Ocean: A Volumetric Census of Temperature and Salinity. Serial Atlas of the Marine Environment, Folio 19, American Geographical Society, New York, 8 pp. + 7 plates.

Wunsch, C. 1978. The North Atlantic general circulation west of 50°W determined by inverse methods. *Reviews of Geophysics and Space Physics*, **16**(4), 583–620.

Wunsch, C., and B. Grant. 1982. Towards the general circulation of the North Atlantic Ocean. *Progress in Oceanography*, **11**(1), 1–59.

Wunsch, G., and J. F. Minster. 1982. Methods for box models and ocean circulation tracers—mathematical programming and nonlinear inverse theory, *Journal of Geophysical Research*, **87**, 5647–5662.

Chapter 10

A Research Plan for Deep-Ocean Disposal of Sewage Sludge off Orange County, California

Norman H. Brooks, Robert G. Arnold, and Robert C. Y. Koh

Environmental Quality Laboratory
California Institute of Technology
Pasadena, California

George A. Jackson

Institute of Marine Resources
Scripps Institute of Oceanography
La Jolla, California

William K. Faisst

Brown and Caldwell
Walnut Creek, California

ABSTRACT

A seven-year research plan has been developed to study the environmental and possible health effects of a proposed outfall for the discharge of sewage sludge into the ocean. The proposed pipeline would be constructed to an unprecedented depth of about 300–400 m off the coast of Orange County in southern California. Anticipated environmental effects associated with marine sludge disposal would be minimal at that depth. Approval to conduct this experiment is being sought through special legislation and regulatory actions at both state and federal levels. Because a sludge outfall serving the County Sanitation Districts of Orange County would be about four or five times deeper than any existing wastewater or sludge outfalls, special research for system engineering and evaluation of environmental effects is necessary. The plan that has been developed to direct this research is designed to examine comprehensively the transport, fate, and effects of sludge discharged at depth off the southern California coast.

10.1. INTRODUCTION

Disposal of sludge presents difficult and costly problems for sewage-treatment agencies throughout the United States. No disposal option is without significant economic and potentially adverse environmental effects. Disposal methods involving landfilling or combustion are permitted if carried out in an appropriate manner; on the other hand, the discharge of sludge to the ocean via an outfall is prohibited by the United States Clean Water Act, even though the Ocean Dumping Act allows permits to be issued for sludge disposal by barge under restrictive conditions (Krier, 1983). Despite existing legal and regulatory impediments, there is renewed interest in using marine waters to dispose of municipal sludge. For many coastal cities, disposal into the ocean is an attractive alternative, not only because it is less expensive than is disposal on land or the combustion of sludge, but also because its potential environmental effects may be less significant than are those associated with other options. National policy for sludge disposal is currently under scrutiny, particularly in regard to marine sludge outfalls and the continuation or expansion of sludge dumping by barges or ships.

The County Sanitation Districts of Orange County (CSDOC) collects, treats, and disposes of wastewater for two million people. The treated effluent is discharged more than 6.5 km offshore at a depth of 60 m within the Southern California Bight. The mean flow rate is approximately 750×10^3 m^3 d^{-1}. The performance of this wastewater effluent outfall has been excellent, with minimal adverse environmental effects. Residual solids from wastewater treatment amount to 136×10^3 kg d^{-1} (dry weight) suspended in 11.4×10^6 liters of wastewater (1.2% solids, by weight).

The County Sanitation Districts of Orange County now proposes an experiment in which sludge would be discharged via a small outfall (45–60 cm in diameter) terminating 12 km offshore at a depth of 300–400 m. The environmental impacts of this activity are expected to be minor, based on what is currently known, and no risk to human health is anticipated. Nevertheless, pathways for possible exposure of humans to trace contaminants or pathogens will be carefully monitored as part of the overall experiment. If the research plan reveals unacceptable health or ecological impacts, the CSDOC would modify or discontinue its marine sludge-discharge operations. Upon termination of the discharge, any sludge-related changes in biological communities should be reversed within only a few years. The CSDOC is currently seeking approval to conduct this experiment through special legislation and regulatory actions.

Because the CSDOC sludge outfall would be about four or five times deeper than existing wastewater or sludge outfalls, special research for system engineering and evaluation of environmental effects must be undertaken. Because of depth-related variation in dissolved-oxygen concentration, temperature, community structure, and many other potentially important environmental variables, monitoring data collected from discharge sites in shallower water probably cannot support the accurate prediction of discharge-related impacts in the 300- to 400-m depth range. Therefore, a comprehensive monitoring and research plan describing 40 individual research tasks in 9 groups has been developed. Highlights of the plan are discussed in Section 10.8. For more details, the reader is referred to a report by Brooks et al. (1982).

Should the deep-water sludge-disposal experiment proceed as proposed, the recommended research and monitoring activities are expected to produce important results that would assist state and national policy-makers in making difficult sludge-disposal decisions. Our primary objective in this chapter is to describe a model program for oceanographic monitoring and research that can be used to assess the impacts, and ultimately the advisability, of marine sludge discharge by Orange County. We will first describe the sludge itself, and then discuss the physical characteristics of the water, sediment, and biota at the proposed discharge location. We will then point out those changes in the marine environment likely to result from the proposed sludge discharge. In general, however, existing information is inadequate for the type of detailed predictions necessary to make informed regulatory decisions: hence the motivation for the proposed experimental discharge and research plan.

10.2. ORANGE COUNTY FACILITIES AND SLUDGE-DISPOSAL PLANNING

All wastewater treated by the CSDOC receives at least primary treatment. About 470×10^3 m^3 d^{-1} of wastewater is also provided with some form of biological treatment (trickling filter or activated sludge). Sludges from these processes are anaerobically digested, mechanically dewatered, and composted prior to disposal as landfill. In 1978, the average disposal rate of digested sludge was 100 metric tons (t) (dry weight) each day.

Marine disposal of sludge is much less expensive than is disposal on land. The estimated cost of oceanic disposal is less than one-fourth that of the parallel land-based alternative (Table 10.1), with annual savings of about $10 million (1981 U.S. dollars). These figures represent only rough estimates and should be refined by further detailed engineering and planning studies.

The proposed deep-water sludge outfall would consist of a single pipeline with an inside diameter of 45–60 cm extending about 12 km from shore to a depth of 300–400 m (Fig. 10.1). Its exact depth and location would be determined during preliminary design. The sludge would be discharged at a rate of approximately 11.4×10^6 liter d^{-1} and would consist of a mixture of digested primary sludge (3.0×10^6 liter d^{-1}) and waste-activated sludge (8.4×10^6 liter d^{-1}). This material may be prediluted with secondary effluent before discharge at a ratio of perhaps one or two parts of effluent per part of sludge. Predilution would both decrease the resulting concentrations of sludge contaminants and also simplify the logistics of maintaining a minimum flow velocity through the pipeline.

10.3. THE PROPERTIES OF SLUDGE

Most municipal sludges, including that from Orange County, contain measurable amounts of trace metals and refractory organic compounds such as pesticides and industrial solvents. Neither dewatering nor anaerobic digestion effectively reduces these contaminants. Marine disposal methods would rely both on source control and on dispersal to ensure that concentrations of such materials are maintained below potentially harmful levels in the ocean.

Sludge chemistry depends on (1) the nature of materials discharged into the sewer system (i.e., the degree of industrialization within the service area), and (2) the processes used to treat wastewater and sludge. Many important contaminants in wastewater are removed by physical, chemical, or biological treatment processes

Table 10.1 Estimated Costs of Alternative Sludge-Disposal Methods for County Sanitation Districts of Orange County[a]

Cost	Disposal by Landfill (1981 U.S. dollars)	Disposal by Deep-Ocean Outfall (1981 U.S. dollars)
Capital costs ($10^6)	35–48	6–11
Annual costs ($10^6 y^{-1})		
Capital[b]	4.0–5.3	0.7–1.2
Operation and maintenance	6.6	0.25
Special research and monitoring program (CSDOC share)	—	0.75–1.25
Total annual cost ($10^6 y^{-1})	10.6–11.9	1.7–2.7
Unit costs ($ t^{-1})[c]	90–100	15–23

[a] Based on preliminary data provided by the County Sanitation Districts of Orange County in February 1981. From Brooks and Krier, 1981.

[b] Based on 10.25% interest, amortized as follows: land, interest only; storage tanks and outfall pipe, 30 y; pumping, dewatering and screening equipment, 10 y; and trucks, 7 y.

[c] Actual digested-sludge discharge is 136 t d^{-1} (dry weight), which is derived from 320 t d^{-1} of raw sludge. For consistency, unit costs are given in costs per ton of original raw sludge.

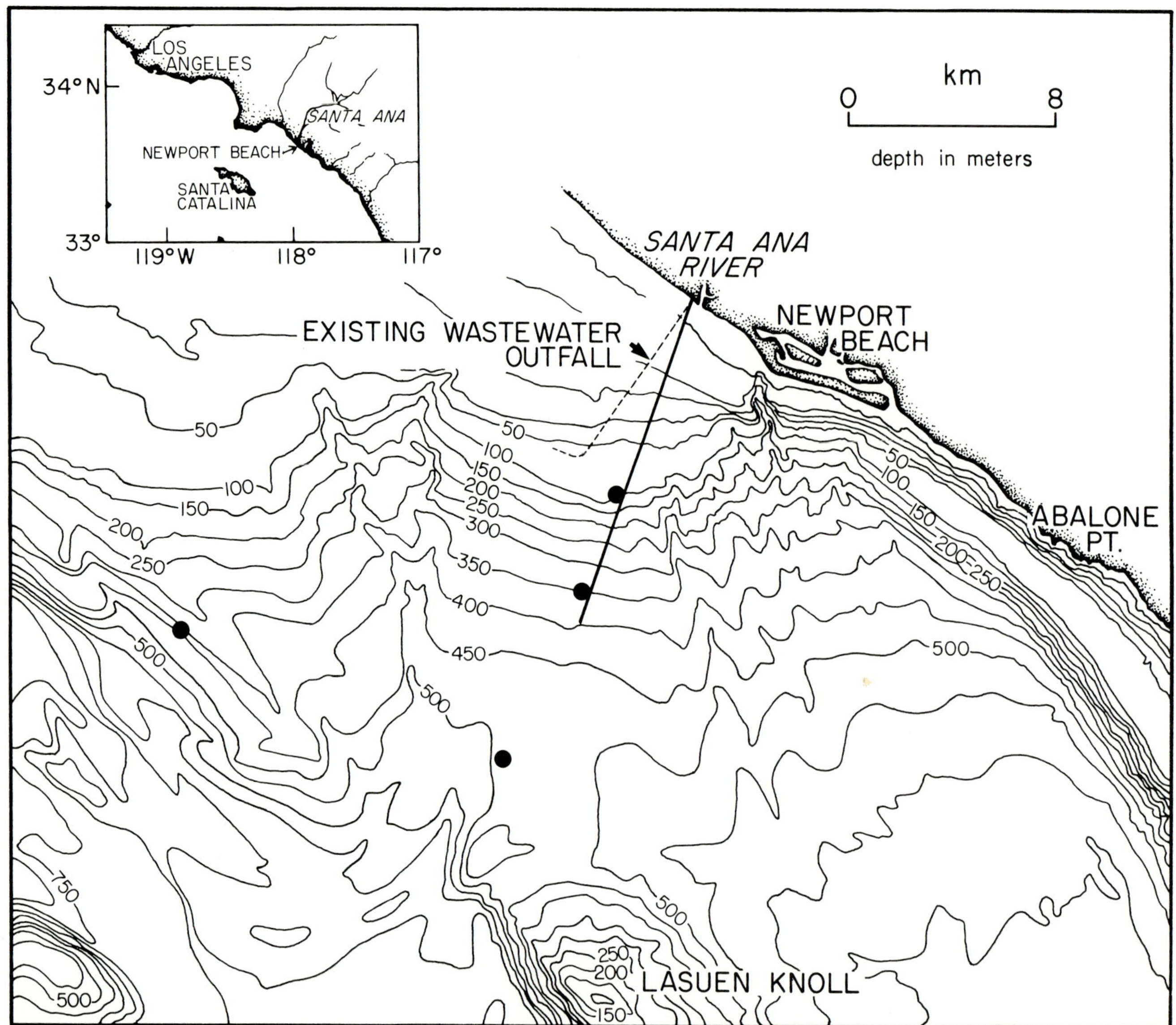

Figure 10.1. Basin Bathymetry off Orange County, California, including the locations of the existing CSDOC wastewater outfall and the proposed sludge outfall. Closed circles denote proposed current-meter moorings, and straight line denotes probable alignment of sludge outfall.

and become concentrated in the sludge. Consequently, a large fraction of some trace metals and trace organic compounds from raw wastewater are found in sludge, mainly in association with particles. Tables 10.2 and 10.3 indicate that, although all pollutants are much more concentrated in sludge than in wastewater effluent, the flux in the sludge volume is about 70 times less in the CSDOC system. Consequently, mass-emission rates for sludge and effluent are approximately equal for many constituents (suspended solids, volatile suspended solids, Ag, Cd, Cr, Cu, and Pb). For some other constituents of sewage, the mass-emission rates from sludge would be fractions of their emission rates in effluent. For biochemical oxygen demand (BOD), polychlorinated biphenyls (PCBs), chlorinated hydrocarbons (CHC), and Ni the mass emissions in sludge

Table 10.2. Projected Pollutant Concentrations in Various CSDOC Sludge Streams[a]

	Concentrations in Sludge (Projected)				Concentrations in Sewage Effluent[c] (1981–1982) (217 MGD) (mg liter^{-1})	Concentration Ratio (Sludge: Effluent)
Substance	Primary[b] (0.80 MGD, Digested) (mg liter^{-1})	Waste-Activated[b] (2.20 MGD, Undigested) (mg liter^{-1})	Trickling-Filter Humus (0.02 MGD, Digested) (mg liter^{-1})	Combined Sludge (3.0 MGD) (mg liter^{-1})		
Suspended solids	25,000	6,800	20,000	11,500	115	100
Volatile suspended solids	13,000	5,100	16,000	7,200	90	81
5-d BOD (20°C)	2,500	1,400	2,000	1,700	150	11
Ag	2.3	0.24	1.5	0.8	0.014	57
Cd[d]	3.1	0.68	8	1.4	0.020	70
Cr	11.5	3	25	5.5	0.074	74
Cu	42	6.6	50	17	0.218	77
Ni	4.8	0.3	2	1.5	0.07	21
Pb	17	2.1	18	6	0.08	75
Zn	58	7	80	21	0.20	105
Polychlorinated biphenyls	0.12	0.028	0.50	0.050	0.00149	34
CHC pesticides	0.0075	0.0022	0.050	0.004	0.000143	28

[a] Projections are based on influent quality and operational parameters from 1981, assuming that planned secondary-treatment capacity at CSDOC Plant No. 2 is in operation. MGD means millions of U.S. gallons per day, with 1 MGD equal to 3.79×10^6 liters d^{-1}.

[b] Figures are based on data representative of the period July 1979 to June 1981.

[c] Figures representative of the 1982 CSDOC average wastewater effluent quality (G. Pamson, personal communication).

[d] Source-control measures planned for 1982–1983 are expected to reduce the average Cd concentration in CSDOC sludge (combined) to approximately 85 mg kg^{-1} (dry weight) before the end of 1983. This represents a reduction of about 23% below the figures for Cd provided here.

Table 10.3. Projected Mass-Emission Rates of Pollutants in Various CSDOC Sludge Streams[a]

| | Mass Emission Rates in Sludge Streams (Projected) | | | | Mass Emission Rate in Sewage Effluent[c] (1981–1982) (217 MGD) (kg d^{-1}) | MER Ratio: Sludge: Effluent (0.014) |
	Primary[b] (0.80 MGD, Digested) (kg d^{-1})	Waste Activated[b] (2.20 MGD, Undigested) (kg d^{-1})	Trickling-Filter Humus (0.02 MGD, Digested) (kg d^{-1})	Combined Sludge (3.0 MGD) (kg d^{-1})		
Suspended solids	75,000	56,000	1,500	133,000	94,500	1.4
Volatile suspended solids	39,000	42,000	1,200	82,000	73,900	1.1
5-d BOD (20°C)	7,500	11,300	200	19,000	123,000	0.15
Ag	7.0	2.0	0.1	9	11.5	0.78
Cd[d]	9.4	5.6	0.6	16	16.4	0.98
Cr	35	25	1.9	62	60.8	1.02
Cu	129	55	3.8	190	179	1.06
Ni	14.4	2.5	0.2	17	57.5	0.29
Pb	51	17.1	1.4	70	65.7	1.07
Zn	175	58	6.1	240	163	1.5
Polychlorinated Biphenyls	0.3	0.2	0.04	0.5	1.22	0.41
CHC Pesticides	0.023	0.018	0.004	0.045	0.12	0.38

[a]See Table 10.2 for footnotes. A dash denotes the absence of data.

would be 0.15, 0.4, 0.4, and 0.3, respectively, times their effluent emissions.

Tables 10.2 and 10.3 reflect the changes in sludge quantity and quality that are expected to take place following construction of additional secondary-treatment facilities for the CSDOC. In general, the projected concentrations of the major trace metals in CSDOC sludges are representative of sludges derived within major U.S. municipalities (U.S. National Research Council, 1984). Source-control and pretreatment efforts within the CSDOC service area have appreciably reduced influent mass-flow rates of several trace contaminants during the past five years: the inputs of Cd, Zn, Pb, Cr, Ni, and Cu have decreased by 40–70% since 1976. Among the trace metals routinely monitored, only Ag remains essentially unaffected by source-control measures.

The physical properties (density, particle size, and settling characteristics) of sludge particles influence their fate subsequent to marine discharge (Kavanaugh and Leckie, 1980; Koh, 1982). The bulk specific gravity of liquid sludge ranges from slightly greater than 1.00 to about 1.02, depending on the concentration of solids and type of sludge. Particulate materials within the sludge have specific gravities ranging from slightly over 1.0 for biological solids to about 2.7 for fine sand, silt, and clay. The size of particles in a sludge sample ranges from colloids 1 μm in diameter to visible material with diameters 50–100 μm. About half of the volume of the

solids in the sludge is comprised of particles <20 μm in diameter; particles 50 μm and smaller comprise 90% of the particle volume.

Settling is affected by particle size, particle density, and degree of coagulation in seawater, and the mean settling velocity decreases with increased dilution due to reduced coagulation effects. Figure 10.2 indicates that 30–35% (by weight) of the sludge particles settle with velocities of less than 10^{-3} cm s^{-1}, or less than ~1 m d^{-1}. This fraction of the solids discharged from a sludge outfall would spread over wide areas prior to reaching the bottom (Section 10.5). It is important to note, however, that two physical mechanisms (gravity settling and downward diffusion) participate in the downward migration of sludge particles. Downward diffusion becomes an important factor in determining the downward migration of particles that fall slowly (<10^{-3} cm s^{-1}) (Koh, 1982).

10.4. CHARACTERISTICS OF THE DISCHARGE SITE

10.4.1. Bathymetry

The underwater topography off the southern California coast is an extension of the rugged mountains onshore. Local bathymetry is characterized by a number of deep basins approximately 1000 m deep. Deeper portions of these basins are completely surrounded by steep slopes below the sill depths. The various sills connecting the basins vary in depth but are generally several hundred meters deep.

The bathymetry off the mouth of the Santa Ana River is shown in Fig. 10.1. Alignments both of the existing Orange County wastewater outfall (a pipe with an inside diameter of 3.05 m) and of a candidate deep-water sludge outfall are also shown. As indicated in the figure, there are important physical differences between the discharge sites of the existing wastewater outfall and proposed sludge outfall. The CSDOC effluent outfall is located on a shelf, and its multi-port diffuser lies at an average depth of 60 m. Because of the relatively flat slope of the shelf (0.75%), the discharge is nearly 8 km offshore. Beyond the 100-m depth, however, the bottom slope becomes much steeper (5%) so that a sludge outfall terminating at a 300- to 400-m depth need only be 12 km long, or about 50% longer than the wastewater outfall.

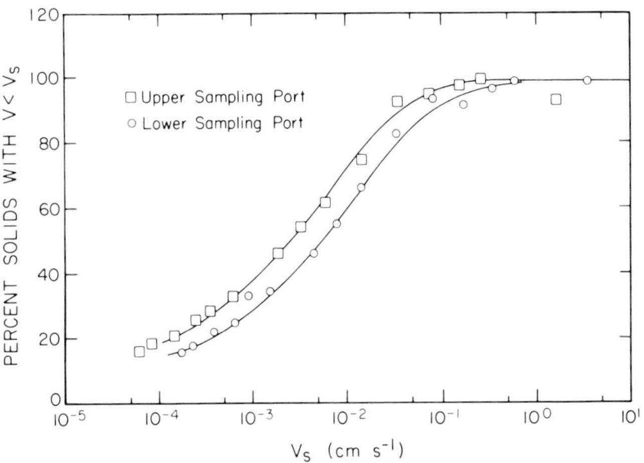

Figure 10.2. Fall-velocity distributions for CSDOC digested primary sludge diluted 100:1 with seawater, as determined from samples at two ports along the settling column. (From Faisst, 1980.)

10.4.2. Profiles of Water Properties

Profiles representing statistical summaries of local receiving-water characteristics (T. Hendricks, personal communication) are presented in Fig. 10.3. Recent oceanographic measurements taken in a preliminary survey conducted by the Sourthern California Coastal Water Research Project (SCCWRP) (1983) in the region surrounding candidate discharge sites generally fall within the ranges shown in this figure. The density stratification in the layer below 150 m is quite small; the typical gradient of relative density is 2×10^{-6} m^{-1} (or 0.2 σ_t units per 100 m) and exhibits little temporal or spatial variation within the basins. The dissolved-oxygen concentration decreases steadily with increasing depth to over 400 m; at depths between 300 and 400 m, concentrations range from 0.5 to 1.5 ml liter^{-1} (0.7–2.1 mg liter^{-1}).

10.4.3. Currents

The rugged bathymetry and the hydrodynamic separation of the southward-flowing California Current at the northern limit of the Southern California Bight combine to make the current structure in the bight very complex (SCCWRP, 1983). Deep drift currents generally run northward through the nearshore basins, whereas local currents in the vicinity of candidate discharge sites are predominantly parallel to the local bottom contours.

The onshore–offshore water motion is much weaker and has a frequency spectrum similar to that of white noise, with only a small peak at the semidiurnal frequency. In contrast, the along-isobath component of the current exhibits a marked diurnal peak, with an apparent spring–neap cycle, and a peak amplitude of approximately 0.1 m s^{-1}. In addition, flow events, presumably topographic waves propagating along the isobaths, occur for a few days and have an amplitude of about 0.3 m s^{-1}. Both the diurnal tide and topographic waves may well be general features of the flow occurring at the depths in which the measurements were taken, and they may be expected to affect the advection and dispersal of sludge at the proposed outfall site. In particular, pollutant transport and dispersion parallel to the bottom contours should be much more intense than is the transport and dispersion normal to them. Measurements taken at the proposed discharge site during the spring of 1982 (T. Hendricks, personal communication; SCCWRP, 1983) show that there is a persistent upcoast drift flow of about 8 cm s^{-1} (7 km d^{-1}) 50 m above the bottom. This indicated that advection will disperse sludge particles over a wide area and minimize buildup in the vicinity of the discharge.

10.4.4. Chemistry and Biology

The chemistry of seawater is closely tied to biological processes. In general, the production of plant matter

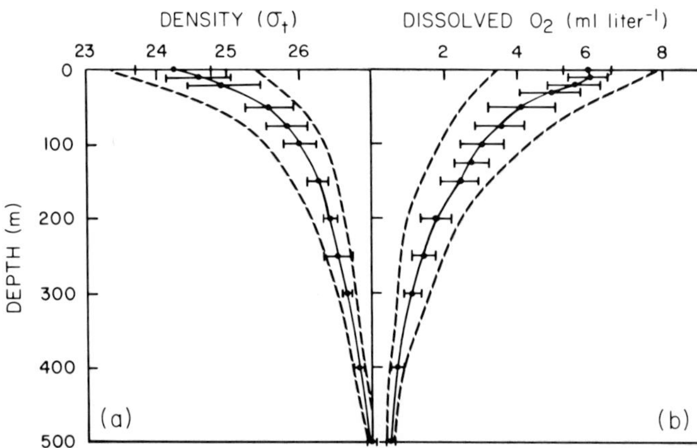

Figure 10.3. Vertical distributions of (a) density and (b) dissolved O$_2$ at CalCOFI station 90.28 (33° 20′ 6″ N, 117° 46′ 6″ W) near the site of proposed deep-water sludge-outfall experiment. (From analysis by T. Hendricks.) Dashed lines indicate the range of observed values, horizontal bars indicate the width of two standard deviations, and the closed circles indicate the mean values.

near the ocean's surface causes a depletion of major nutrients and trace metals and releases oxygen into the water. The settling of organic matter and its biodegradation in deeper waters produce opposing effects; concentrations of metals such as Cu, Cd, and Ni generally increase with depth, whereas concentrations of oxygen generally decrease.

In the southern California region, there is an oxygen minimum at depths between 400 and 800 m (Emery, 1960). Oxygen concentrations at the minimum are about 0.4 ml liter^{-1} (0.6 mg liter^{-1}), as compared to typical surface concentrations of 6 ml liter^{-1} (8.5 mg liter^{-1}) (Fig. 10.3). Some of the submarine basins in southern California, such as the Santa Monica, San Pedro, and Santa Barbara basins, draw their deep water from the oxygen-minimum layer, have low exchange rates with outside waters, and have a high natural oxygen demand. Consequently, their dissolved-oxygen concentrations are at times nearly zero. Waters at the proposed Orange County sludge-discharge sites are less topographically constrained, leading to higher local exchange rates and oxygen concentrations [typically 0.5–1.0 ml liter^{-1} (0.7–1.4 mg liter^{-1}) at 400 m and 1.0–1.5 ml liter^{-1} (1.4–2.1 mg liter^{-1}) at 300 m].

Preliminary data from SCCWRP (1983) show no simple depth-related trend in data for epibenthic organisms collected at depths between 300 and 627 m. Although the benthic fauna in the area of the proposed outfall are not depauperate, measures of benthic invertebrate biomass are completely dominated by two species of sea urchin. There are no major fisheries at the depth and location of the proposed outfall site; however, there is a small sable fishery in the area, focused primarily at Lasuen Knoll to the south.

10.5. OCEANIC PROCESSES DETERMINING THE TRANSPORT, FATE, AND EFFECTS OF SLUDGE

Material presented in this section is intended to provide a qualitative overview of the various processes that may materially affect marine sludge disposal at a depth of 300–400 m. The research plan outlined in Section 10.8 consists of 40 tasks designed either to measure directly the environmental impacts of the CSDOC sludge disposal at 300–400 m or to characterize and model the physical, chemical, and biological processes that affect these impacts. Figure 10.4 illustrates both the mecha-

nisms described and the interrelationships among the processes of interest.

Initial dilution is the immediate entrainment of ambient seawater into the turbulent, buoyant jet issuing from the end of the outfall. The process is normally considered to be completed when the initially buoyant jet either achieves neutral buoyancy and ceases to rise in the water column, as shown in Fig. 10.5, or reaches the surface. Plume submergence (trapping below the surface) depends on the ambient density stratification. Both the magnitude of initial dilution and the maximum height of plume rise can be predicted with reasonable accuracy by using discharge and receiving-water characteristics as inputs to existing mathematical models (Fisher et al., 1979). At the completion of the initial dilution process, the sludge–seawater mixture would be a neutrally buoyant drifting cloud, referred to as a wastefield.

Passive advection due to oceanic currents is the primary means by which the wastefield is transported away from the discharge site following the initial dilution of the sludge. The strength and persistence of currents at the maximum height of plume rise have profound effects on the ultimate fate of sludge and, in turn, on the viability of the proposed disposal scheme.

Lateral spreading and additional wastefield dilution accompany the advection process. Spreading and diffusion of the wastefield is caused by (1) gravitational effects (the cloud gets thinner as it spreads horizontally at the level of neutral buoyancy), (2) shear dispersion (the current speed and direction vary with depth, tending to pull the wastefield apart), and (3) diffusion due to turbulent eddies. These three effects are not easily separated, but they all contribute to the dilution and dispersal of the waste.

Sedimentation of wastefield particles occurs simultaneously with passive advection and lateral spreading. Whereas sedimentation further reduces the wastefield concentration of particle-associated contaminants, the process also leads to the accumulation of those materials among marine sediments. The particles essentially rain out of the dilute wastefield as it drifts downcurrent. Under the conditions anticipated, 30–50% (by weight) of the discharged sludge solids would have fall velocities of less than 10^{-3} cm s^{-1} (<1 m d^{-1}); therefore, it is likely that the process would occur over a long period during which particles might be transported tens of kilometers. Although sedimentation of sludge particles is expected to occur over a large area, sediment accu-

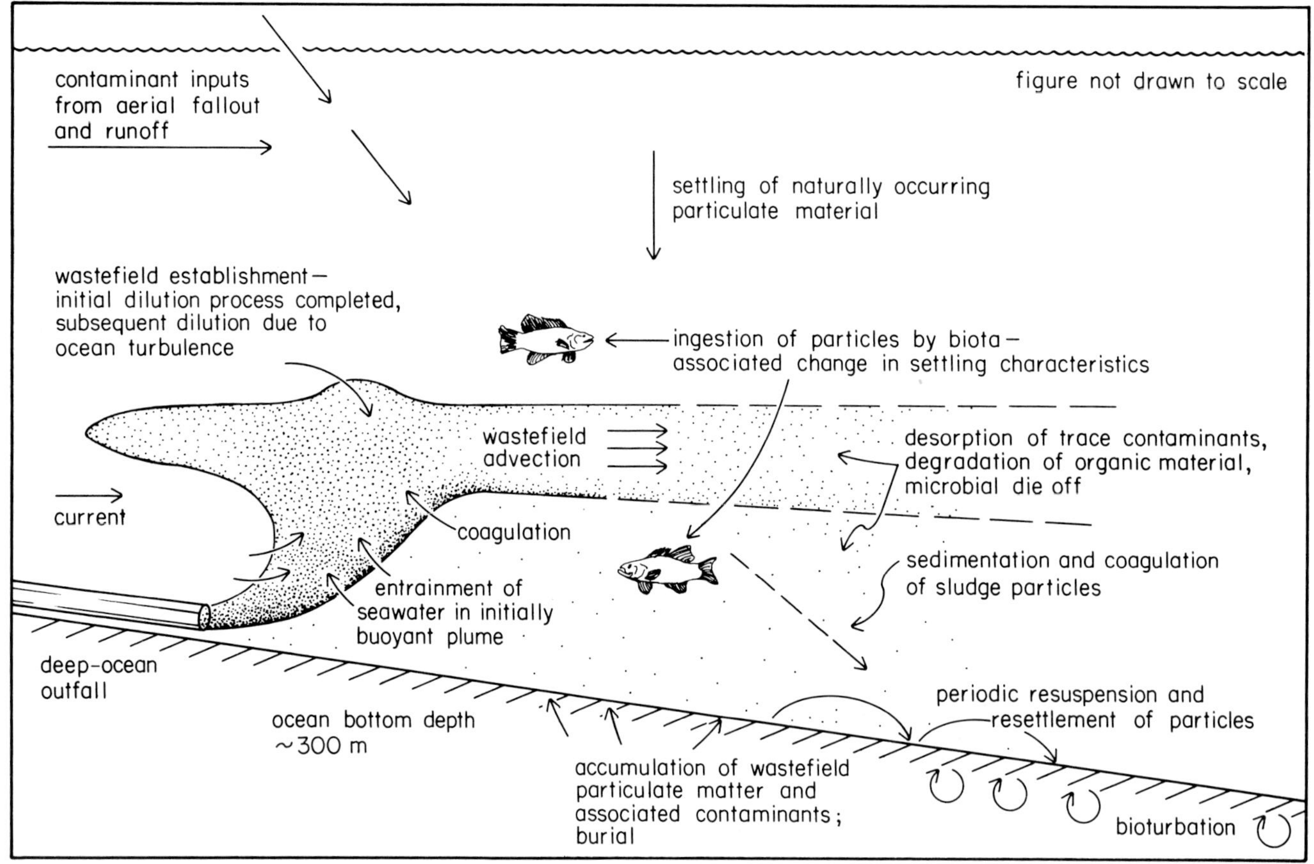

Figure 10.4. Factors affecting fluxes of particulate matter and fates of contaminants in marine waters.

mulation rates would be low outside a small region surrounding the outfall terminus. Within this region (perhaps 1 km in radius), the small percentage of relatively large, fast-settling particles (greater than about 0.5 cm^{-1}, or 500 m d^{-1}) would collect. Any particles or oil droplets lighter than seawater would rise toward the surface. There is a small fraction (as yet not determined) of floatable material in sludge, but efforts will be made in the treatment plants to eliminate such material from the sludge to the maximum extent possible.

Coagulation (or flocculation) of sludge particles, or of sludge with naturally occurring marine particles, is a process whereby small particles coalesce into larger, faster-settling solids. Coagulation is poorly understood in the ocean, but shear currents and small-scale turbulence enhance the process, whereas high dilution impedes it because of lower particle concentration and less frequent collisions in dilute solutions (Faisst, 1980). The

increased ionic strength resulting from the dilution of sludge with seawater probably favors coagulation and accelerates particle sedimentation.

Microbial die-off (or disappearance) can occur in marine waters due to the combined effects of sedimentation, wastefield dispersion, and bacterial or viral mortality. Because the proposed discharge site is about 12 km offshore and the plume would not immediately affect the surface, pathogens could reach shore only by gradual upwelling and mixing over periods of days or weeks. During such a period, pathogen concentrations would be reduced to very low levels. Microorganisms attached to particles would be cotransported, gradually sinking for the most part. If there is buoyant material (i.e., oil and grease) in the sludge discharge, associated pathogens could reach the ocean surface and then be advected toward the shore by wind-driven currents.

Zooplankton feeding on sludge particles in the subsur-

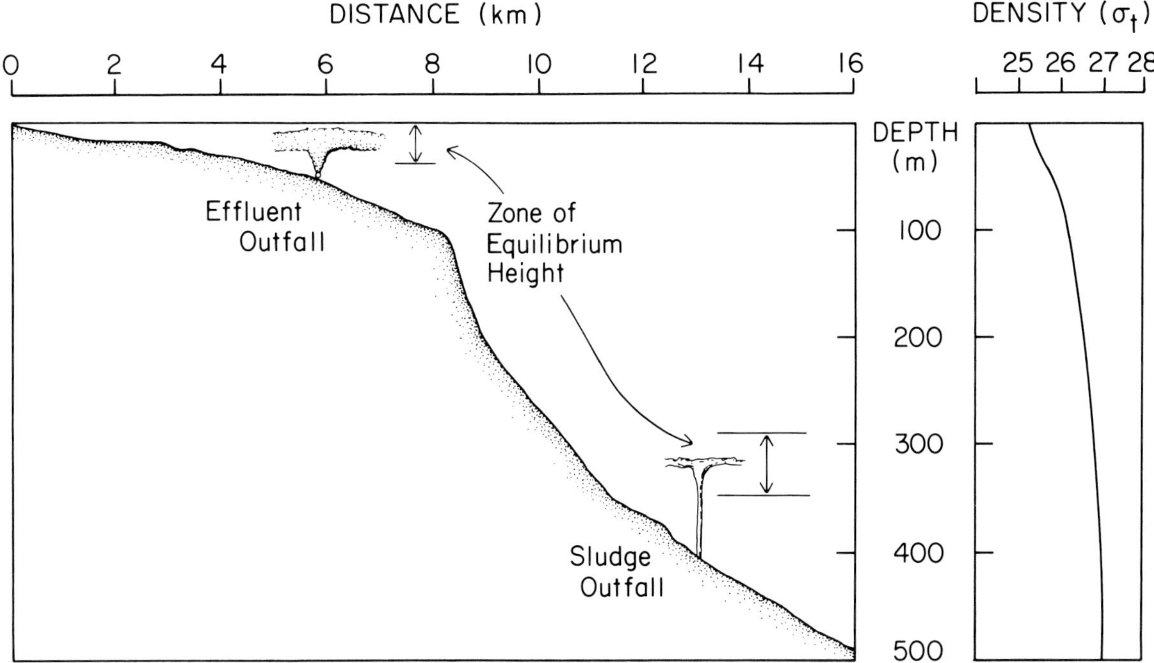

Figure 10.5. Cross-section of the water off Orange County, showing depth and distance from shore both for the proposed sludge outfall and for the diffuser for the existing effluent outfall. Density stratification (typical profile shown) causes the height of plume rise to be restricted to the ranges shown.

face wastefield may affect the plankton community or associated food chain in ways that are as yet unidentified. Zooplankton feeding may cause some sludge components to be incorporated into rapidly settling fecal pellets, thereby accelerating deposition. The resultant sedimentation pattern for these constituents would be smaller in area than in the absence of such activity, but peak fluxes of particulate matter and outfall-related contaminants would be higher.

Bacterial degradation of particles and dissolved organic material from sludge will consume dissolved oxygen in the water column. If currents carry the sludge into the deeper basins, particles may have to settle several hundred meters in order to reach the bottom. During this long settling period (about 100 d for particles settling at 10^{-3} cm s^{-1}), a significant fraction of the particulate organic carbon discharged may be oxidized to CO_2 with attendant oxygen consumption (Jackson, 1982).

The oxygen-consumption kinetics at the relatively cold temperatures ($<10°C$) typical of the proposed discharge depth is not yet characterized. Laboratory tests will yield the data needed to develop predictions for dissolved oxygen in the water column. The proposed discharge depth (300–400 m) represents a compromise. Based on previous analyses (Jackson, 1982), this is the deepest water in which ambient dissolved-oxygen concentrations assure satisfaction of discharge-related oxygen demand without risking unacceptably low oxygen levels in the water column.

Mobilization of tracer contaminants from sludge particles into the water column would probably result from the dilution of sludge with seawater. The oxidation of some environmentally important trace metals (present in anoxic sludge in their reduced forms) favors dissolution of metal–sulfides and some metal–hydroxide compounds. Release of the metals and refractory organic compounds are also promoted by waste dilution and breakdown of particulate organic material.

Sediment resuspension and *bioturbation* also play important roles in determining the sediment character in the vicinity of the sludge discharge. Bottom currents at the discharge site periodically reach speeds capable of resuspending surface sediments (Koh, 1982). Such events tend to destroy the apparent chronology of sediment deposition, to oxygenate surface sediments, to deplete dissolved-oxygen concentrations in bottom water, and gradually to transport sludge materials away from the discharge point toward deeper sediments. The increase in sediment porosity through bioturbation reduces the degree of vertical stratification among sediments and permits sediment oxygenation and other exchanges with the water column. Together, resuspension and bioturbation may greatly accelerate the release of contaminants from the sediments to overlying waters.

Bioaccumulation of dichlorodiphenyltrichloroethane (DDT) and polychlorinated biphenyl compounds (PCBs) is anticipated among marine organisms exposed to sludge contaminants; however, we expect that the relatively low concentrations of critical contaminants in CSDOC sludge will not result in concentrations within the tissue of fish in excess of limits or standards set by the State or by the Food and Drug Administration. Furthermore, we do not expect adverse ecological effects due to biomagnification in the marine food chain. Before the industrial inputs were eliminated from the sewer system, emissions of DDT from the Whites Point outfall system (located several kilometers upcoast from the existing Orange County wastewater outfalls), which endangered local pelican populations a decade ago, were about 6000 times greater (Goldberg, 1979) than the anticipated mass-emission rates of the proposed sludge outfall (Table 10.3).

10.6. ANTICIPATED EFFECTS OF THE PROPOSED SLUDGE DISCHARGE

With the qualitative discussion of processes as background, we will provide a semiquantitative description of environmental conditions predicted to follow initiation of the proposed sludge discharge, with emphasis on near-field behavior. Treatment of the far-field effects is limited to the results of order-of-magnitude calculations. For this purpose it is assumed that a sludge outfall would be constructed along the candidate alignment shown in Fig. 10.1 and would terminate at a depth of 300–400 m. Furthermore, the sludge flow rate is assumed to be 11.4×10^6 liter d^{-1} of unthickened, digested primary and undigested waste-activated sludge. Treated wastewater may be mixed with the sludge to maintain minimum flow velocity in the outfall and to prevent the deposition of solids.

The bulk density of the sludge–wastewater mixture would be slightly greater than that of freshwater, perhaps 1.005–1.010 g cm^{-3}; therefore, it would rise in the midst of the denser seawater at the discharge depth as a buoyant plume. Due to ambient density stratification (Fig. 10.3), the plume would stop rising after reaching its neutral buoyancy level (Fig. 10.5). Because of ambient density stratification, the dilute sludge cloud would always be trapped in deep water, never intruding into the shelf waters (100 m deep and within 8 km of shore). The terminus of the CSDOC wastewater outfall and its rising plume are also shown in Fig. 10.5 for comparison.

Assuming a one-to-one predilution of wastewater and sludge, the mixture will rise in the receiving water to depths of 50–110 m above the discharge point or, equivalently, to depths of 350–290 m for a discharge at 400 m (Koh, 1982). The dilution ratios (seawater: discharge) of the plume will vary between 450 and 750, or between 900 and 1500 if dilution prior to discharge is considered. If we pick a plume dilution of 600 (1200 overall) and assume that secondary effluent has a negligible BOD, whereas the sludge has a 5-d BOD of 1700 mg liter^{-1} (measured at 20°C), then the discharge-related BOD in the layer would be 1.4 mg liter^{-1} ($\sim$1 ml liter^{-1}). Given unlimited time, total oxygen consumption might be 50% greater than the measured 5-d BOD, or 2.1 mg liter^{-1} (1.5 ml liter^{-1}). As indicated in Fig. 10.3, the ambient dissolved-oxygen content at a depth of 350 m is $\sim$1 ml liter^{-1}, indicating that local impacts might be appreciable. On the other hand, at the low ambient temperature (8–10°C) at the depth of the plume, the rate of oxygen consumption will be slowed considerably. Further dilution of the drifting plume may then be fast enough to prevent serious oxygen depletion. Other sludge constituents would also be reduced by a factor of 1200 (factor of 2 for predilution $\times$ 600 for initial dilution with seawater); for example, suspended solids would be reduced from about 12,000 to 10 mg liter^{-1}, a value probably several times that of the background.

Following initial dilution, further mixing occurs as the plume drifts with the current. Such motion would be largely horizontal and parallel to local bottom contours. Turbulent diffusion at those depths is likely to be slower than it is near the surface. If it is assumed that the horizontal diffusion coefficient at depth is 10% of its value in the surface layer, then by using results from Brooks (1960) it can be calculated that turbulent mixing would provide additional dilution by a factor of 2 or 3 within 12 h, with a further reduction of BOD to 0.5–0.7 mg liter^{-1}, or ultimate oxygen consumption to 0.7–1.1 mg liter^{-1}.

Jackson et al. (1979) estimated the rates of sludge degradation at 8–10°C to be 1% per day. If the ambient oxygen concentration were conservatively estimated to be 0.5 ml liter^{-1} (0.7 mg liter^{-1}), it would be 50% depleted after 0.35 mg liter^{-1} is consumed (neglecting replenishment and further mixing after 12 h); thus, sludge degradation would halve the oxygen concentration of the wastefield within 40–70 d. In addition, particle removal by settling and mixing processes would decrease this impact on oxygen concentrations.

Based on these calculations, it seems possible that the naturally low concentrations of ambient dissolved oxygen would be slightly depressed locally in the diluted sludge field near the discharge point. The sludge-discharge system, however, would be specifically designed to prevent appreciable oxygen depletion. Among the possible methods of controlling dissolved-oxygen depletion are: (1) the BOD mass-emission rate could be reduced by digesting the waste-activated sludge component, (2) the outfall depth could be reduced to move the plume into water containing more dissolved oxygen, and (3) the sludge could be diluted with increased amounts of effluent.

The benthic community structure would change in response to a sludge-enriched substrate; a few species would be favored at the expense of less pollution-tolerant organisms. Echinoderms and crustaceans are expected to be negatively affected in the vicinity of the outfall terminus (Thompson et al., 1984). Total biomass may be enhanced significantly and species diversity reduced over a considerable area. Local patches of anoxia may develop in the most highly affected bottom areas if the transport of oxygen from the overlying water to the sediment surface cannot match the requirements of aerobic respiration. Under such circumstances, hydrogen sulfide, which is toxic to most marine species, may be generated in the sediments.

The fluxes of various trace contaminants in CSDOC sludge are expected to be modest (Table 10.2), and mobilized metals and organic toxicants are likely to be transported from the Southern California Bight to the open ocean before unacceptable concentrations (relative to standards provided in the California State Ocean Plan) develop in the vicinity of the outfall. Acute toxic effects are not expected because of high dilution and wide dispersal of the sludge solids. Nevertheless, sludge constituents may be a source of chronic toxic effects among populations of organisms that live or feed on the bottom near the discharge site. Signs of chronic toxicity attributed to southern California marine outfalls in the

past include the incidence of fin erosion and liver enlargement among fish (Sherwood, 1982).

The areal extent of discharge-related impacts within the local benthos would depend on many factors, only some of which are now understood. At present, we can state that sludge particles will be widely dispersed in the environment. The bulk of this material may be dispersed throughout the Southern California Bight since typical particle fall velocities are on the order of 1 m d^{-1}, whereas characteristic rates of net horizontal advection at the depth to which the plume rises are 7 km d^{-1}. Hence, we expect benthic impacts to be minor except in a relatively small region near the discharge point; however, there is not enough information to make quantitative predictions regarding changes in sediments due to sludge-particle deposition.

In summary, we can estimate the short-term fate of sludge near the discharge point with some confidence; however, oceanographic processes that affect the long-term fates and effects of sludge contaminants are sometimes poorly understood. For instance, the roles of coagulation and biological activity in accelerating particle deposition cannot be quantified at this time. In addition to deposition rate, the sediment quality would undoubtedly be affected by biological activity and periodic sediment resuspension, both of which are poorly characterized. Water-quality predictions depend on such poorly understood factors as contaminant partitioning between the liquid and the solid phases, temperature effects on oxygen demand, and the effectiveness of advection in clearing sludge-related pollutants from the vicinity of the discharge and from the Southern California Bight as a whole.

10.7. GOALS OF THE RESEARCH PLAN

A research plan has been constructed to clarify aspects of the behavior and effects of the proposed deep-ocean sludge discharge and to provide information with which to make future decisions regarding sludge discharge to the ocean. Its specific goals are: (1) to characterize the physical, chemical, and biological properties of the sludge; (2) to recommend the depth and location of the end of the sludge outfall, processes for in-plant treatment of sludge before discharge, and possible modifications in industrial waste pretreatment (or source control) programs for trace contaminants; (3) to make predischarge predictions of sludge-induced environmental impacts, including changes in oxygen concentrations in the water column and in benthic sedimentation rates;

(4) to observe changes in water-column and benthic communities and to measure the distribution of important chemicals in the water, in the sediment, and in organisms; (5) to find what risks (if any) to human health are associated with deep-water sludge discharge, either directly through human contact with seawater or indirectly through consumption of seafood; (6) to determine the mechanisms by which ecological and chemical changes occur in the marine environment; (7) to assess the predictability of the effects of deep discharge of sludge in this and other settings by comparing model predictions with monitoring data; (8) to summarize and recommend methodology for designing systems for the discharge of sludge into the ocean; and (9) to summarize the effects of the Orange County sludge discharge, and to evaluate their significance with respect to marine resources.

10.8. TASK GROUPS IN THE RESEARCH PLAN

With these goals in mind, we have developed a research plan consisting of 40 tasks arranged into 9 groups. These groups, which are interrelated as shown in Fig. 10.6, are briefly described below, by using the same classification and numbering system for the tasks as in the detailed description presented in Brooks et al. (1982).

Task Group 1: Survey for Discharge-Related Effects

1.1. Measurement of Water-Quality Characteristics
1.2. Water-Column Biological Monitoring
1.3. Inventory of Local Sport and Commercial Fishing Activities
1.4. Trawl Monitoring Program
1.5. Benthic Surveys
1.6. Age Dating and Chemical Analyses of Sediment Cores by Layers
1.7. Measurement of Sedimentation Rate via Sediment-Trap Deployment
1.8. Bioaccumulation of Trace Contaminants
1.9. Designation of Control Sites for Water-Column and Sediment Monitoring

Comparison of important water-quality and benthic parameters measured near the outfall and at control stations, both before and after discharge begins, would be the basic method for detecting and documenting the effects of the discharge. With this monitoring data, predictive models can be calibrated and improved. Predischarge monitoring data would also contribute to the

design of the sludge-discharge system. In terms of total effort and cost, this group would be by far the largest.

Task Group 2: Site Characteristics

2.1. Measurement of Local Currents
2.2. Hydrographic Surveys for Determination of Geostrophic Currents
2.3. Site Bathymetry

In addition to the usual water-quality monitoring data, a predischarge survey of the bathymetry and currents of the area is necessary for siting the outfall and the network of sampling stations and for interpreting and modelling the observed discharge-related effects.

Task Group 3: Sludge Characterization

3.1. Measurement and Projection of Sludge-Quality Parameters
3.2. Evaluation of In-Plant Unit Operations Affecting Sludge Characteristics
3.3. Measurement of Sludge-Particle Size Distribution, Settling Characteristics, and Coagulation Effects
3.4. Mobilization of Trace Contaminants within the Water Column

Experiments described in this task are designated to characterize sludges from the CSDOC treatment plants in terms of selected environmentally important parameters, to determine how alternatives for sewage treatment and industrial-waste pretreatment (source control) might affect those sludge characteristics, and to investigate the behavior of sludge particles and associated trace contaminants when placed into marine waters. Of special interest are the settling properties of sludge particles and the fractionation of trace contaminants such as metals and pesticides between the dissolved and the particulate phases.

Task Group 4: Modelling

4.1. Initial Dilution and Height of Plume Rise
4.2. Estimation of Water-Column Dissolved Oxygen Concentrations
4.3. Estimation of Particulate and Trace-Contaminant Fluxes to Local Sediments
4.4. Sediment Concentrations of Dissolved Oxygen and Sludge-Related Contaminants
4.5. Biological Models

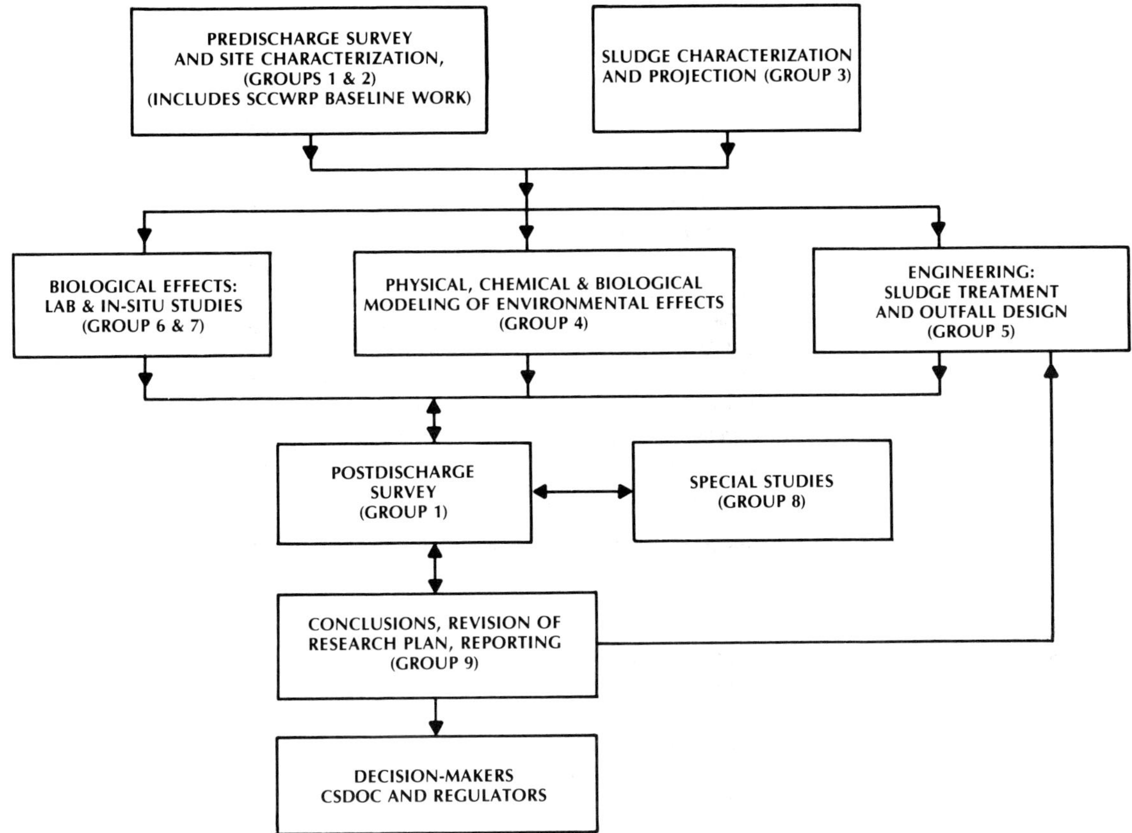

Figure 10.6. Relationships among generalized tasks comprising the overall sludge-disposal experiment.

The behavior of the waste after discharge can be predicted by a series of conceptual and computer models; each model focuses on particular processes and scales. For example, the initial-dilution model covers only the first few minutes following discharge, during which the buoyant plume rises from the end of the outfall, producing initial dilutions of several hundred to one; on the other hand, particle-deposition models describe processes that take days or weeks. Such models would form a basis for predicting effects, not only for this project but also for future projects. Predictions would be used both in the outfall design and in the selection of appropriate sludge-treatment operations and pretreatment measures for industrial waste.

Task Group 5: Preliminary Design

5.1. Evaluation of In-Plant Design Alternatives
5.2. Selection of Discharge Site and Submarine Pipeline Technology

Characteristics both of the discharged sludge and of the initial-mixing process in the ocean would be within the control of the design engineer. Preliminary modelling by Jackson et al. (1979) favors discharge at a depth of approximately 300–400 m through a pipe 45–60 cm in diameter. Available alternatives, both in the treatment plant and in the submarine discharge structure, would be examined in terms of their effects on sludge quality, initial mixing, and sludge behavior subsequent to discharge. Capital and operating costs would be estimated for each alternative.

Task Group 6: Biological Impacts—Laboratory Studies

6.1. Effect of Sludge Particles and Organics on Zooplankton Feeding
6.2. Dependence of Zooplankton Respiration on Ambient O_2 Concentration
6.3. Microbial Decomposition of Water-Column Organics
6.4. Sea Urchin Toxicology: Lab Study

Specific laboratory experiments would be carried out

to investigate biologically mediated pollutant transformations and biological sensitivities to pollutants. These data would be needed for developing good predictive models of sludge effects.

Task Group 7: Biological Impacts—*In Situ* Studies

7.1. Benthic Oxygen Consumption
7.2. Rates of Particle Ingestion by Macrobenthos
7.3. Effect of Sludge Particles on the Feeding Habits of Marine Fauna: Baited Camera Experiments
7.4. Field Observation of Zooplankton Feeding in the Presence of Sludge Constituents

Field studies would allow one to observe the effects of sludge on a community of organisms as opposed to individuals or single species. Direct *in situ* manipulation of the kind recommended in this plan would permit ecologists to study the mechanisms of ecological change.

Task Group 8: Special Studies

8.1. Identification of Sludge Tracers
8.2. Remote Tracking of the Wastefield
8.3. Extrapolation from the Hyperion Sludge-Disposal Experience
8.4. Assessment of Potential Public Health Impacts

This group includes special studies that do not fit in other categories. Potential health impacts, tracer studies, and extrapolation from the Hyperion sludge-outfall experience (at a 100-m depth) are examples.

Task Group 9: Integration, Analysis, and Interpretation

9.1. Analysis and Interpretation of Survey Data
9.2. Comparison of Model Predictions with Observations
9.3. Effects of Organic Enrichment on the Dynamics of Midwater Plankton Communities
9.4. Revision of the Research Plan
9.5. Annual Summaries of Project Results, Conclusions, and Recommendations

The results of individual tasks and task groups do not stand alone, but must be woven into an overall picture to fulfill the goals of the project and summarize the results. This group of tasks includes a cross-disciplinary interpretation of modelling data, a comparison of model predictions with data, and the revision of the research plan. Summaries and recommendations would be prepared annually to assist the regulatory agencies and the CSDOC in making decisions about the continuation of the sludge discharge.

10.9. RESEARCH ADMINISTRATION, SCHEDULING, AND COST

The organization for research administration would be separate from the CSDOC and its normal operating functions. Research tasks should be funded through a joint administrative board representing the sponsoring agencies. The board would appoint a scientist or engineer, preferably with experience both in oceanographic research and in waste-disposal problems, to be the project research administrator. Using funds provided by the sponsoring agencies as a single pool, the project research administrator would carry out the research plan by awarding grants and contracts. This person would also play a key role in disseminating the research results.

A research review committee of scientists and engineers would be appointed by the research administrator with the concurrence of the administrative board. Membership of this committee would be based on professional qualifications, rather than on organizational affiliation or geographical location. The functions of this research committee, in collaboration with the research administrator, would be: (1) to approve the research plan and all revisions; (2) to establish priorities among research tasks; (3) to approve procedures for the solicitation of proposals and the awarding of grants and contracts; (4) to evaluate research proposals submitted by organizations wishing to perform the work, and to recommend awards; and (5) to prepare an annual summary of the findings and evaluate their significance.

When processing proposals, the administrator and the research review committee would be assisted by a system of external peer review by individual experts. For renewal of grants or contracts, peer review would include evaluation of the proposer's performance under previous agreements for the sludge-disposal experiment. Investigators would be required to submit annual progress reports as well as a final report to the research administrator. Publication in peer-reviewed professional journals should be encouraged.

It is expected to take 2.5 y from approval of the project to the start of discharge. During this period, the research project would be formally started, and 2 y of predischarge work would be conducted. After discharge 5 y of observation are planned, making a total of 7 y for the research project. The estimated costs for each of these years (although subject to substantial uncertainty) range from about $1.5 to 2.0 million (U.S. 1982 dollars).

ACKNOWLEDGMENTS

Work leading to the preparation of this chapter was supported by the National Oceanic and Atmospheric Administration (Grant No. NA81RACOO153) and the County Sanitation Districts of Orange County. Many of the ideas represented in the recommended scientific program originated within a broad-based Research Planning Committee, established to advise on content of the research plan. A complete list of committee members is available within a report by Brooks et al. (1982).

REFERENCES

Brooks, N. H. 1960. Diffusion of sewage effluent in an ocean current. *In*: Proceedings of an International Conference on Waste Disposal in the Marine Environment, E. A. Pearson (Ed.). Pergamon Press, New York, pp. 246–267.

Brooks, N. H., and J. E. Krier. 1981. Alternative strategies for ocean disposal of municipal wastewater and sludge. *In*: Use of the Ocean for Man's Waste: Engineering and Scientific Aspects. Proceedings of a Symposium (23–24 June 1981), Lewes, Delaware. Marine Board, National Research Council, Washington, D.C., Appendix A, pp. 271–295.

Brooks, N. H., R. G. Arnold, R. C. Y. Koh, G. A. Jackson, and W. K. Faisst. 1982. Deep Ocean Disposal of Sewage Sludge off Orange County, California: A Research Plan. EQL Report Number 21, Environmental Quality Laboratory, California Institute of Technology, Pasadena, 116 pp.

California State Water Resources Control Board. 1983. Final Environmental Impact Report: Amendment of the Water Quality Control Plan, Ocean Waters of California. California State Water Resources Control Board, Sacramento, California, 2 vols., paginated separately.

Emery, K. O. 1960. The Sea off Southern California. John Wiley & Sons, New York, 366 pp.

Faisst, W. K. 1980. Characterization of particles in digested sludge. *In*: Particulates in Water: Characterization, Fate, Effects and Removal, M. C. Kavanaugh and J. O. Leckie (Eds.). American Chemical Society, Washington, D.C., pp. 259–282.

Fisher, H. B., E. J. List, R. C. Y. Koh, J. Imberger, and N. H. Brooks. 1979. Mixing in Inland and Coastal Waters. Academic Press, New York, 483 pp.

Goldberg, E. D. (Ed.). 1979. Proceedings of a Workshop on Assimilative Capacity of U.S. Coastal Waters for Pollutants, Crystal Mountain, Washington. Environmental Research Laboratories, Boulder, Colorado, 284 pp.

Jackson, G. A., R. C. Y. Koh, N. H. Brooks, and J. J. Morgan. 1979. Assessment of Alternative Strategies for Sludge Disposal into Deep Ocean Basins off Southern California. EQL Report Number 14, Environmental Quality Laboratory, California Institute of Technology, Pasadena, California, 121 pp.

Jackson, G. A. 1982. Sludge disposal in southern California basins. *Environmental Science and Technology*, **16**, 746–757.

Kavanaugh, M. C., and J. O. Leckie (Eds.). 1980. Particulates in Water: Characterization, Fate, Effects and Removal. American Chemical Society, Washington, D.C., 401 pp.

Koh, R. C. Y. 1982. Initial sedimentation of waste particulates discharged from ocean outfalls. *Environmental Science and Technology*, **16**, *757–763*.

Krier, J. E. 1983. Ocean discharge of municipal wastes: legal and institutional aspects. *In*: Ocean Disposal of Municipal Wastewater: Impacts on the Coastal Environment, Vol. 2, E. P. Myers (Ed.). Sea Grant College Program, Massachusetts Institute of Technology, Cambridge, Massachusetts, pp. 659–705.

Myers, E. P. (Ed.). 1983. Ocean Disposal of Municipal Wastewater: Impacts on the Coastal Environmental. Sea Grant College Program, Massachusetts Institute of Technology, Cambridge, Massachusetts, 1115 pp.

Sherwood, M. J. 1982. Fin erosion, liver condition, and trace contaminant exposure in fishes from three coastal regions. *In*: Ecological Stress and the New York Bight: Science and Management, G. F. Mayer (Ed.). Estuarine Research Foundation, Columbia, South Carolina, pp. 359–377.

Southern California Coastal Water Research Project (SCCWRP). 1983. A Survey of the Slope off Orange County, California: First Year Report to the County Sanitation Districts of Orange County. Southern California Coastal Water Research Project, Long Beach, California, 178 pp.

Thompson, B. E., J. D. Laughlin, and D. T. Tsukada. 1984. Ingestion and oxygen consumption by slope echinoids. *In*: Southern California Coastal Water Research Project Biennial Report 1983–1984. Southern California Coastal Water Research Project, Long Beach, California, pp. 93–107.

U.S. National Research Council. 1984. Ocean Disposal Systems for Sewage Sludge and Effluent. National Academy Press, Washington, D.C., 126 pp.

Chapter 11

Cadmium Pollution Associated with a Coastal Lead-Smelting Plant

J. Michael Bewers, Douglas H. Loring, Kate Kranck, and Gerald H. Seibert

Atlantic Oceanographic Laboratory
Bedford Institute of Oceanography
Dartmouth, Nova Scotia, Canada

Rene Levaque Charron

Brunswick Mining and Smelting Corporation Ltd.
Belledune, New Brunswick, Canada

John F. Uthe, Chiu L. Chou, and Douglas G. Robinson

Fisheries Research Branch
Halifax Fisheries Research Laboratory
Halifax, Nova Scotia, Canada

ABSTRACT

This chapter describes the extent of Cd contamination of water, sediment, and lobsters in a coastal embayment that is adjacent to, and which received discharges from, a lead-smelting plant. A survey of the extent of contamination of water, sediment, and suspended particulate matter was carried out in 1980. At that time, the severity of contamination of lobsters in the area justified the imposition of increased controls both on the discharges from the plant and on the procedures for marketing lobsters from the area. The nature of improved discharge controls and changes in lobster contamination in the area between 1980 and 1982 are described.

11.1. INTRODUCTION

A Pb- and Zn-smelting and refining plant has been in operation on the southern shore of the Gulf of St. Lawrence (Fig. 11.1) at Belledune, New Brunswick, since 1966. Cadmium is present in the ore concentrates that are smelted at the plant and limits the extent of byproduct recycling within the plant. Cadmium is of environmental concern in the area due to the presence of populations of edible shellfish that are known to accumulate this element (Topping, 1973; Freeman and

Uthe, 1974). In 1977 the smelting company reported the results of its own monitoring program for Pb and Cd in lobsters (*Homarus americanus*) (Dugdale and Hummel, 1977). The concentrations that this study found were not high enough to warrant concern; however, in April 1980 the company supplied additional results, obtained in the period 1977–1979, indicating that Cd concentrations in claw and tail muscles of lobsters captured during the summer within Belledune Harbor had risen substantially. Cadmium concentrations in the hepatopancreas (digestive gland, mid-gut gland, or tomalley) in the 1979 samples were also judged high enough to pose a potential health hazard to humans consuming relatively large quantities of lobster tomalley (Uthe and Zitko, 1980). As a result, a field survey was undertaken in April through May of 1980 to investigate the concentrations of Cd in sediments, suspended particulate matter, water, and lobsters from Belledune Harbor and the adjacent coastal zone. Salinity and temperature profiles and current-meter measurements were also made at this time to determine the major physical processes that might be responsible for the dispersal and movement of waterborne contaminants in the area. Studies of lobsters

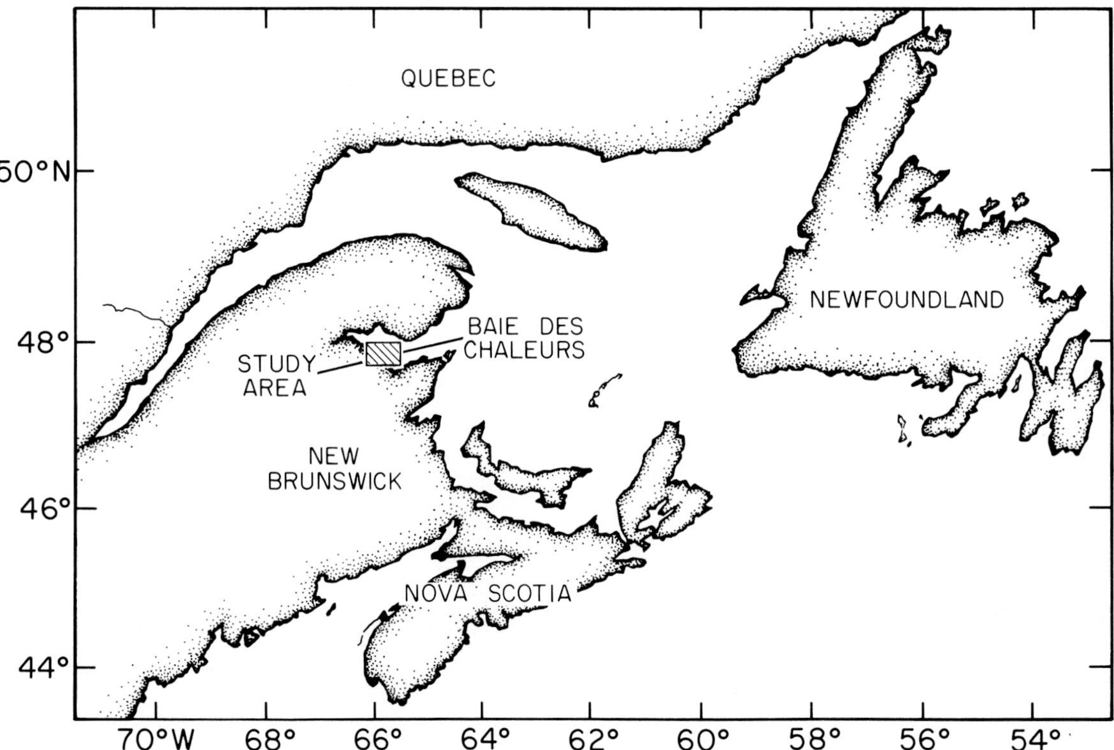

Figure 11.1.   Location of the study area in the Baie des Chaleurs, Gulf of St. Lawrence, Canada.

have continued on an annual basis since 1980 (Uthe et al., 1983). This chapter describes the results of these investigations and the consequences of more stringent control measures introduced by the company in 1980 to limit Cd discharges.

11.1.1. Industrial Discharges into Belledune Harbor

11.1.1a. Lead-smelter plant operations

The Pb-smelting and refining process at the Brunswick Mining and Smelting Corporation Ltd. plant at Belledune comprises three main stages: sintering, smelting, and refining. The raw material for the plant is a Pb concentrate derived from a Pb–Zn–Cu–Ag ore. The smelter receives 225,000 metric tons (t) of Pb concentrate annually plus minor amounts of additional concentrate and Cottrell precipitator dusts from other NORANDA (NORANDA Mines Ltd., Toronto, Ontario, Canada) company operations. Typically, the Pb concentrate and the Cottrell dusts contain 33% and 47% Pb and 0.013% and 0.8% Cd, respectively. The plant has a nominal capacity of 65,000 t y^{-1} of refined Pb, 3500 t y^{-1} of Cu matte, and lesser amounts of Pb–Sb slag and Pb–Bi alloys. About 200,000 t of H_2SO_4, an additional byproduct, is passed by pipeline to an adjacent plant for the production of diammonium phosphate fertilizer (Neumann and Schnarr, 1971).

The concentrate and residual materials are mixed with fluxes and fed to a sintering machine, where the sulphur is ignited and burned. This process releases SO_2, other oxides, and volatile substances, including any Cd not fixed in the sinter. The gases and dust pass through an electrostatic precipitator, which removes 98% of the dust prior to further gas cleaning and the production of H_2SO_4. The sinter is subsequently broken into pieces, which are mixed with coke and fed into a blast furnace while the fine materials are recycled. The products of the blast furnace are Pb bullion, slag, and baghouse dust. The baghouse dust is normally recycled, but when it approaches or exceeds 10% Cd, it is stockpiled for future sale; in 1979, 67% of the baghouse dust produced (500 t) was sold commercially. The slag, containing silicates of the Fe, Ca, Mg, and most of the Zn present in the original concentrate, is granulated and stockpiled. The Pb bullion is transferred to the refining stage, which comprises a sequence of drossing (skimming) operations, and is finally cast into ingots. The H_2SO_4 produced from the sulphur oxides released at the sintering stage is transported by pipeline to a neighboring fertilizer plant and used to produce H_3PO_4 from phosphate rock, of which 127,000 t was consumed in 1970. The phosphoric acid is then reacted with NH_3 to yield diammonium phosphate fertilizer.

11.1.1b. Control measures introduced by Brunswick Mining and Smelting Corporation Ltd. in 1980

Before 1980, Cd escaped from the smelting plant into the air, as dust, into surface runoff, and as drainage from the slag pond. Cadmium-containing dust also became wind-borne from the delivery yard and the waste-material storage area, and Cd was transported from the site in storm water. All of these processes resulted in contamination of the surrounding environment (Wixson et al., 1975; Gale and Wixson, 1979). Although runoff from the smelting plant drained into the sea at several points along the property, the predominant routes were through the slag pond and via a contaminated ditch on the eastern side of the plant. The large flow of non-contact saline cooling water should have contributed little to the influx of metals to Belledune Harbor; however, before entering the harbor, the cooling water was combined with the overflow from the slag pond. Similarly, freshwater, used for slag granulation and other purposes inside the plant, and the saltwater, used to slurry the slag and convey it to the slag pile, were also combined in the slag pond, which in turn discharged into the harbor.

Since 1980, clean runoff has been collected from the south, west, and east of the plant property and discharged into coastal waters via two uncontaminated diversion ditches (Fig. 11.2). The slag is now granulated with recycled water, pumped into a cyclone, and dewatered. The granulation and conveyance water flows to a holding pond from which it is again returned to the granulation stage. The dewatered slag is hauled by truck to a nearby disposal area that is monitored so that if the drainage water is found to be unduly contaminated it can be rerouted to the plant for treatment. Treatment facilities for the wastewater were installed in November 1980 to deal with (1) surface drainage from within the site; (2) wastewater from sewage, laundry, and laboratory facilities; and (3) overflow from the recycled process-water stream. Water from all of these sources are now collected in a holding pond and then pumped to the water-treatment plant. The treatment

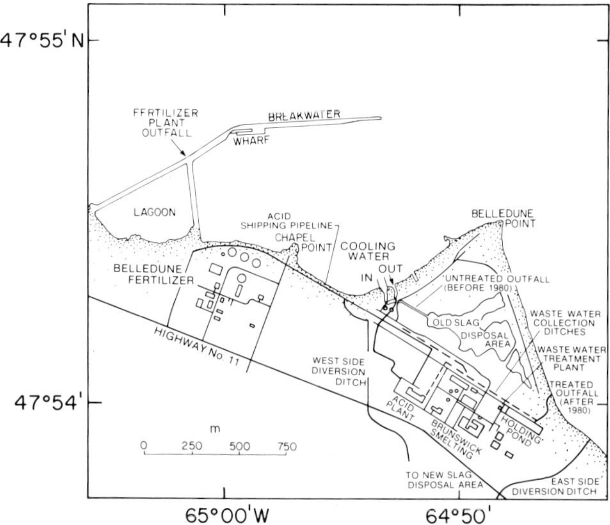

Figure 11.2. Belledune Harbor with a detailed plan of the smelting plant.

process comprises conventional lime neutralization for metal removal, addition of $FeCl_3$ for As precipitation, addition of flocculating agents, and settling. The treated effluent is then discharged to the east of Belledune Harbor (Fig. 11.2), but there are facilities to recirculate the effluent to the holding pond for further treatment if the quality of the final effluent is found to be unsatisfactory. The sludge from the treatment plant is recycled to the sintering stage of the smelting process. Since 1980, seawater has been used only as non-contact cooling water in the acid-production plant. Since there is no longer a discharge from the slag pond, the reduced volume of cooling water is discharged directly into Belledune Harbor without treatment, and the aqueous effluent from the fertilizer plant is discharged outside Belledune Harbor, northwest of the breakwater (Fig. 11.2).

11.1.1c. *Environmental studies*

As part of its environmental surveillance program, the smelter has conducted monitoring studies for the incidence of metals in the vicinity of its plant since the early 1970s. Measurements in the marine environment have revealed an area of contamination extending as far as 15 km to the east of the plant (Levaque Charron, 1979). Furthermore, yearly comparisons have shown a gradual increase in Cd concentrations, particularly in blue mussels (*Mytilus edulis*) and lobsters (Dugdale and Hummel, 1977; Prairie and Levaque Charron, 1981;

Prairie and Trudel, 1983). The processing of Cottrell dusts since the mid 1970s may have resulted in these increased Cd emissions.

11.2. SAMPLING AND ANALYTICAL METHODS

11.2.1. Water and Suspended Particulate Matter

A vertical profiling current meter and a salinity–temperature profiler were deployed at a number of locations both inside Belledune Harbor and in the coastal area outside the harbor. Data from these devices was used to determine water circulation and vertical density stratification in the harbor and its environs. Water samples were collected with Teflon®-coated (E.I. duPont de Nemours and Company, Wilmington, Delaware), 12-liter General Oceanics (Miami, Florida) sampling bottles at 6 harbor and 12 coastal-zone stations. The water was sampled at a single depth (7 m) since physical oceanographic measurements indicated that the water column was vertically well mixed. The water samples were filtered through preweighed 47-mm-diameter 0.4-μm Nuclepore® (Nuclepore Corporation, Pleasanton, California) filters to remove and collect the suspended particulate matter. The filters were placed in closed plastic containers and returned wet to the laboratory. The filtrate was then acidified with high-purity HCl to a pH of <2 and stored in pre-cleaned, 2-liter conventional (soft) polyethylene bottles.

Water samples were analyzed for Cd by graphite furnace atomic absorption spectrophotometry (GFAAS) following sample preconcentration by chelation and solvent extraction (Bewers et al., 1976). These methods for analyzing seawater have been recently compared with those used by other laboratories in an intercalibration exercise conducted under the auspices of the International Council for the Exploration of the Sea (ICES) (Bewers et al., 1981; Bewers and Windom, 1982; Berman et al., 1983). Filters were washed, dried, and weighed to determine the quantities of suspended particulate matter per unit volume of water.

11.2.2. Sediments

Samples of sediment were obtained by using a Van Veen grab at 22 locations within Belledune Harbor (Fig. 11.3) and from 13 locations outside (Fig. 11.4). Samples of the surface layer of the sediments were

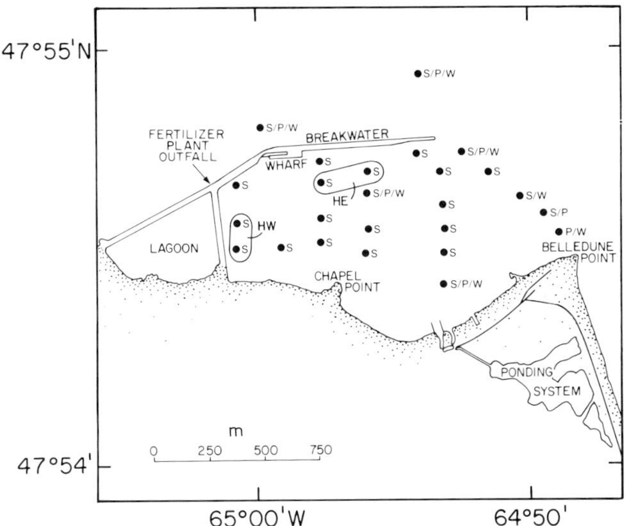

Figure 11.3. Locations of sampling sediment (S), suspended particulate matter (P), water (W), and lobster from the Harbor East (HE) and the Harbor West (HW) sites in Belledune Harbor.

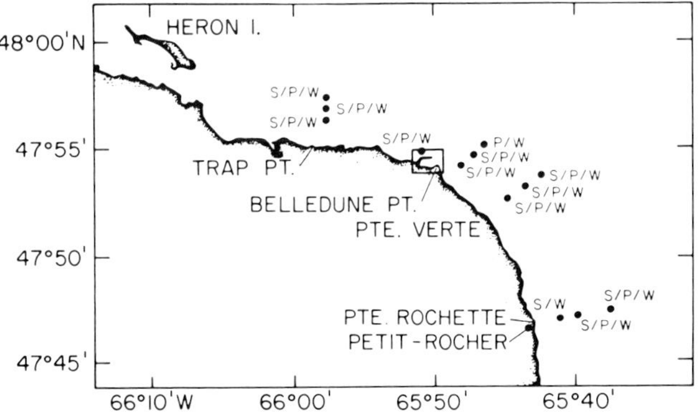

Figure 11.4. Locations of sampling stations for sediment (S), suspended particulate matter (P), and water (W) in the coastal zone.

collected from these grabs and returned to the laboratory in polyethylene bags. In the laboratory, a representative portion of each sample was used to determine, by wet sieving, the percentage (by weight) of sand-sized (2.0–0.053 mm in diameter) and mud-sized (<0.053 mm in diameter) material in each sample. This information was used to establish the textural characteristics of the sediment according to the nomenclature used by Loring and Rantala (1977) and to estimate the textural partitioning of Cd. Another portion of each sample was dried in an oven for 24 h at 60°C for chemical analysis.

The mud fraction was retained and oven-dried for chemical analysis.

Total Cd concentrations were determined on a dry-weight basis by the digestion and GFAAS analytical techniques described by Rantala and Loring (1980) in a 10- to 500-mg portion of the total sediment sample, in a portion of the <0.053-mm fraction of the sediments, and in the samples of suspended particulate matter. A separate 2-g portion of the total sample was leached for 24 h with aqueous acetic acid (25%, by volume) to partition Cd into its weak-acid soluble and insoluble fractions. This technique is used to separate the proportions of Cd that are weakly bound from those that are strongly bound to the sediment particles (Loring, 1981). Cadmium solubilized by leaching with acetic acid is defined here as the non-detrital fraction, whereas the Cd retained in the residue after leaching with acetic acid and subsequently released digestion with HF is defined here as belonging to the detrital fraction.

11.2.3. Biological Samples

In April through May of 1980, 1981, and 1982, lobster fishing was carried out at the sites shown in Figs. 11.3 and 11.5 by commercial fishermen using standard lobster traps. Lobsters were transported live to Halifax, packed individually in open polyethylene bags over ice. Following the determination of length, weight, and sex, the intact hepatopancreas was removed, weighed, and stored frozen in polyethylene bags. Prior to analysis by atomic absorption spectrophotometry, the gland was homogenized by hand-kneading the tissue within the plastic bag. Fresh-frozen claw and tail muscles from

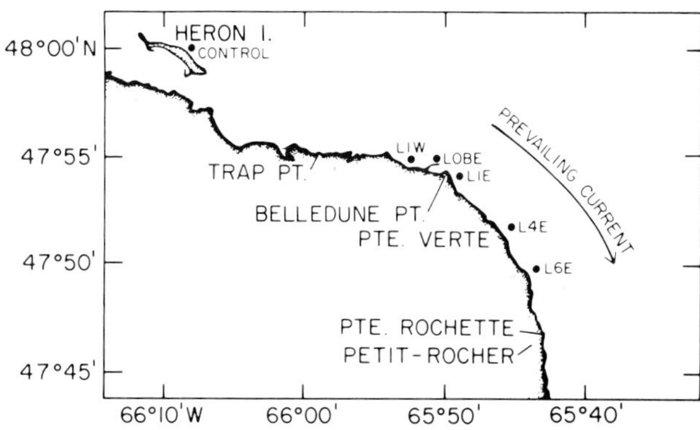

Figure 11.5. Sampling sites used for the annual geographical assessment of Cd concentrations in lobster hepatopancreas. The direction of the prevailing current is also shown.

animals collected in 1980 and 1981 were sampled for analysis by dissection of a small cube of frozen tissue from within the muscle (Uthe et al., 1980). All Cd results for lobsters are expressed on a wet-weight basis. The geometric means of the Cd concentrations were calculated for each sample set since probit analysis (Laitinen, 1960) and comparisons of arithmetic and geometric means with median values showed satisfactory data normalization (Uthe et al., 1983).

In order to assess potential dangers to humans consuming lobsters from the Belledune area, the concentrations of Cd were determined in cooked lobster meat. The tail and claws were removed from some of the animals selected for study from the annual geographical surveys carried out in 1981 and 1982 and were immediately steam-cooked over boiling, glass-distilled water in an all-glass system for 10 min. The cooked meat was removed by shucking, weighed, and homogenized with an equal weight of glass-distilled water by using a Polytron® (Kinematica, Lucerne, Switzerland) homogenizer and frozen. The analytical methods have been used satisfactorily in intercomparison studies carried out within ICES (Topping, 1982).

11.3. RESULTS AND DISCUSSION

11.3.1. The 1980 Survey

11.3.1a. *Physical oceanographic conditions*

A previous oceanographic survey by the Canadian Hydrographic Service (1965) indicates that the steady surface circulation near Belledune is composed of an easterly coastal flow of 10–20 cm s^{-1} (Fig. 11.5). This is consistent with the general cyclonic surface circulation in the Gulf of St. Lawrence and the flows induced by freshwater discharge at the head of the Baie des Chaleurs. The width (d) of such a coastal flow can be approximated by the expression $d = R/F$ (Leblond, 1980), where R is the Rossby radius and F is the internal Froude number. By using representative values for density and thickness of the flow, d is 9 km. The inshore portions of the coastal flow at Belledune can be countered by a wave- or wind-induced westerly drift. Reddy (1968), through the analysis of aerial photographs of Belledune Point prior to the construction of the harbor, showed that there had been a small westward displacement (75 m) of the point between the years 1933 and 1963 as a result of an inshore westerly drift.

The distribution of dissolved substances is determined primarily by transport processes, including advection and turbulent diffusion. Under normal conditions, the coastal flow past the mouth of Belledune Harbor is of a scale that prevents it from actually being able to penetrate the harbor, resulting in the harbor being a relatively isolated system. A possible circulation system induced in the harbor by such a coastal flow is shown in Fig. 11.6. Exchange between the harbor and the coastal flow would therefore be limited to diffusion or tidal flushing, except during periods of strong easterly and northeasterly winds. A lower bound on the flushing time can be calculated by defining it to be the volume of the harbor divided by the entering tidal flux. This is equal to 4.2 semi-diurnal lunar tidal cycles, or about 2 d, assuming complete mixing inside the harbor and no exchange between the water parcels entering and leaving the harbor. Under less than these ideal conditions, we expect the flushing time to be considerably longer than 2 d. Correspondingly, it might be anticipated that only limited amounts of non-conservative contaminants escape the harbor, especially if internal sediment movement is relatively sluggish.

11.3.1b. *Sediment composition*

Analysis of the sediments showed that the sediments contained 14–73% sand-sized (2.0–0.053 mm in diameter) particles, with the remaining material being < 0.053

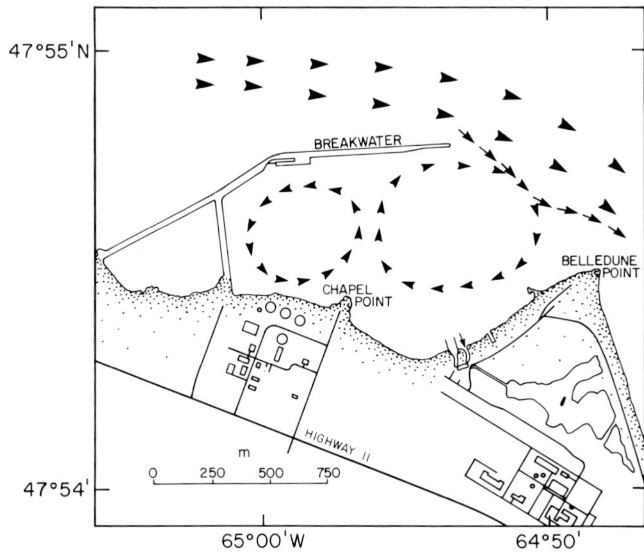

Figure 11.6. Hypothetical water circulation induced by coastal flow.

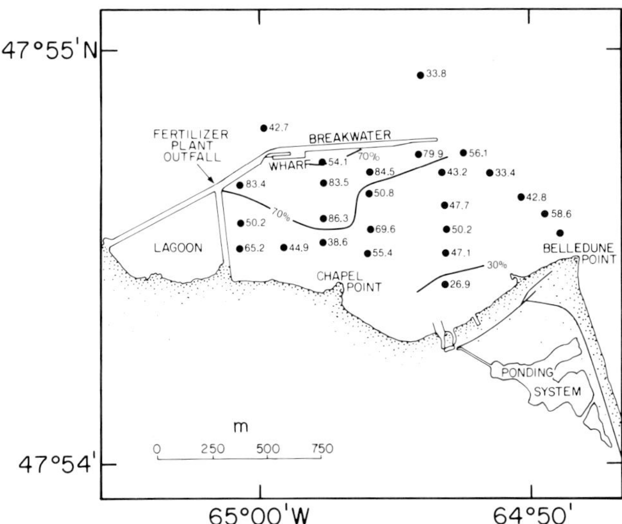

Figure 11.7. Percentage of fine-grained (<0.053 mm in diameter) material in Belledune Harbor. The 30% (by weight) and 70% (by weight) contours for the fine-grained material are indicated by the solid lines.

mm in diameter. Within Belledune Harbor, the highest percentages (>70%) of fine-grained material were found in the sediment near the outfall and the wharf (Fig. 11.7), and the lowest concentrations were found in the sediment along the southern side of the harbor. Texturally, this resulted in a wide variety of sediment types that range in classification from muddy sands (>70% sand, by weight) to the sandy muds (5–30% sand, by weight) that were found alongside of the lagoon and wharf breakwater. Sediments of a comparably mixed nature were also found in the adjacent coastal area.

Belledune Harbor sediments contained high concentrations of Cd ranging from 9 to 61 mg kg^{-1} (dry weight) (Fig. 11.8a). The highest concentrations (30–61 mg kg^{-1}) occurred in the inner harbor and decreased to 9–17 mg kg^{-1} off, and to the east, of the plant outfall. The Cd enrichment of these sediments was found to be 100–200 times that of the background level in the adjacent coastal sediments ($\sim$0.3 mg kg^{-1}). The total concentrations of Cd both in the mud (measured) and sand-sized (calculated) fractions are presented as absolute values in Fig. 11.8b. The calculated total Cd in the sand-sized fraction of the sediments varied from 4 to 91 mg kg^{-1}. The highest absolute concentrations (88–91 mg kg^{-1}) of Cd occurred just off the wharf and formed part of the Cd-rich (31–91 mg kg^{-1}) sand fraction in the northwestern corner of the harbor. Seaward, over a

distance of 1.5 km, the Cd concentrations in the sand-sized fraction decreased sharply toward the mouth of the harbor, where values were in the range of 5–52 mg kg^{-1}, following a dispersal pattern similar both to Cd in the whole sediment (Fig. 11.8a) and to Cd in the mud fraction (Fig. 11.8b). The sand fraction at the mouth of the harbor was enriched by a factor of about 150 as compared with the concentration (0.05 mg kg^{-1}) in the offshore sands.

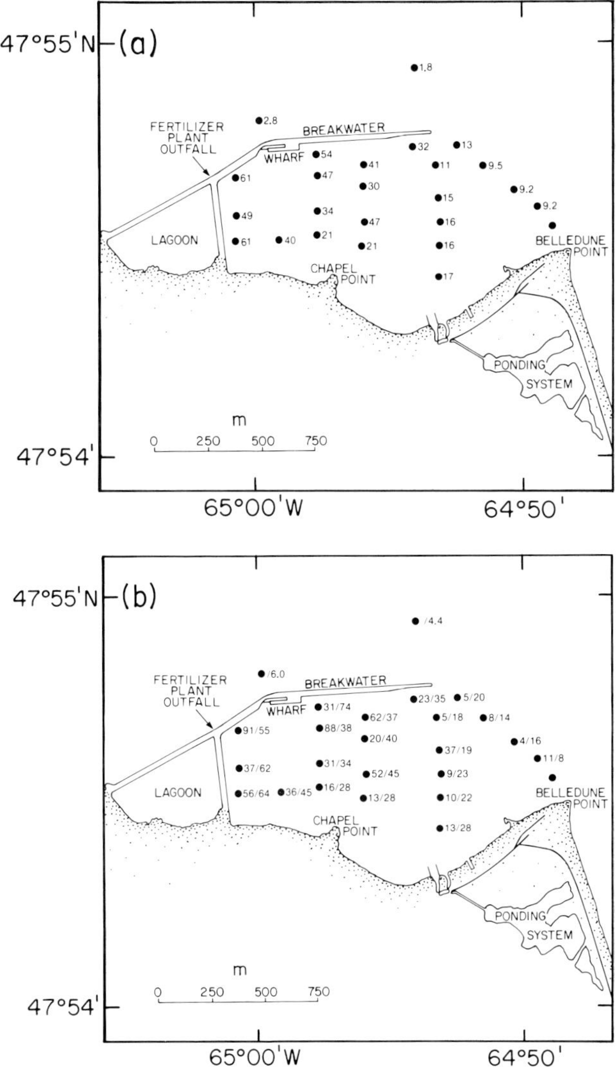

Figure 11.8. (a) Total concentration of Cd (mg kg^{-1}) in Belledune Harbor sediments. (b) Absolute Cd concentrations (mg kg^{-1}) in the sand–mud fractions of harbor sediments. The first number refers to the Cd concentration of the sand-size fraction; the second number refers to the mud-size fraction.

In the mud fraction, absolute Cd concentrations varied from 8 to 74 mg kg^{-1} (Fig. 11.8b) and had a comparable distribution pattern to that found for Cd in the whole sediment. High Cd concentrations (28–74 mg kg^{-1}) in the mud fraction were found in the western part of the harbor and decreased seaward to values between 8 and 28 mg kg^{-1}. The Cd-rich mud fractions found within the harbor were enriched by factors of between 50 and 150 as compared with the background Cd level (0.5 mg kg^{-1}) in the mud fraction of the coastal sediments.

Analysis of the contribution that each size fraction makes to the total Cd in the sediments of Belledune Harbor showed that between 13 and 56% of the Cd occurred in the sand fraction and that the balance was associated with the mud fraction. The highest proportion of the Cd contributed by sands occurred in the western part of the harbor, and, similarly, the sand fraction having the highest Cd concentrations was found just off the lagoon breakwater. Concentrations decreased with distance from the breakwater toward the mouth of the harbor, where the concentration was ~5 mg kg^{-1}. This pattern paralleled the distribution of Cd in the mud fraction. The results show that both coarse and fine-grained, Cd-rich material occurred in the sediments. Most of this material appeared to be trapped in the harbor, with little of it having been transported more than 1.5 km seaward.

In order to determine the particle size with which the Cd was associated, a sample of bottom sediment from the harbor was suspended in filtered, deionized water and fractionated by settling as described by Kranck (1980). After progressively longer times, subsamples were withdrawn by pipette from the settling suspension, and the grain size and Cd contents of the suspended material were determined. Results show that the modal particle size in subsequent samples decreased, whereas there was no concomitant change in Cd concentrations (Fig. 11.9). Thus, Cd was distributed throughout the range of particle sizes in the sample rather than being concentrated in a particular size interval. This distribution pattern was confirmed by size analysis of three subsamples of Cd-rich ore concentrate from the Belledune smelter; among them, these samples contained all grain sizes from 1 to 500 μm (Fig. 11.9). The lack of size differentiation between Cd-containing particles indicates that the decline in Cd concentrations from the harbor to the offshore region was due to dilution by uncontaminated material from outside sources rather

than to a decrease in grain size in Cd-rich material during transport from the harbor.

The coarse Cd-rich particles were probably derived from spills of ore concentrate at the unloading wharf. The source of the fine material was probably the slag pond since the highest concentrations of Cd in suspended matter were found at the outfall in the southeastern corner of the harbor. Due to the internal circulation pattern of Belledune Harbor (Fig. 11.6), the Cd-rich particles in the discharge from the slag pond were presumably trapped in the mixed sediments at the western end of the harbor. Outside the harbor, most of the Cd (70–99%) associated with the sediments was found in the mud fraction. Coarser sedimentary particles contributed little to the total Cd concentration.

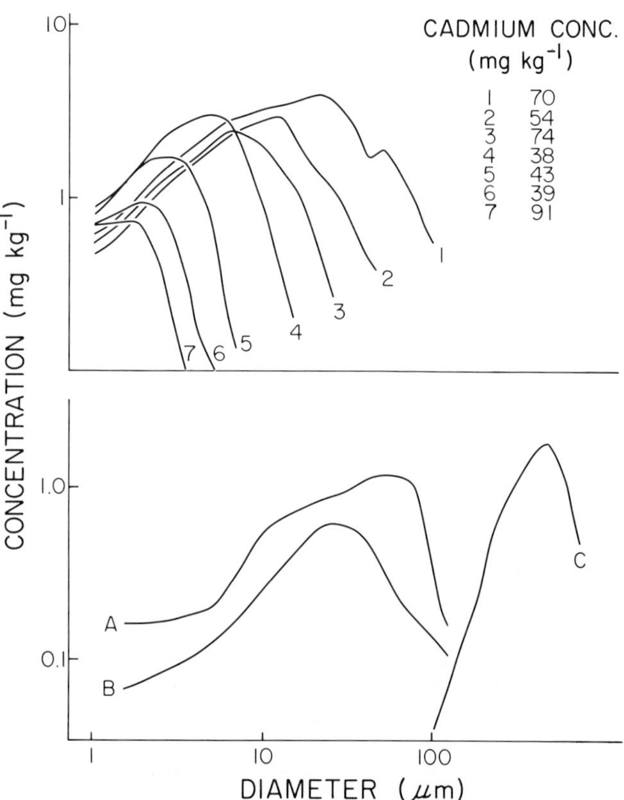

Figure 11.9. Grain-size analyses of and Cd concentrations in sediment from the Belledune area. Labels 1–7 refer to subsamples of settling-fractionated bottom sediment from station 27. Lable A refers to dust from the stockpile, B refers to Pb concentrate, and C refers to blast-furnace slag from the stockpile.

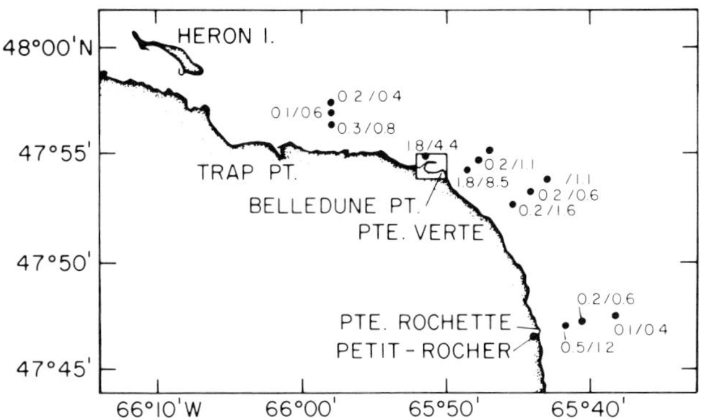

Figure 11.10. Cadmium concentrations (mg kg^{-1}) in the total and mud fractions of the coastal sediments. The first number refers to the whole sediment; the second number refers to the mud fraction.

At stations to the east of the harbor (Fig. 11.10), concentrations of total Cd varied between 0.1 and 1.8 mg kg^{-1}. Except for sediments at the station about 3 km seaward of Belledune Point and the two stations close to the northern edge of the breakwater, which had Cd concentrtions of 1.8, 2.8, and 1.8 mg kg^{-1}, respectively, all of the sediment samples collected outside the harbor contained total Cd concentrations <0.5 mg kg^{-1}.

The distribution of Cd in the mud fraction of the sediments outside the harbor reflected the movement of fine particulate material from within the harbor and its seaward dilution with uncontaminated particulate material. The highest concentration (8.5 mg kg^{-1}) was found just to the east of the harbor, and concentrations declined with distance along the coast in the direction of the prevailing longshore current. The concentrations on a transect to the west of the harbor probably reflect the background concentrations of Cd in coastal sediments, although even these samples would, to some extent, be affected by the metal-concentrate shipping activities that occur in the Dalhousie–Baie des Chaleurs area. The sediments of the Baie des Chaleurs contain relatively high concentrations of Cd (0.9 mg kg^{-1}) in comparison with other fine-grained sedimentary basins in the Gulf of St. Lawrence (0.3 mg kg^{-1}) (Loring, 1981).

The acetic acid leaching procedure indicated that between 24 and 71% of the total sedimentary Cd was in the non-detrital phase (Table 11.1). Within the harbor, the lowest percentages of non-detrital Cd occurred adjacent to the breakwater, the wharf, and the causeway (Fig. 11.11). The proportion of non-detrital Cd increased towards the center of the harbor (50–60% of the total) and even further towards the mouth of the harbor, where the proportion of non-detrital Cd reached 60–70% of the total. This pattern appears to represent the seaward migration of fine-grained Cd-rich material that has become associated with the sediment particles by absorption or flocculation.

Offshore, where the Cd concentrations were much lower, the non-detrital fraction contributed 35 to 82% of the total Cd in the sediments. The highest proportions of non-detrital Cd were found in the fine-grained offshore sediments. This probably reflects the normally greater proportion of weak-acid leachable metals in fine-grained, compared to coarse-grained, sedimentary particles (Loring, 1981).

11.3.1c. Composition of suspended particulate matter

Five samples of suspended particulate matter obtained from the harbor contained 6–96 mg kg^{-1} of Cd (Fig. 11.12). The highest concentration occurred near the outfall of the smelting plant, but high concentrations (24 mg kg^{-1}) persisted in the center of the harbor and at the seaward end of the breakwater (15 mg kg^{-1}). The sharp gradient in concentration near Belledune Point was probably a consequence of a contaminated, easterly, nearshore flow close to the point being adjacent to an inflow of relatively clean water containing naturally occurring suspended particulate matter.

Samples of suspended particulate matter obtained from outside the harbor contained from 2.1 to 6.2 mg

Table 11.1. Detrital and Non-detrital Cd Concentrations in the Sediments of Belledune Harbor and the Adjacent Coastal Zone

Location	Number of Samples	Average Detrital Cd Concentration (mg kg^{-1})	Average Non-detrital Cd Concentration (mg kg^{-1})	Average Percentage of Non-detrital Cd	Range (%)
Belledune Harbor	22	14.7	13.1	47	24–71
Coastal zone	13	0.38	0.27	57	35–82

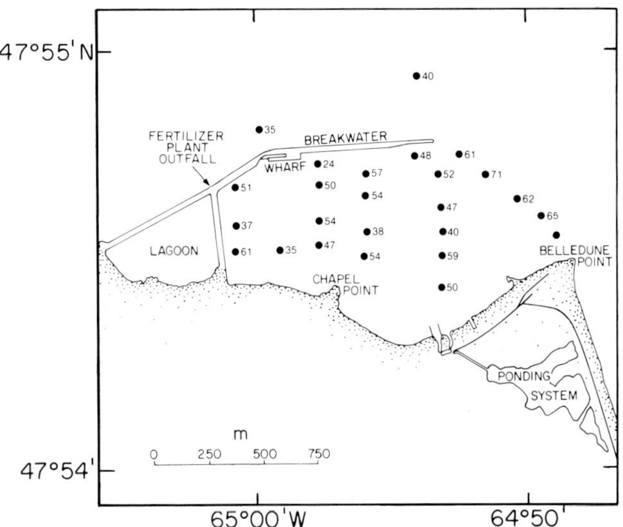

Figure 11.11. Non-detrital Cd concentrations expressed as a percentage by weight of the total Cd concentrations in harbor sediments.

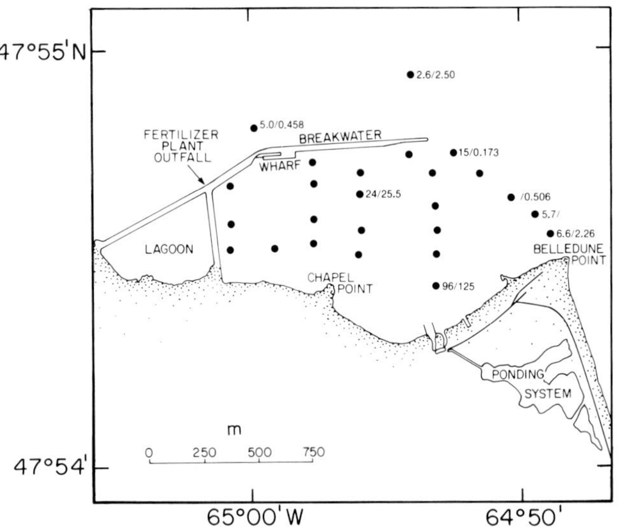

Figure 11.12. Cadmium in suspended particulate matter (mg kg^{-1}) and water (μg liter^{-1}) in Belledune Harbor. The first number refers to suspended particulate matter; the second number refers to water.

kg^{-1} of Cd (Fig. 11.13). Although these values were much higher than typical for coastal sediments, they did not appear to reflect a high degree of contamination when the fine-grained nature of the suspended particulate matter was considered. Overall, there was little indication that the sediments, their fine-grained frac-

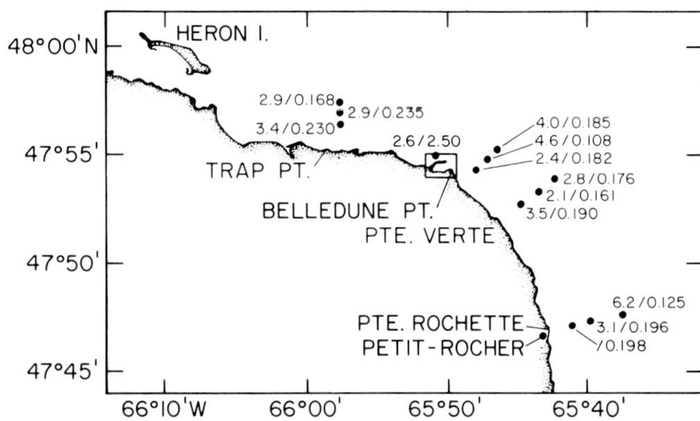

Figure 11.13. Cadmium concentrations in suspended particulate matter (mg kg^{-1}) versus Cd in water (μg liter^{-1}) in the coastal zone. The first number refers to suspended particulate matter; the second number refers to water.

tions, or the suspended particulate matter were contaminated by Cd farther than 3 km seaward of Belledune Harbor.

11.3.1d. *Water composition*

The concentrations of Cd in filtered water samples taken from the coastal zone are also shown in Fig. 11.13. In comparison to the concentrations of Cd found in the water of the Gulf of St. Lawrence (Yeats et al., 1978), the Cd concentrations in the vicinity of Belledune Harbor were quite high. It must be recognized, however, that the coastal longshore flow is a relatively low-salinity current containing the constituents of fresh-water discharges into the western and southwestern Gulf of St. Lawrence (El Sabh, 1975). Since the region of the upper Baie des Chaleurs may be regarded as a source of metals (including Cd), one might expect the metal concentrations close to the coast at Belledune to be somewhat higher than those in other areas of the gulf. The water within Belledune Harbor was clearly contaminated with concentrations of up to 125 μg liter^{-1} near the smelting-plant outfall (Fig. 11.12). Nevertheless, rapid dilution was clearly occurring; consequently, the concentrations declined very rapidly toward and outside the mouth of the harbor. There is some evidence of contamination of the coastal zone in the results of the analyses of seawater from stations just outside the harbor, north of the breakwater; however, the high concentrations at these stations may have been a consequence of discharges from the fertilizer plant. The general concentrations of Cd in seawater outside the harbor were 125–230 ng liter^{-1} (Fig. 11.13), which are probably typical of the local inshore region since this represents only minor augmentation of the Cd concentration compared with open surface water in the Gulf of St. Lawrence (Loring, 1981). Only the water in and around the harbor appeared to be detectably contaminated from local sources.

11.3.1e. Lobsters

The analyses of hepatopancreatic tissue from lobsters collected within and around the harbor in 1980 showed that very high concentrations of Cd were present in lobsters in the harbor (Table 11.2), especially in those taken from within the harbor and close to the breakwater [the Harbour West (HW) site (Fig. 11.3)]. This pattern is similar to that found in the abiotic component of this study. Due to the circulation within the harbor (Loring et al., 1980), the HW site, although not situated exactly at the former smelter-effluent outfall, would be susceptible to contamination from this source. The geometric mean of the Cd concentration (176 mg kg^{-1}, wet weight) in the hepatopancreas in 1980 was substantially higher than that for the Harbor East (HE) site (62.3 mg kg^{-1}) in the same year. This difference is somewhat surprising since the distance between the two sampling sites is less than 800 m, and it suggests that only limited mixing of the lobster populations between the two areas was occurring. It is possible, however, that the lobsters sampled at the HE site contained a number of animals that had only recently wandered into the harbor from uncontaminated areas outside.

The mean concentrations of Cd in the hepatopancreas of lobsters taken from sites outside Belledune Harbor in 1980 were distributed geographically (Table 11.2) as one would expect, given the source of the Cd and the nature of the prevailing coastal circulation. The mean concentration at the L1W site (Fig. 11.5) was substantially lower than was the mean concentration at stations up to 10 km east of the harbor, but it was elevated compared with those for the control site (Heron Island). The mean concentration reported for the Heron Island site (3.85 mg kg^{-1}) was similar to the mean Cd level in lobsters taken at Petit Rocher in 1973 (4.30 mg kg^{-1}) and well below the average mean concentrations of Cd in the hepatopancreas from lobsters from other areas of the Atlantic Coast of Canada (Uthe and Freeman, 1980). However, this value from 1973 was significantly lower ($p < 0.05$) than the value from the Petit Rocher site in 1980, suggesting that Cd contamination was fairly widespread within the Bay of Chaleur. Thus, the spatial extent of Cd contamination in lobsters was greater than that detected in the sediments, the suspended particulate matter, or the water. Since the Food and Agriculture Organization–World Health Organization (1972) recommend that human beings ingest no more than 1.0 μg kg^{-1} body weight of Cd per day, then the concentration of Cd in lobster hepatopancreatic tissue was judged to be sufficiently

Table 11.2. Geometric Mean of the Cd Concentration (μg g^{-1}, Wet Wt) in the Hepatopancreas of Lobsters Captured within or around Belledune Harbor

Distance from the Harbor	Sample Site[a]	1980			1981			1982	
		$\overline{X}_g$	Number of Animals	ANCOVA[b] 1980–1981	$\overline{X}_g$	Number of Animals	ANCOVA[b] 1981–1982	$\overline{X}_g$	Number of Animals
30 km west	Heron Island	3.85	30	*	4.83	27	n.s.	4.69	23
1.6 km west	L1W	11.9	29	n.s.	20.0	28	n.s.	14.3	33
Within Belledune Harbor	Harbor East	176.0	29	*	210.0	46	n.s.	147.0	35
	Harbor West	62.3	28	n.s.	68.5	44	n.s.	76.3	35
Outside the Harbor									
Breakwater	L0BE	21.4	29	*	65.7	27	n.s.	46.7	27
3.1 km east	L1E	28.1	31	*	40.6	36	n.s.	40.4	36
7.8 km east	L4E	28.0	26	*	42.7	48	*	23.8	38
10.0 km east	L6E	17.3	19	*	25.5	31	*	15.6	36
	Limestone Point[c]	—	—	—	13.6	26	n.s.	13.9	32
17.4 km east	Petit Rocher	11.6	31	n.s.	15.7	41	n.s.	12.9	36

[a] See Figs. 11.4 and 11.5 for sample locations.

[b] ANCOVA means analysis of covariance; an asterisk indicates that the data is significant at $p < 0.05$.

[c] Limestone Point was not sampled in 1980.

high to justify action to reduce Cd discharges from local industry and to impose some restrictions on lobster fishing, processing, and marketing procedures in the area.

As a result of the 1980 study, Belledune Harbor was closed to commercial lobster fishing in 1980. In addition, all lobsters taken from between the L1W and L4E sampling sites (Fig. 11.5) were processed only after the removal and destruction of the lobster carapace, which included the raw hepatopancreas. This meant that only the meat from the tail and claws of the lobsters were processed for human consumption. Product inspection ensured that no contaminated meat reached the commercial market. These controls have remained in place since 1980.

11.3.2. Lobster Surveys in 1981 and 1982

The results of the annual geographical surveys of Cd concentrations in lobster hepatopancreatic tissue are summarized in Table 11.2. No attempt was made to

select animals in the catch; instead, analysis of covariance (ANCOVA), with lobster weight as the covariant, was used to estimate the significance of the differences among the means from year to year.

Significant ($p < 0.05$) differences were found between mean concentrations recorded in 1980 and those in 1981, with the higher concentrations being found in 1981 in the HW, L0BE, L1E, L4E, L6E, and Heron Island sites. This general increase in mean concentration of Cd in the vicinity of the harbor between 1980 and 1981 is not surprising since the new plant-control measures were not put into operation until November 1980, well after the major annual feeding and activity period of the animals (Levaque Charron and Eljarbo, 1981). The small but significant increase in mean concentrations of Cd in animals at the control site (Heron Island) between 1980 and 1981 may be sampling anomaly or a real increase caused by some unknown factor. The absence of significant changes in the concentrations at the HE site is somewhat surprising but may be related to the presence of lobsters that have wandered into the harbor to this site from less contaminated areas shortly before the

Table 11.3. Concentration (μg g^{-1}, Wet Wt) of Cd in the Muscle[a] of Lobsters from Sites HW and HE

Date of Collection	Laboratory	Number of Samples	Tissue Analyzed	Mean	Range
July 1975	Brunswick Mining and Smelting[b]	8	Tail	0.40	0.18–0.86
July 1976	Brunswick Mining and Smelting[b]	5	Tail	0.60	0.22–1.11
July 1977	Brunswick Mining and Smelting[b]	30	Tail and claws (pooled)	0.46	0.10–3.60
July 1978	Brunswick Mining and Smelting[b]	10	Tail and claws (pooled)	1.49	0.36–4.08
July 1979	Brunswick Mining and Smelting[b]	28	Tail and claws (pooled)	2.70	0.40–11.0
April 1980	Department of Fisheries and Oceans[c]	12	Tail and claws (pooled)	—[d]	0.20–1.85
April 1981	Department of Fisheries and Oceans[c]	24	Tail and claws (pooled)	0.22	0.03–0.64

[a] The muscle was fresh-frozen

[b] From Levaque Charron and Eljarbo, 1981.

[c] Unpublished data.

[d] Muscle tissue was only analyzed in lobsters from HW displaying a selected range of Cd concentrations in the hepatopancreas; therefore, a mean concentration for Cd in muscle cannot be calculated.

sampling took place. Overall, in 1982 the trend toward increasing concentrations in hepatopancreatic tissue with time, apparent up until 1981, appears to have ceased or have been slightly reversed, with decreased concentrations evident at stations L4E and L6E (to the east of Belledune Harbor).

The concentrations of Cd in uncooked lobster muscle from animals captured at the HW site in 1980 and 1981 were determined. These results, and data from earlier studies of lobster muscle carried out by the Brunswick Smelting Corporation, are presented in Table 11.3. It is clear from these data that the concentration of Cd in lobster muscle peaked in 1979 and decreased during the subsequent two years. This trend contrasts with that found in hepatopancreatic Cd concentrations, which appeared to have peaked two years later, in 1981 (Table 11.2).

We have also studied Cd concentrations in pooled, cooked tail and claw muscles from individual lobsters from each of the sampling sites both in 1981 and in 1982. The mean concentrations and the measurements of the corresponding concentrations of Cd in the hepatopancreas are shown in Table 11.4. The results for both years reflect the same geographical distribution as was found for hepatopancreas; however, the concentrations of Cd in the cooked meat were lower by factors between 150 and 400 than were the concentrations in uncooked hepatopancreas. In general, the mean concentrations of Cd in pooled meat from lobsters from all areas were lower in 1982 than in 1981. This could be a sampling anomaly since, with the exception of samples taken from the L1E site (Fig. 11.5), the mean concentrations of Cd in hepatopancreas were also lower in 1982. An analysis of variance, however, showed no statistically significant decrease in any individual sample pair.

Table 11.4. Concentration (mg kg^{-1}, Wet Wt) of Cd in Pooled Samples of Cooked Muscle and Raw Hepatopancreas of Lobsters from the Area around Belledune Harbor[a]

Sample Site	Year	Number of Samples	Cooked Meat		Hepatopancreas	
			$\overline{X}_g$	Range	$\overline{X}_g$	Range
Heron Island	1981	10	0.04	0.03–0.09	5.80	4.36–12.0
	1982	9	0.03	0.02–0.03	5.32	3.51–10.1
L1W	1981	10	0.12	0.03–1.82	29.60	8.30–135.0
	1982	6	0.03	0.03–0.06	8.77	5.76–17.5
HW	1981	8	0.89	0.16–3.06	218.00	62.70–572.0
	1982	21	0.74	0.11–3.26	165.00	33.30–728.0
HE	1981	11	0.35	0.07–1.28	67.90	17.00–355.0
	1982	12	0.23	0.04–0.762	62.50	14.70–470.0
L0BE	1981	10	0.28	0.05–1.24	79.60	13.50–308.0
	1982	13	0.16	0.02–1.14	41.00	6.70–190.0
L1E	1981	10	0.21	0.05–0.68	34.30	8.15–130.0
	1982	12	0.18	0.04–0.33	47.80	10.30–134.0
L4E	1981	10	0.19	0.05–0.43	56.60	7.67–127.0
	1982	11	0.11	0.03–0.20	23.40	11.50–47.0
L6E	1981	10	0.11	0.05–0.28	30.70	12.20–50.9
	1982	8	0.08	0.02–0.14	19.50	5.80–54.3
Beach Point, Prince Edward Island	1981	10	0.03	0.02–0.09	11.80	4.30–49.9
Petit Rocher	1982	10	0.06	0.04–0.10	14.00	7.10–32.3

[a]Sample numbers for lobster hepatopancreas differ from those shown in Table 11.2.

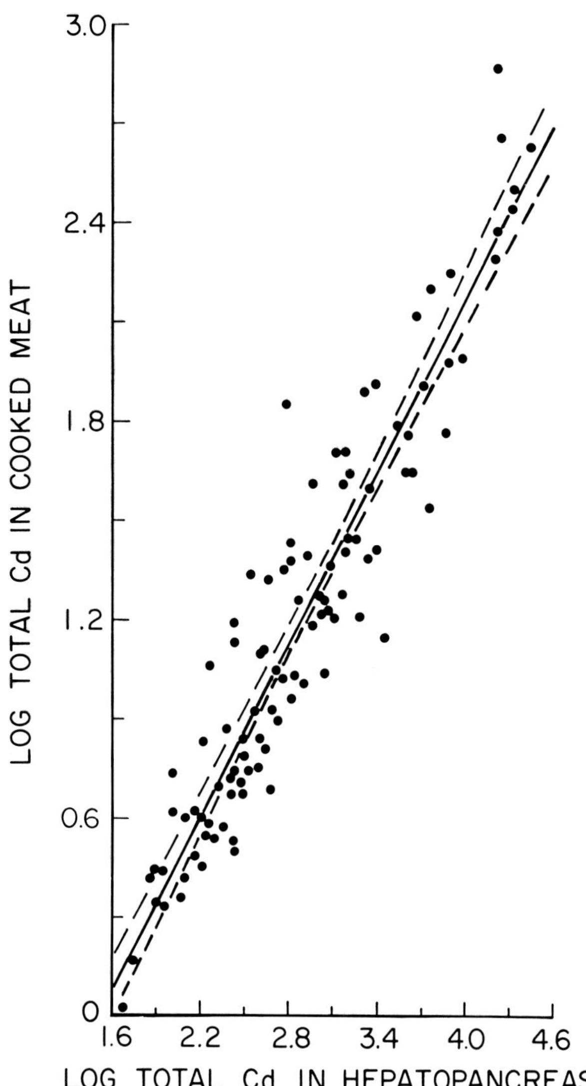

Figure 11.14. The relationship between the total amount (μg) of Cd in cooked, pooled lobster tail and claw meat and the amount (μg) in the hepatopancreas of the same animal (log base 10).

It should be noted that Cd concentrations in cooked meat from lobsters from the HW site (Table 11.4) were higher than the concentrations in uncooked muscle (Table 11.3), probably due to shrinkage and the inclusion of the parenchyma tissue in the cooked meat. Ray et al. (1981) have shown that parenchyma tissue has a higher Cd concentration than does muscle tissue.

Finally, we have also studied the relationship between the total amounts of Cd in the raw hepato-

pancreas and in the pooled, cooked muscle. Each sampling site gave a similar relationship, which suggests that we were looking at a general relationship. All of the data were then combined in Fig. 11.14, the relationship for which is

$$\mathrm{Log_{10}Cd}_{Cooked\ Meat}$$

$$= -1.321 + 0.876\ \mathrm{Log_{10}Cd}_{Raw\ Hepatopancreas} \qquad 1$$

with a correlation coefficient of 0.939 ($n = 100$). This analysis provides some confidence in routinely monitoring cooked lobster meat in order to assess the effects of more stringent discharge controls and to determine the need for the introduction and maintenance of processing and marketing restrictions, without having to resort to continuous field monitoring, which involves much larger costs.

11.4. CONCLUSIONS

Until 1980, the operation of the Brunswick Mining and Smelting Corporation Ltd. lead-smelting and refining plant at Belledune, New Brunswick, had given rise to Cd contamination of sediment, water, and lobsters in the vicinity of Belledune Harbor. By 1980, lobsters caught within the harbor contained such high concentrations of Cd in their tissues that some controls on both fishing and processing procedures for lobsters had to be introduced, and further efforts were made to reduce the flux of Cd into the local marine environment. The hepatopancreas of lobsters captured immediately outside the harbor contained substantially lower concentrations of Cd than did the hepatopancreas of lobsters inside the harbor. These concentrations decreased further as the distances between the harbor and the location at which the animals were captured increased. The Cd distribution in the sediments and water in the area was similar to that for the lobster tissues, with marked contamination of both water and sediments being evident at sites within and close to the harbor.

In 1980, additional discharge-control measures were introduced by the smelting plant, resulting in some stabilization of the concentrations of Cd in lobsters by 1982. It is likely that the Cd concentrations in lobster tissues will decline in the future, but it is too early to predict what the new equilibrium values might be. Given the extent of sediment contamination in the area, it is likely that it will take a number of years for the system to approach some new equilibrium condition.

Methods of exploiting the marine coastal zone as a receiving medium for the wastes from coastal heavy industry and for food production can often be in conflict. The regular monitoring of fisheries likely to be affected by disposal activities is a useful approach to the early detection of such interferences. When monitoring results dictate, regulatory controls can be imposed, or other action can be taken, to prevent conflicting uses of the coastal zone from becoming wholly incompatible.

REFERENCES

Berman, S. S., A. P. Mykitiuk, P. A. Yeats, and J. M. Bewers. 1983. ICES Fifth Round Intercalibration for Trace Metals in Seawater: ICES 5/TM/SW (Section 3) Preliminary Report. Paper No. CM 1983/E:24, unpublished report, International Council for the Exploration of the Sea, Copenhagen, Denmark, 33 pp.

Bewers, J. M., and H. L. Windom. 1982. Comparison of sampling devices for trace metal determinations in seawater. *Marine Chemistry*, **11**, 71–86.

Bewers, J. M., J. Dalziel, P. A. Yeats, and J. L. Barron. 1981. An intercalibration for trace metals in seawater. *Marine Chemistry*, **10**, 173–193.

Bewers, J. M., B. Sundby, and P. A. Yeats. 1976. The distribution of trace metals in the western North Atlantic off Nova Scotia. *Geochimica et Cosmochimica Acta*, **40**, 687–696.

Canadian Hydrographic Service. 1965. Data Compilation: Belledune Point Current Survey, 1964 Season. Data Series Report No. 65-1, Bedford Institute of Oceanography, Dartmouth, Nova Scotia, Canada, 60 pp.

Dugdale, P. J., and B. L. Hummel. 1977. Cadmium in the lead smelter at Belledune: its association with heavy metals in the ecosystem. *In*: Proceedings of the First International Cadmium Conference (January 1978), San Francisco. Metal Bulletin Ltd., London. pp. 53–73.

El Sabh, M. I. 1975. Transport and Currents in the Gulf of St. Lawrence. Bedford Institute of Oceanography Report Series No. BI-R-75-9, Bedford Institute of Oceanography, Dartmouth, Canada, 180 pp.

Food and Agriculture Organization–World Health Organization. 1972. Evaluation of Certain Food Additives and the Contaminants Mercury, Lead and Cadmium. Technical Report Series No. 505, World Health Organization, Geneva, Switzerland, 32 pp.

Freeman, H. C., and J. F. Uthe. 1974. Geographical Distribution of Cadmium and Arsenic in Lobster Digestive Gland (Hepatopancreas). Paper CM 1974/E:16, unpublished report, International Council for the Exploration of the Sea, Copenhagen, Denmark, 3 pp.

Gale, N. L., and B. G. Wixson. 1979. Cadmium in forest ecosystems around lead smelters in Missouri. *Environmental Health Perspectives*, **28**, 23–37.

Kranck, K. 1980. Experiments on the significance of flocculation in the settling of fine-grained sediment in still water. *Canadian Journal of Earth Sciences*, **17**, 1517–1526.

Laitinen, H. A. 1960. Chemical Analysis. McGraw-Hill, New York, 611 pp.

Leblond, P. H. 1980. On the surface circulation in some channels of the Canadian Arctic Archipelago. *Arctic*, **33**, 189–197.

Levaque Charron, R. 1979. Marine Impact Survey of the Belledune Harbor Area, New Brunswick, for the Period May, 1979, to August, 1980. Internal Report No. 389, unpublished report, Centre de Récherche NORANDA, Pointe Claire, Quebec, Canada, 46 pp.

Levaque Charron, R., and I. Eljarbo. 1981. Lobster Growth Study of the Belledune, New Brunswick, Area, 1973–78. Technical Memorandum 110, Centre de Recherche NORANDA, Pointe Claire, Quebec, Canada, vii + 47 pp.

Loring, D. H. 1981. Potential bioavailability of metals in eastern Canadian estuarine and coastal sediments. *Rapports et Procès Verbaux de la Conseil Internationale de l'Exploration de la Mer*, **181**, 93–101.

Loring, D. H., and R. T. T. Rantala. 1977. Geochemical Analysis of Marine Sediments and Suspended Particulate Matter. Report No. 700, Canadian Technical Report of Fisheries and Aquatic Sciences, Ottawa, 56 pp.

Loring, D. H., J. M. Bewers, G. H. Siebert, and K. Kranck. 1980. A preliminary survey of circulation and heavy metal contamination in Belledune Harbor and adjacent areas. *In*: Cadmium Pollution of Belledune Harbor, New Brunswick, Canada, J. F. Uthe and V. Zitko (Eds.). Report No. 700, Canadian Technical Report of Fisheries and Aquatic Sciences, Ottawa, 56 pp.

Neumann, G. W., and J. R. Schnarr. 1971. "The Brunswick Story": Concentrator operation at Brunswick Mining and Smelting Corporation's no. 12 mine. *The Canadian Mining and Metallurgical Bulletin*, **64**, 51–61.

Prairie, R., and R. Levaque Charron. 1981. Marine Environmental Impact Survey of the Belledune Harbor Area, New Brunswick, for the Period July–September, 1980. Internal Report 402, Centre de Récherche NORANDA, Pointe Claire, Quebec, Canada, 41 pp.

Prairie, R., and L. Trudel. 1983. Heavy Metal Accumulation by Blue Mussels (*Mytilus edulis*) in Belledune Area, New Brunswick, June–September, 1981. Internal Report No. 411, Centre de Récherche NORANDA, Pointe Claire, Quebec, Canada, 46 pp.

Rantala, R. T. T., and D. H. Loring. 1980. Direct determination of cadmium in silicates from a fluoroboric acid matrix by graphite furnace atomic absorption spectrophotometry. *Atomic Spectroscopy*, **1**, 163–165.

Ray, S., D. W. McLeese, and L. E. Burridge. 1981. Cadmium in tissues of lobster captured near a lead smelter. *Marine Pollution Bulletin*, **12**, 383–386.

Reddy, M. P. M. 1968. Wave Conditions and Littoral Drift near Belledune Point, Chaleur Bay. Report No. AOL 68-2, Department of Energy, Mines and Resources, Marine Sciences Branch, Ottawa, 25 pp.

Topping, G. 1973. Heavy metals in shellfish from Scottish waters. *Aquaculture*, **1**, 379–384.

Topping, G. 1982. Report on the Sixth ICES Trace Metal Intercomparison Exercise for Cadmium and Lead in Biological Tissue. Cooperative Research Report No. 111, International Council for the Exploration of the Sea, Copenhagen, Denmark, 27 pp.

Uthe, J. F., and H. C. Freeman. 1980. Cadmium in hepatopancreas of American lobster (*Homarus americanus*) from eastern Canada. *In*: Cadmium Pollution of Belledune Harbor, New Brunswick, Canada, J. F. Uthe and V. Zitko (Eds.). Report No. 963, Canadian Technical Report of Fisheries and Aquatic Sciences, Ottawa, pp. 73–76.

Uthe, J. F., and V. Zitko (Eds.). 1980. Cadmium Pollution of Belledune Harbor, New Brunswick, Canada. Report No. 963, Canadian Technical Report of Fisheries and Aquatic Sciences, Ottawa, v + 107 pp.

Uthe, J. F., C. L. Chou, and D. G. Robinson. 1980. Cadmium in American lobster (*Homarus americanus*) from the area of Belledune Harbor, New Brunswick, Canada. *In*: Cadmium Pollution of Belledune Harbor, New Brunswick, Canada, J. F. Uthe and V. Zitko (Eds.). Report No. 936, Canadian Technical Report of Fisheries and Aquatic Sciences, Ottawa, pp. 65–72.

Uthe, J. F., C. L. Chou, D. G. Robinson, and R. Levaque Charron. 1983. Cadmium in American Lobster (*Homarus americanus*) from the Area of a Lead Smelter. Report No. 1163, Canadian Technical Report of Fisheries and Aquatic Sciences, Ottawa, pp. 194–197.

Wixson, B. G., N. L. Gale, and K. Downey. 1975. Control of environmental contamination by cadmium, lead, and zinc near a new lead belt smelter. *In*: Proceedings of the University of Missouri's Eleventh Annual Conference on Trace Substances in Environmental Health, Vol. 11, D. D. Hempkill (Ed.). University of Missouri, Columbia, 455 pp.

Yeats, P. A., J. M. Bewers, and A. Walton. 1978. Sensitivity of coastal waters to anthropogenic trace metal emissions. *Marine Pollution Bulletin*, **9**, 264–268.

PART III: CHEMICAL PROCESSES AND TRANSPORT

Chapter 12

Adsorption Kinetics of Waste-Generated Arsenate and Phosphate at the Iron Oxyhydroxide–Seawater Interface

Geoffrey E. Millward, John G. Marsh, and Stuart A. Crosby

Department of Marine Science
Plymouth Polytechnic
Plymouth, Devon, United Kingdom

Michael Whitfield and William J. Langston

The Laboratory
Marine Biological Association
Plymouth, Devon, United Kingdom

Peter O'Neill

Department of Environmental Science
Plymouth Polytechnic
Plymouth, Devon, United Kingdom

ABSTRACT

This chapter reports kinetic analyses of the uptake of arsenate and phosphate onto fresh and aged Fe(II)-derived precipitates in seawater. Uptake by fresh Fe(II)-derived material in seawater had two kinetic regimes.

At temperatures of 15 and 20°C and a pH < 7.5, and a temperature of 2°C and a pH < 7.9, the adsorption followed first-order kinetics that were dependent on the slow oxidation of Fe(II). The rate constants were in good agreement with other values obtained from experiments on the oxidation of Fe(II). At temperatures of 15 and 20°C and a pH > 7.5, and a temperature of 2°C and a pH > 7.9, the data fitted a reversible reaction of the type $2A \rightleftharpoons X + Y$ (where A is the dissolved species, and X and Y are the adsorbed species). These analyses suggested that the rate-determining step was the adsorption process rather than the oxidation of Fe(II). Conditional rate constants were obtained as a function of pH and temperature. Similar adsorption experiments were carried out with precipitates obtained from natural sources such as sediment pore waters and acidic mine streams. In general, the adsorption behavior of the natural material was similar to that of the synthetic precipitates. Aged Fe(II)-derived material did not adsorb phosphate at all in the pH range of 7.3–8.2, and about 50% of the arsenate remained dissolved at equilibrium. The arsenate adsorption was inversely dependent on pH at a temperature of 20°C but was independent of pH at 2°C.

12.1.   INTRODUCTION

An urgent need exists for the development of quantitative geochemical models that can be used to predict the consequences of the disposal of toxic wastes into the sea (United Nations IMCO/FAO/UNESCO/WMO/ QHO/IAEA/UN/UNEP Joint Group of Experts on the Scientific Aspects of Marine Pollution, 1982). Ideally, these models should include not only the time constants of the physical processes (advection and mixing) but also the time constants of the important chemical processes. The fate of dissolved elements in seawater is largely controlled by heterogeneous processes involving the surfaces of fine particles and highly reactive colloidal Fe and Mn oxyhydroxides. The application of the equilibrium concept to these systems is limited to the relatively fast sorption reactions or to the very slow reactions, in which case it can be neglected (Pankow and Morgan, 1981). Reactions that equilibrate over intermediate time scales are often in the non-steady state for periods over which considerable advection may occur, thus leading to the lateral dispersion of the waste material. Many trace constituents may show this relatively slow transfer to the particulate phase, and the rate constants for the sorption reactions are therefore of key importance in any time-dependent modelling.

In this chapter the role of amorphous and poorly crystallized Fe oxyhydroxides is examined because these form an important fraction of the total surface available for adsorption. The Fe precipitates found in seawater can be derived from either Fe(III) or Fe(II) salts, and Crosby et al. (1983) have shown that there are considerable differences in the characteristics of the surfaces of the two materials. Precipitates formed by the relatively rapid hydrolysis of Fe(III) have been used by many workers to demonstrate their sorption capacity for both anions (Yates and Healy, 1975; Lijklema, 1980; Pierce and Moore, 1980; Crosby et al., 1981; Pierce and Moore, 1982) and cations (Kinniburgh et al., 1976; Davis and Leckie, 1978; Inoue and Munemori, 1979; Swallow et al., 1980; Millward and Moore, 1982). The Fe(II)-derived precipitates, however, are regarded as being potentially more important in seawater because they have a natural source in the anoxic zone of sediment pore waters and also industrial sources, such as acidic mine wastes (Boyden et al., 1979; Hoff et al., 1982) and iron-rich wastes from the titanium pigment industry (Gerlach, 1981). The Fe(II) must undergo oxidation prior to hydrolysis and precipitation (Sung and Morgan, 1980).

Phosphate and arsenate were chosen as the adsorbates in this investigation because they are important byproducts of human activity and because Fe oxyhydroxides appear to dominate the solution chemistry of these Group V elements. While arsenic does have two oxidation states in naturally occuring waters, As (V) predominates, and it can be compared to the behavior of phosphate. The time dependence of the adsorption has been followed with a view to obtaining kinetic data, and the chemical systems studied are analogous to the oceanic situation in which the concentration of dissolved toxic wastes is controlled by the presence of fresh and aged Fe(II)-derived precipitates.

12.2.   EXPERIMENTS

12.2.1.   Procedure

The adsorption reactions were carried out in 1-liter reactors, which gave the optimum practical ratio of volume:surface area (Crosby, 1982). For experiments involving synthetically produced Fe precipitiates, 1 ml of a stock solution of $FeCl_2$ was injected into 1 liter of seawater (with a salinity of 34 ‰) to give a final Fe concentration of 5×10^{-5} moles liter^{-1}. The precipitation experiments were carried out at 15 and 2°C for phosphate runs and at 20 and 2°C for arsenate runs. Fresh precipitates were formed in the presence of the adsorbate while the solutions were stirred and open to the air. Aged precipitates were formed in seawater at the appropriate temperature and open to the air, and they were stirred for 24 h prior to the addition of the adsorbates. Separate experiments were carried out for phosphate and for arsenate. The phosphate was added in the form of NaH_2PO_4 to give a final concentration of 1×10^{-6} moles liter^{-1}, and the arsenate was added as Na_3AsO_4 for a final concentration of 1.33×10^{-6} moles liter^{-1}. In experiments with fresh precipitates, the pH was continuously monitored during the adsorption reaction. Immediately after the injection of the Fe solution, the pH decreased by about 0.5 units but returned to the required value within 2 min, and further pH changes were limited to ± 0.2 units throughout the duration of the run. The uptake of the adsorbates onto fresh and aged precipitates was followed over several hours by extraction and filtering 25-ml quantities of the suspension at selected intervals. The dissolved fraction was defined as that fraction passing through a 0.45-μm Millipore® (Millipore U. K. Ltd., London) filter.

Dissolved Fe from natural sources was sampled in relatively high concentrations (up top 80×10^{-6} moles liter^{-1}) in water samples from acidic mine streams (which had a pH <5.0 and a salinity of 0‰) and interstitial waters of stream sediments located in the mineralized area of southwestern England. The interstitial waters were sampled by using a dialysis-bag collection technique developed by Knox et al. (1981). Before use, the water samples were filtered sequentially through glass-fiber/cellulose and 0.45-μm Millipore® filters. The procedure of Fe precipitation from the interstitial waters was similar to that just described, in that a 20-ml aliquot of the Fe-rich interstitial water was injected directly into the seawater. In the case of the acidic mine stream waters, the adsorbate was added, followed by a concentrated mixture of seawater salts, giving a final salinity of 34‰. The Fe already in solution was then precipitated by rapid adjustment of the pH to the required value.

12.2.2. Analytical Methods

The concentration of phosphate in the filtrate was determined using the molybdenum blue complex single-reagent method described by Murphy and Riley (1962). The absorption at 885 nm was measured by using 10-cm cells in a Pye Unicam® (Pye Unicam Ltd., Cambridge, U.K.) SP500 spectrophotometer. The coefficient of variation for replicate samples at the level of 1×10^{-6} moles liter^{-1} was $\pm 2\%$.

The concentration of arsenate remaining in solution was determined by the analysis of its gaseous hydride generated in a closed reaction vessel by the addition of $NaBH_4$ to an aliquot of the sample in $6M$ HCl. The arsine was swept into an Ar/H_2 flame by an auxiliary argon flow, and it was detected at 193.7 nm by using an IL® 151 (Instrumentation Laboratory Ltd., London, U.K.) spectrophotometer with background correction. The coefficient of variation of arsenate at the level of 1.33×10^{-6} moles liter^{-1} was $\pm 5\%$.

The total Fe and Fe(II) in all samples was determined by using the Ferrozine® (Hach Chemical Co., London, U.K.) complex, which absorbs at 562 nm (Gibbs, 1979). The quantity of Fe(II) was analytically defined as that yielding absorbance within 5 min of the addition of the complexing agent without addition of HCl or the reducing agent (Crosby, 1982).

12.3. RESULTS AND DISCUSSION

12.3.1. Adsorption onto Fresh Fe(II)-Derived Precipitates

Precipitate-characterization studies carried out at Plymouth Polytechnic (Crosby et al., 1983) and elsewhere (Sung and Morgan, 1980) have shown that the Fe(II)-derived precipitate is poorly crystallized lepidocrocite (γ-FeOOH). This product appears over time scales of hours and has a surface area of about 120 m^2 g^{-1}, as determined by N_2-gas adsorption. Investigations involving the phosphate adsorption onto the fresh Fe(II)-derived precipitate showed that two kinetic regimes were evident.

12.3.1a. At 15 and 20°C and a pH < 7.5, and at 2°C and a pH < 7.9

Under these conditions the adsorption reaction was relatively slow, and phosphate removal appeared to follow the appearance of the solid phase. If this were the case, the rate-determining step would be the slow oxidation of the Fe(II):

$$O_2 + 4Fe^{2+} + 4H^+ \xrightarrow{\text{slow}} 4Fe^{3+} + 2H_2O \qquad 1$$

This would then be followed by a fast hydrolysis and precipitation reaction:

$$Fe^{3+} + 3H_2O \xrightarrow{\text{fast}} Fe(OH)_3 + 3H^+ \qquad 2$$

Under conditions in which the adsorption process itself is also relatively fast, the rate of uptake of the adsorbate must be equated to the rate of removal of Fe(II):

$$\frac{-d\left[\,\mathrm{Fe(II)}\,\right]}{dt} = \frac{-d\left[\,\mathrm{PO_4}\,\right]}{dt} \qquad\qquad 3$$

Sung and Morgan (1980) showed that Fe oxidation followed first-order kinetics, whereby

$$\left[\mathrm{Fe(II)}\right]_t = \left[\mathrm{Fe(II)}\right]_0 \exp(-k_1 t) \qquad\qquad 4$$

with $k_1 = k[\mathrm{OH}^-]^2 P_{\mathrm{O_2}}$; k is the homogeneous oxidation rate constant; $[\mathrm{OH}^-]$ is the hydroxyl ion concentration calculated from the pH and $K_w(T)$, where T is temperature; and $P_{\mathrm{O_2}}$ is the partial pressure of oxygen (0.2 atm in systems open to the atmosphere). Combining equations 3 and 4

$$\left[\mathrm{PO_4}\right]_t = \left[\mathrm{PO_4}\right]_0 \exp(-k_1 t) \qquad\qquad 5$$

indicates that phosphate (or arsenate) adsorption kinetics should reveal the iron-oxidation rate constant.

Plots of $\log_{10}$ ($[x]_t/[x]_0$) versus time (where x is phosphate or arsenate) are shown in Fig. 12.1. The rate constants derived from those plots are listed in Table 12.1. For phosphate adsorption at 15°C and a pH of 7.3, the derived k value is 1.2×10^{13} M^{-2} atm^{-1} min^{-1}. Arsenate adsorption at 20°C and a pH of 7.4 gave a k value of 7.3×10^{12} M^{-2} atm^{-1} min^{-1}. These values compare well with those obtained from Fe (II)-oxidation experiments. Sung and Morgan (1980) found k to be 1.8×10^{12} M^{-2} atm^{-1} min^{-1} in 0.5 M NaCl at 25°C and a pH of 7.2, and Murray and Gill (1978) found k to be 8.9×10^{11} M^{-2} atm^{-1} min^{-1} in Puget Sound seawater at 15°C and a pH of 8.0.

Further experiments were carried out at 2°C and were fitted to the first-order rate law. For the phosphate adsorption at 2°C and a pH of 7.9, k was 3.1×10^{13} M^{-2} atm^{-1} min^{-1}, and for arsenate adsorption at 2°C and a pH of 7.5, k was 1.9×10^{13}. The closest comparison to this data is that at 5°C, a pH of 6.76, and an ionic strength of 0.11 M NaClO$_4$, k is 4.8×10^9 M^{-2} atm^{-1} min^{-1} (Sung and Morgan, 1980). Thus, from this agreement between rate constants determined directly from Fe(II) oxidation and those determined indirectly from phosphate and arsenate adsorption, it can be

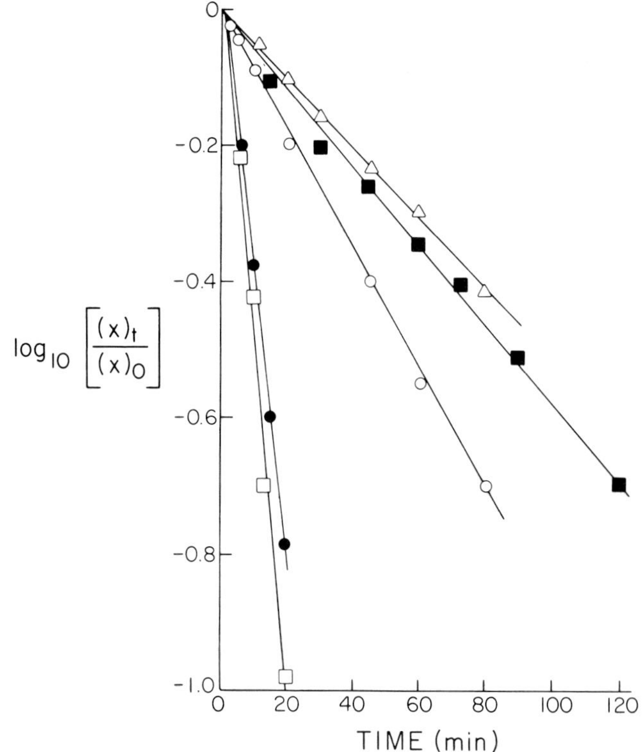

Figure 12.1. First-order kinetic plot for phosphate and arsenate uptake onto freshly forming Fe(II)-derived precipitates in seawater: ($\triangle$) $(\mathrm{PO_4})_0 = 1 \times 10^{-6}$ moles liter^{-1} at 15°C and a pH of 7.1; ($\bigcirc$) $(\mathrm{PO_4})_0 = 1 \times 10^{-6}$ moles liter^{-1} at 15°C and a pH of 7.3; ($\square$) $(\mathrm{PO_4})_0 = 1 \times 10^{-6}$ moles liter^{-1} at 2°C and a pH of 7.9; ($\bullet$) $(\mathrm{AsO_4})_0 = 1.33 \times 10^{-6}$ moles liter^{-1} at 20°C and a pH of 7.4; and ($\blacksquare$) $(\mathrm{AsO_4})_0 = 1.33 \times 10^{-6}$ moles liter^{-1} at 2°C and a pH of 7.5.

concluded that there is a strong coupling between the Fe(II)-oxidation process and the adsorption of the anions. These adsorption reactions were dependent on pH and temperature, and the half-life of the phosphate uptake varied from 57 min at 15°C and a pH of 7.1 to 5.3 min at 2°C and a pH of 7.9. Similar behavior was observed for the arsenate adsorption, and the arsenate data (Table 12.1) indicated an approximate activation energy of 92 kJ mole^{-1}, which is an overall value for the oxidation, hydrolysis, and adsorption processes.

12.3.1b. At 15 and 20°C and a pH > 7.5, and at 2°C and a pH > 7.9

At the higher pHs, the rate of adsorption of the anions increased, and the first-order rate equation was no longer applicable. Attempts to fit the phosphate adsorp-

Table 12.1. Variation of Rate Constant $(k)^a$ with pH and Temperature

Solution Composition[b] (moles liter^{-1})	pH	Temperature (°C)	k (M^{-2} atm^{-1} min^{-1})	$t_{1/2}$[c] (min)
Phosphate				
1×10^{-6}	7.1	15	1.7×10^{13}	57.0
1×10^{-6}	7.3	15	1.2×10^{13}	34.4
1×10^{-6}	7.9	2	3.1×10^{13}	5.3
Arsenate				
1.33×10^{-6}	7.4	20	7.3×10^{12}	7.7
1.33×10^{-6}	7.5	2	1.9×10^{13}	51.6

[a] Derived from plots (Fig. 12.1) yielding k_1 and from the relationship described by Sung and Morgan (1980): $k_1 = k[OH^-]^2 P_{O_2}$.
[b] All solutions were 34 ‰ in salinity and contained Fe at a concentration of 50×10^{-6} moles liter^{-1}.
[c] Half-life.

Table 12.2. Variation of the Rate Constant (k_2) for Phosphate Adsorption^a from Seawater

Solution Composition (moles liter^{-1} of PO$_4$)	pH	Temperature (°C)	k_2 (liter M^{-1} min^{-1})	$t_{1/2}$ (min)
Synthetic Fe[b]				
1×10^{-6}	7.7	15	2.0×10^5	5
1×10^{-6}	7.9	15	5.4×10^5	2
1×10^{-6}	8.1	15	1.5×10^6	< 1
1×10^{-6}	7.9	2	5.6×10^4	18
1×10^{-6}	8.1	2	8.4×10^4	13
2×10^{-6}	7.9	2	1.1×10^4	45
Natural Fe[c]				
$1 \times 10^{-6\,d}$	7.5	15	1.3×10^4	78
$1 \times 10^{-6\,d}$	7.8	15	1.1×10^5	9
$1 \times 10^{-6\,e}$	7.9	15	3.2×10^5	3
$1 \times 10^{-6\,f}$	8.1	15	9.9×10^5	< 1
$1 \times 10^{-6\,f}$	7.6	2	5.6×10^3	178

[a] Seawater with a salinity of 34 ‰.
[b] All solutions contained Fe at a concentration of 50×10^{-6} moles liter^{-1}.
[c] Interstitial water samples.
[d] Fe (Total) $= 52 \times 10^{-6}$ moles liter^{-1}; Fe(II) $= 63\%$ Fe (Total).
[e] Fe (Total) $= 38 \times 10^{-6}$ moles liter^{-1}; Fe(II) $= 63\%$ Fe (Total).
[f] Fe (Total) $= 65 \times 10^{-6}$ moles liter^{-1}; Fe(II) $= 63\%$ Fe (Total).

tion data to the integrated rate equation for the heterogeneous oxidation of Fe(II) (Sung and Morgan, 1980) also proved unsuccessful (Crosby, 1982). In this case the oxidation step is rapid, and the autocatalytic effect of the newly formed Fe(III) surface complicates the kinetics such that the adsorption reaction becomes rate determining. In order to summarize the data at the higher pH, various rate equations for reversible reactions (Laidler, 1965; Swinbourne, 1971) were tested against the data. From many analyses (Crosby, 1982; Marsh, 1984) it has been shown that for these conditions a reaction of the following type applies:

$$2A \underset{k_2'}{\overset{k_2}{\rightleftharpoons}} X + Y \qquad\qquad 6$$

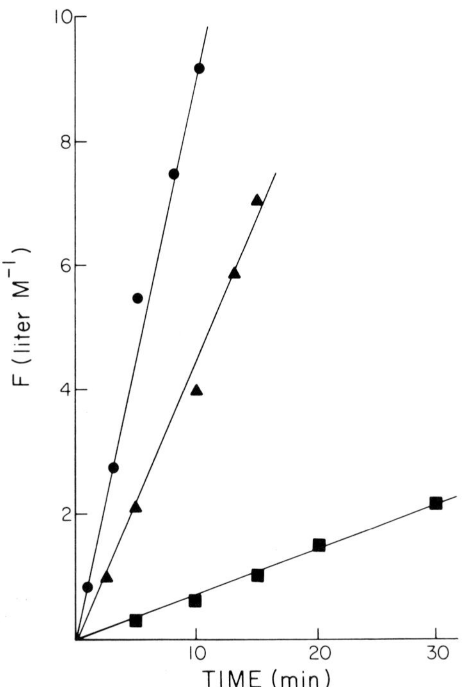

Figure 12.2. Plot of the integrated rate function F (equation 7) versus time for phosphate adsorption $[(PO_4)_0 = 1 \times 10^{-6}$ moles liter$^{-1}]$ onto fresh Fe(II)-derived precipitates in seawater: ($\bullet$) at 15°C and a pH of 7.9, ($\blacktriangle$) at 15°C and a pH of 7.7, and ($\blacksquare$) at 2°C and a pH of 8.1.

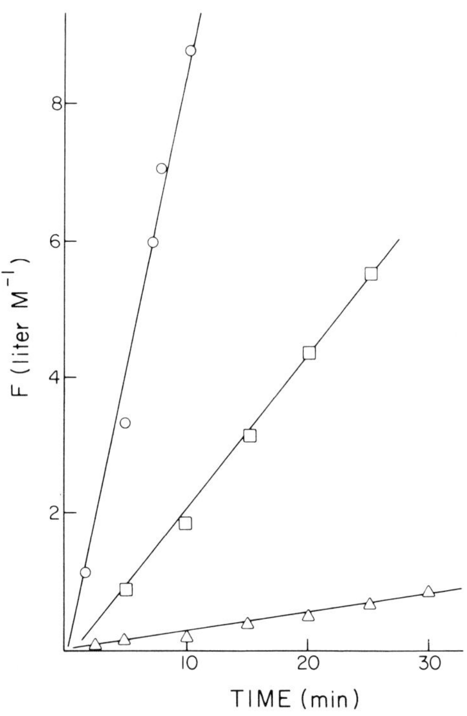

Figure 12.3. Plot of the integrated rate function F (equation 7) versus time for phosphate adsorption $[(PO_4)_0 = 1 \times 10^{-6}$ moles liter$^{-1}]$ onto Fe precipitates derived from sediment interstitial waters: ($\bigcirc$) at 15°C and a pH of 8.1, ($\square$) at 15°C and a pH of 7.9, and ($\triangle$) at 15°C and a pH of 7.5.

where A is the adsorbate (either phosphate or arsenate), and X and Y are the adsorbed species, possibly sited in different ways on the oxyhydroxide surface, The mechanistic implications of this reaction are not clear, but the adsorption mechanism may not be ion exchange since the amount adsorbed is considerably greater than is the number of positive exchange sites (Balistrieri and Murray, 1979). Even though the mechanism requires further investigation, the adoption of the reaction in equation 6 does allow for the determination of conditional rate constants useful in pollutant-dispersion calculations. The integrated form of the above reaction is

$$k_2 t = \left(\frac{X_0 - X_\infty}{2X_0 X_\infty} \right) \ln \left[\frac{X_0 X_\infty - 2X_t X_\infty + X_t X_0}{X_0(X_t - X_\infty)} \right] \qquad 7$$

where X_0, X_t, and X_∞ are the concentrations of either phosphate or arsenate at time = 0, time = t, and time = ∞, respectively. Under the conditions of these experiments (15 or 20°C, and a pH > 7.5; and 2°C and a pH > 7.9) almost all of the adsorption profiles showed that about 10% (the range being 7–16%) of the initial adsorbate remained in solution at equilibrium. The plots of the integrated-rate function, designating the right-hand side of equation 7 as F, versus time for some of the phosphate-adsorption profiles is shown in Fig. 12.2 for precipitates derived from reagent-grade Fe(II) and in Fig. 12.3 for Fe(II) from natural sources. The rate constants are summarized in Table 12.2. The precipitates from the synthetic sources show rapid uptake at 15°C at a pH of 7.7–8.1, with half-lives of a few minutes; however, under the same conditions, at 2°C, the half-lives increase by about an order of magnitude, and the activation energy at a pH of 7.9 was estimated

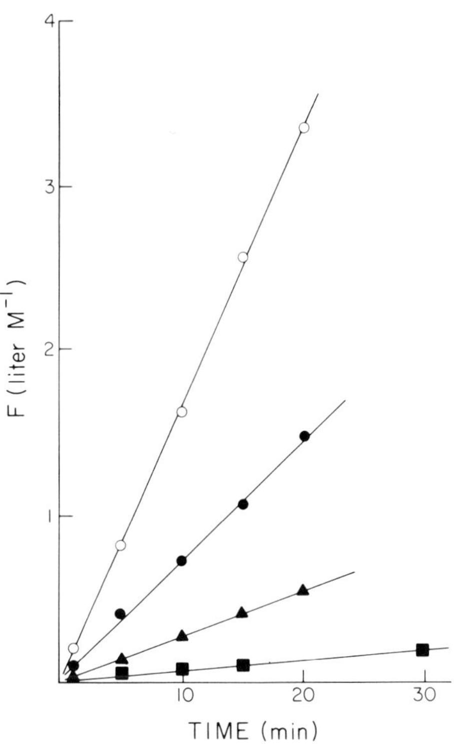

Figure 12.4. Plot of the integrated rate function F (equation 7) versus time for arsenate adsorption onto freshly forming Fe(II)-derived precipitates at 2°C and a pH of 8.0 with concentrations of ($\bigcirc$) $(AsO_4)_0 = 1.33 \times 10^{-6}$ moles liter^{-1}, ($\bullet$) 2.66×10^{-6} moles liter^{-1}, ($\blacktriangle$) 6.65×10^{-6} moles liter^{-1}, and ($\blacksquare$) 13.3×10^{-6} moles liter^{-1}.

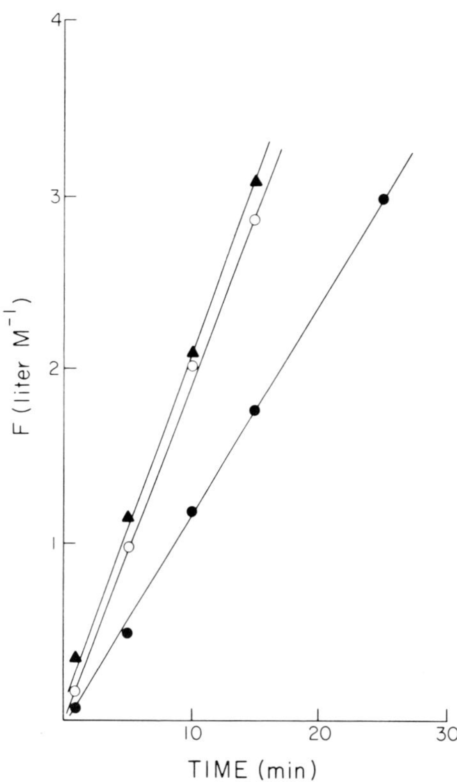

Figure 12.5. Plot of the integrated rate function F (equation 7) versus time for arsenate adsorption [$(AsO_4)_0 = 1.33 \times 10^{-6}$ moles liter^{-1}] onto Fe precipitates derived from natural sources: ($\blacktriangle$) mine stream water at 2°C and a pH of 8.3, ($\bigcirc$) interstitial water of sediment at 2°C and a pH of 8.0, and ($\bullet$) mine stream water at 2°C and a pH of 7.7.

to be 115 kJ mole^{-1}. The natural samples did not always give clean adsorption curves, which reduced the amount of analyzable data; however, the trends shown by the precipitates from natural sources are generally in agreement with the synthetic ones, suggesting that complicating factors such as the presence of organic matter in the natural samples exerts a minor influence over the adsorption process. Similar experiments and analyses were carried out for the arsenate–iron oxyhydroxide interaction, and equation 7 was applied to the adsorption data both for the synthetic precipitates (Fig. 12.4) and the natural precipitates (Fig. 12.5). Again, the trends between the results for the two precipitate sources were compatible (Table 12.3); however, both the phosphate and arsenate data showed that the rate constants in this kinetic regime were dependent on the

initial concentration. No satisfactory explanation can be provided for this observation, and further investigation into the concentration effects is required.

12.3.2. Adsorption onto Aged Fe(II)-Derived Precipitates

The adsorption profiles for phosphate and arsenate uptake onto aged Fe(II)-derived material are shown in Fig. 12.6. The most prominent feature of these plots is that the aged Fe(II)-derived precipitate does not react with phosphate in seawater. Phosphate was not adsorbed at any pH in the range of 7.3 to 8.2 (in seawater and in freshwater) even when the experiments were extended for periods of 60 h (Crosby et al., 1981). In

Table 12.3. Variation of the Rate Constant (k_2) for Arsenate Adsorption[a] from Seawater

Solution Composition (moles liter^{-1} of AsO$_4$)	pH	Temperature (°C)	k_2 (liter M^{-1} min^{-1})	$t_{1/2}$ (min)
Synthetic Fe				
1.33×10^{-6}	8.0	20	9.5×10^5	< 1
1.33×10^{-6}	8.0	2	1.7×10^5	4
2.66×10^{-6}	8.0	2	1.3×10^5	3
6.65×10^{-6}	8.0	2	3.2×10^4	5
13.3×10^{-6}	8.0	2	6.9×10^3	11
Natural Fe				
$1.33 \times 10^{-6\,c}$	7.7	20	9.8×10^5	< 1
$1.33 \times 10^{-6\,c}$	7.7	2	1.2×10^5	6
$1.33 \times 10^{-6\,d}$	8.0	2	1.9×10^5	4
$1.33 \times 10^{-6\,c}$	8.3	2	1.9×10^5	4

[a] Seawater with a salinity of 34 ‰.
[b] All solutions contained Fe at a concentration of 50×10^{-6} moles liter^{-1}.
[c] Fe (Total) in mine stream water $= 65 \times 10^{-6}$ moles liter^{-1}; Fe(II) = 95% Fe (Total).
[d] Fe (Total) in interstitial water samples $= 45 \times 10^{-6}$ moles liter^{-1}; Fe(II) = 100% Fe (Total).

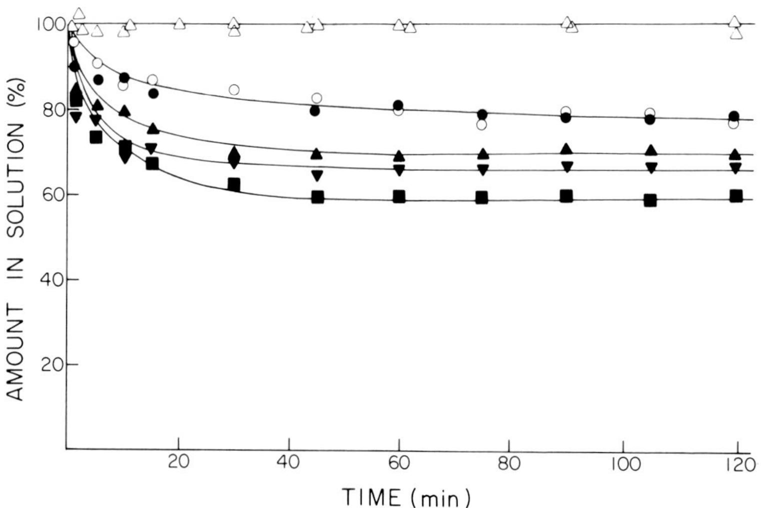

Figure 12.6. Adsorption of phosphate $[(PO_4)_0 = 1 \times 10^{-6}$ moles liter$^{-1}]$ at 15°C and arsenate $[(AsO_4)_0 = 1.33 \times 10^{-6}$ moles liter$^{-1}]$ at 20°C onto aged Fe(II)-derived precipitates in seawater. The graph shows the percentage of dissolved species in the filtrate as a function of time. (△) Phosphate adsorption at a pH of 7.3–8.2; and arsenate adsorption at: (○) a pH of 8.5, (●) a pH of 8.2, (▲) a pH of 7.9, (▼) a pH of 7.6, and (■) a pH of 7.3.

contrast, some adsorption of arsenate onto this material did occur, and the equilibrium value of the adsorption was inversely dependent on pH. The explanation for these observations may be that the adsorption is controlled by the speciation of the anion and the surface charge.

Crosby et al. (1981) have suggested an electrostatic-type model for phosphate adsorption onto Fe(II)-derived material. The product of Fe(II) precipitation has been identified as lepidocrocite, which has a pH_{ZPC} (zero point of change) of 6.2 (Cabrera et al., 1977). For arsenate in seawater at a pH of 7.0, the major species are $HAsO_4^{2-}$ (83%) and $H_2AsO_4^-$ (16%), whereas at a pH of 8.1 the composition is 98% $HAsO_4^{2-}$ and about 1% AsO_4^{3-} (Turner et al., 1981). If electrostatic interactions are important, then increases in the fraction of species with higher negative charges would lead to increases in the electrostatic repulsion and a decrease in adsorption with pH increases. In contrast, the speciation of phosphate is such that at a pH of 8.2 the composition is 70% HPO_4^{2-} and 30% PO_4^{3-}, so considerable electostatic repulsion exists between the anion and the surface, and therefore no adsorption takes place. The major problem with this argument is the assumption that the anions do not form complexes with the major cations in seawater. Equilibrium calculations for phosphate in seawater at a pH of 8.2 show that up to 40% could be present as $NaHPO_4^-$ and 13% as $MgHPO_4$ (Turner et al., 1981). There are no comparable formation constants available for corresponding arsenate species. While the physical-adsorption model may not be correct in every detail, it is suggested that the differences in the dissolved species, rather than differences in surface character, is the controlling factor. This has to be the case since the solids were prepared in a consistent manner, and the adsorption runs for arsenate and phosphate were carried out under identical conditions.

The profiles for arsenate adsorption onto aged precipitates fitted the integrated-rate equation 7 as shown in Fig. 12.7, and the rate constants are presented in Table 12.4. The variation of the rate constant k_2 with pH is given by

$$\log_{10} k_2 = 6.99 - 0.37 \, \text{pH} \qquad 8$$

with a correlation coefficient (r^2) of 0.91. In view of the uptake of arsenate onto these precipitates further experiments were carried out at a temperature of 2°C. Adsorption of arsenate was independent of pH, with about

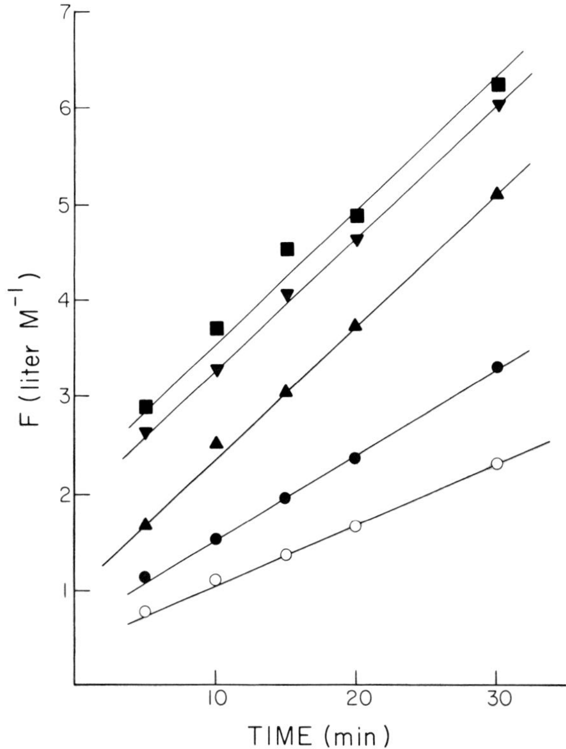

Figure 12.7. Plot of the integrated rate function F (equation 7) versus time for arsenate adsorption [$(AsO_4)_0 = 1.33 \times 10^{-6}$ moles liter^{-1}] onto aged Fe(II)-derived precipitates in seawater at 20°C and ($\bigcirc$) a pH of 8.5, ($\bullet$) a pH of 8.2, ($\blacktriangle$) a pH of 7.9, ($\blacktriangledown$) a pH of 7.6, and ($\blacksquare$) a pH of 7.3.

Table 12.4. Adsorption of Arsenate[a] onto Aged Fe(II)-Derived[b] Precipitates in Seawater

pH	k_2 (liter M^{-1} min^{-1})
7.3	1.8×10^4
7.6	1.6×10^4
7.9	1.5×10^4
8.2	1.0×10^4
8.5	0.6×10^4

[a] Initial concentration was 1.33×10^{-6} moles liter^{-1}.

[b] Initial concentration contained Fe at a concentration of 50×10^{-6} moles liter^{-1}.

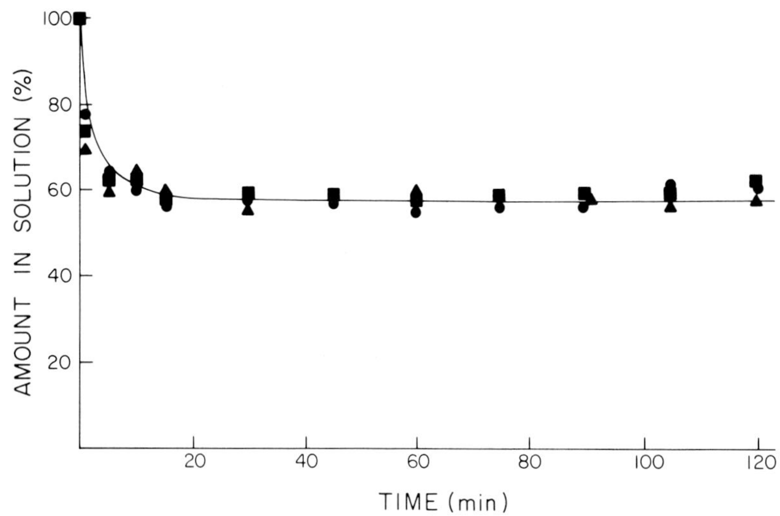

Figure 12.8.   Adsorption of arsenate $[(AsO_4)_0 = 1.33 \times 10^{-6}$ moles liter$^{-1}]$ onto aged Fe (II)-derived precipitates in seawater at 2°C and (●) a pH of 8.3, (■) a pH of 7.3, and (▲) a pH of 7.0.

50% adsorption at equilibrium over the pH range of 7.0–8.3. The adsorption at equilibrium at 2°C (Fig. 12.8) is greater than at 20°C (Fig. 12.6). The explanation for the enhanced uptake at the lower temperature lies in the much larger surface areas of these precipitates: Marsh et al. (1984) have shown that Fe(II)-derived precipitates formed at 20 and 2°C have surface areas of 121 m^2 g^{-1} and 299 m^2 g^{-1}, respectively. The adsorption profiles shown in Fig. 12.8, however, did not fit any of the kinetic analyses attempted here, and no rate constants could be evaluated.

12.4.   CONCLUSIONS

The uptake of phosphate and arsenate onto fresh and aged Fe(II)-derived precipitates in seawater has been studied at appropriate pHs. The kinetic analysis of the adsorption profiles showed two distinct regimes dependent on pH and temperature. When the temperature was 15–20°C and the pH < 7.5, and when the temperature was 2°C and the pH < 7.9, the adsorption followed a first-order reaction related to the oxidation of Fe(II) in solution. At higher pHs, the adsorption profiles fit a reversible reaction of the second order, which is dependent on the initial and equilibrium concentrations of the adsorbates. The adsorption rate constants derived from this analysis showed good agreement between those Fe precipitates obtained from natural and

those from synthetic sources; thus, chemical analogues of this type are probably a reasonable representation of actual processes in seawater.

The usefulness of the kinetic analysis is that the data can be incorporated into quantitative physicochemical models involving phosphate and arsenate scavenging by Fe precipitates. When phosphate and arsenate wastes are disposed of in warm (15–20°C), temperate ocean waters with a pH of 7.9–8.1, the interactions with freshly precipitating Fe(II)-derived oxyhydroxides would take place rapidly with half-lives of 1–3 min; thus, lateral dispersion of phosphate and arsenate would be limited to the vicinity of the dumpsite. However, at temperatures of 15 or 20°C and a pH < 7.5, in which case the adsorption can be first-order, the half-life can be on the order of 60 min. In this case there exists a greater potential for the transport of the dissolved waste, which may be redeposited bound to Fe oxyhydroxides some distance from the site. Where the wastes are disposed of in cold (2°C), polar ocean waters with a pH of 7.9–8.1, adsorption onto fresh Fe(II)-derived precipitates have first-order half-lives of about 5 min and second-order half-lives with a maximum of about 45 min. Here again there exists a potential for the greater dispersal of the material from the dumpsite. Aged Fe(II)-derived precipitates would not absorb dissolved phosphate waste under normal oceanic conditions although these precipitates removed < 50% of dissolved arsenate in about 20 min. Thus, in the presence of aged Fe(II)-derived precipitates, there exists the possibility for considerable transport and mixing in the dissolved phase.

ACKNOWLEDGMENTS

Stuart A. Crosby and John G. Marsh were supported by Natural Environment Research Council research studentships. The authors thank E. Ian Butler and David R. Turner of the Marine Biological Association of the United Kingdom for helpful discussions.

REFERENCES

Balistrieri, L., and J. W. Murray. 1979. The surface of goethite (α-FeOOH) in seawater. *In*: Chemical Modeling in Aqueous Systems: Speciation, Sorption, Solubility and Kinetics, E. A. Jenne (Ed.). American Chemical Society Symposium, Series 93, American Chemical Society, Washington, D.C., pp. 275–298.

Boyden, C. R., S. R. Aston, and I. Thornton. 1979. Tidal and seasonal variations of trace elements in two Cornish estuaries. *Estuarine, Coastal and Marine Science*, **9**, 303–317.

Cabrera, F., L. Madrid, and P. De Arambarra. 1977. Absorption of phosphate by various oxides: Theoretical treatment of the adsorption envelope. *Journal of Soil Science*, **28**, 306–313.

Crosby, S. A. 1982. The Interaction of Phosphate with Iron Oxyhydroxides in Simulated Estuarine Conditions. Ph.D. Dissertation, Department of Marine Science, Plymouth Polytechnic, Plymouth, U.K., 281 pp.

Crosby, S. A., E. I. Butler, D. R. Turner, M. Whitfield, D. R. Glasson, and G. E. Millward. 1981. Phosphate adsorption onto iron oxyhydroxides at natural concentrations. *Environmental Technology Letters*, **2**, 371–378.

Crosby, S. A., D. R. Glasson, A. H. Cuttler, E. I. Butler, D. R. Turner, M. Whitfield, and G. E. Millward. 1983. Surface areas and porosities of Fe(II)- and Fe(III)-derived oxyhydroxides. *Environmental Science and Technology*, **17**, 709–713.

Davis, J. A., and J. O. Leckie. 1978. Surface ionisation and complexation at the oxide/water interface. *Journal of Colloid and Interface Science*, **67**, 90–107.

Gerlach, S. A. 1981. Marine Pollution: Diagnosis and Therapy. Springer-Verlag, Berlin, 218 pp.

Gibbs, M. M. 1979. A simple method for the rapid determination of iron in natural waters. *Water Research*, **13**, 295–297.

Hoff, J. T., J. A. J. Thompson, and C. S. Wong. 1982. Heavy metal release from mine tailings into sea water—a laboratory study. *Marine Pollution Bulletin*, **13**, 283–286.

Inoue, Y., and M. Munemori. 1979. Coprecipitation of mercury (II) with iron (III) hydroxide. *Environmental Science and Technology*, **13**, 443–445.

Kinniburgh, D. G., M. L. Jackson, and J. K. Syers. 1976. Adsorption of alkaline earth, transition and heavy metal cations by hydrous oxide gels of iron and aluminium. *Soil Science Society of America Journal*, **40**, 796–799.

Knox, S., D. R. Turner, A. G. Dickson, M. I. Liddicoat, M. Whitfield, and E. I. Butler. 1981. Statistical analysis of estuarine profiles: application to manganese and ammonia in the Tamar Estuary. *Estuarine, Coastal and Shelf Science*, **13**, 357–371.

Laidler, K. J. 1965. Chemical Kinetics. McGraw-Hill, London, 566 pp.

Lijklema, L. 1980. Interaction of orthophosphate with iron (III) and aluminium hydroxides. *Environmental Science and Technology*, **14**, 537–541.

Marsh, J. G. 1984. The Removal of Arsenic from Aquatic Systems by Iron Oxyhydroxides. Ph.D. Dissertion, Department of Marine Science, Plymouth Polytechnic, Plymouth, U.K., 258 pp.

Marsh, J. G., S. A. Crosby, D. R. Glasson, and G. E. Millward. 1984. BET nitrogen adsorption studies of iron oxides from natural and synthetic sources. *Thermochimica Acta*, **82**, 221–229.

Millward, G. E., and R. M. Moore. 1982. The adsorption of Cu, Mn and Zn by iron oxyhydroxide in model estuarine solutions. *Water Research*, **16**, 981–985.

Murphy, J., and J. P. Riley. 1962. A modified single solution method for the determination of phosphate in natural waters. *Analytica Chimica Acta*, **27**, 31–36.

Murray, J. W., and G. Gill. 1978. The geochemistry of iron in Puget Sound. *Geochimica et Cosmochimica Acta*, **42**, 9–19.

Pankow, J. F., and J. J. Morgan. 1981. Kinetics for the aquatic environment. *Environmental Science and Technology*, **15**, 1306–1313.

Pierce, M. L., and C. B. Moore. 1980. Adsorption of arsenite on amorphous iron hydroxide from dilute aqueous solution. *Environmental Science and Technology*, **14**, 214–216.

Pierce, M. L., and C. B. Moore. 1982. Adsorption of arsenite and arsenate on amorphous iron hydroxide. *Environmental Science and Technology*, **16**, 1247–1253.

Sung, W., and J. J. Morgan. 1980. Kinetics and product of ferrous iron oxygenation in aqueous systems. *Environmental Science and Technology*, **14**, 561–568.

Swallow, K. C., D. N. Hume, and F. M. M. Morel. 1980. Sorption of copper and lead by hydrous ferric oxide. *Environmental Science and Technology*, **14**, 1326–1331.

Swinbourne, E. S. 1971. The Analysis of Kinetic Data. Nelson, London, 126 pp.

Turner, D. R., M. Whitfield, and A. G. Dickson. 1981. The equilibrium speciation of dissolved components in freshwater and seawater at 25°C and 1 atm pressure. *Geochimica et Cosmochimica Acta*, **45**, 855–881.

United Nations IMCO / FAO / UNESCO / WMO / IAEA / UN / UNEP Joint Group of Experts on the Scientific Aspects of Marine Pollution. 1982. An Oceanographic Model for the Dispersion of Wastes Disposed of in the Deep Sea. Unpublished report of the GESAMP Working Group, unpaginated.

Yates, D. E., and T. W. Healy. 1975. Mechanism of anion adsorption at the ferric and chromic oxide/water interfaces. *Journal of Colloid and Interface Science*, **52**, 222–228.

Chapter 13

Metal Binding in Heterogeneous Multicomponent Systems: Mathematical and Experimental Modelling

R. Scott Altmann and James O. Leckie

Department of Civil Engineering
Stanford University
Stanford, California

ABSTRACT

Experimental observations of the adsorption and complexation characteristics of trace metals with humic substances and with hydrous oxides have revealed variation in binding intensity as a function of degree of site occupancy and that the binding intensities for the same solid will depend on the cation. A model is proposed for partitioning trace metals between solids and solution in complex, heterogeneous, aqueous systems. It is based on two observations: (1) that the natural ligands of interest (humic matter, hydrous oxide, clays, and bacterial surfaces) are structurally complex and therefore unlikely to be characterized by models employing either simple invariant stoichiometries or constant reaction energetics and (2) that the variable nature of metal speciation as a function of system composition will require the use of information from experimentally determined behavior in predefined systems. The model incorporates composite isotherms to represent the functional dependence of metal partitioning on the composition of the system. The applicability of the model is demonstrated by comparison with experimental observations of Cu binding in various systems containing polyligands.

13.1.   INTRODUCTION

Modern technology is currently introducing various trace metals into the biosphere at rates comparable to those due to the natural sedimentary cycle (Cartwright et al., 1977; Morgan, 1977; Wood and Goldberg, 1977). Extensive research defining the fate of some of these elements (e.g., Cu, Cd, and Pb) in aquatic, terrestrial, and atmospheric environments reflects a general concern for their toxic properties. Studies by Sunda and

Guillard (1976), Giesy et al. (1977), Anderson and Morel (1978), and others have demonstrated that the toxicity of Cu, Pb, and other metals to a variety of organisms is correlated with the activity of the aquo metal cation and not with the total metal concentration. Development of (1) an understanding of the mechanisms that determine metal speciation in complex natural environments and (2) a means of predicting speciation, given accessible information, therefore become important goals if environmentally sound decisions are to be made concerning the management of such diverse activities as disposing of wastes into the ocean, dredging, acid-rain production, the operation of fossil-fuel power plants, and mining.

The speciation of a particular element in an aqueous environment is determined through the resolution of a wide array of competing reaction mechanisms, including complexation with both inorganic and organic soluble ligands, changes in oxidation state, precipitation and dissolution, solid-phase adsorption and desorption, and ion exchange. The importance of the association between metals and particulate matter in aquatic environments has been thoroughly demonstrated (Martin, 1970; Turekian et al., 1973; Brewer, 1975; Turekian, 1977). Surficial sediments are also likely to be important sinks for trace metals through complexation with particle-bound organic compounds (in particular, humic substances) and through adsorption onto mineral solids (in particular, hydrous metal oxides and clays). This chapter presents a conceptual and mathematical model describing trace-metal partitioning between a defined bulk solution phase and a heterogeneous solid phase consisting of macromolecular, organic-complexing ligands and mineral surfaces.

13.2. BACKGROUND

Naturally occurring solid phases found in oxidized environments can be classified as either organic or mineral. Important mineral constituents include amorphous to crystalline hydrous metal oxides (e.g., $FeOOH$, MnO_2, TiO_2, Al_2O_3, and SiO_2), the layered aluminosilicate clays (e.g., montmorillonite), and carbonates. Organic materials are represented by the variably oxidized, refractory, moderate- to very high-molecular-weight, complex polymers operationally classified as humic and fulvic compounds, macrophyte exudates, and bacterial cell surfaces with associated exopolymers. Although diverse, compounds within this mélange possess certain common characteristics that are expected to influence the overall nature of their interaction with trace metals. Of particular note are (1) the near-universality of bound oxygen as an available electron donor, (2) the high number of functional groups per compound or particle (polyfunctionality), and (3) the structural complexity and heterogeneity of ligand sites at the microscopic scale. Observations indicate that humic material and the hydrous oxides of Fe and Mn are particularly important in trace-metal binding in environments ranging from seawater to soils (Neihof and Loeb, 1972; Jenne, 1976; Suarez and Langmuir, 1976; Hunter and Liss, 1979; Hunter, 1980).

The adsorption and complexation of trace metals such as Cu, Cd, and Pb with humic and fulvic acids have been examined (Mantoura and Riley, 1975; Buffle et al., 1977; Stevenson, 1977; Gamble et al., 1980; Saar and Weber, 1980). The adsorption of these metals onto a number of amorphous-to-crystalline hydrous oxides has been quantified over a wide range of adsorbate: metal ratios as a function of the pH of the solution and the ionic strength (Davis and Leckie, 1978; Benjamin and Leckie, 1981). Observations of the metal-binding characteristics of humic material (Gamble et al., 1980), hydrous oxides (Benjamin and Leckie, 1980, 1981), and natural sediments (Lion et al., 1982) have revealed behavioral similarities, including (1) a variation in binding intensity with degree of site occupancy and (2) the binding intensities for the same solid being dependent on the cation.

Metal binding by oxides and humic matter has generally been interpreted by assuming a single or several reaction mechanism(s). A series of variously sophisticated and useful models, including ion exchange (Dugger et al., 1964), surface complexation (Stumm et al., 1970), physical adsorption (James and Healy, 1972; MacNaughton and James, 1974), and surface complexation with electrostatic interactions (Davis and Leckie, 1978), are available for hydrous oxides and similar surface–solution interfaces. In all of these models the intrinsic energies of interaction between the site and the metal are assumed to apply uniformly to all sites. In most instances, for both inorganic and organic polyligands, experimental observations at low surface coverage and over a relatively narrow range of metal:solid ratios, support this inherently homogeneous polyligand assumption. This has not been observed to be an acceptable assumption, however, when systems are examined over a wider range of surface coverage. This dependence on surface coverage has been reported for oxide binding of Cd, Zn, Cu, and Pb (Benjamin and Leckie, 1980), for a fulvic acid with Cu (Gamble et al., 1980), and we have observed it for various oxides, a humic acid, and natural sediment samples. This chapter focuses on modelling this variable binding behavior by a wide range of polyligand types.

13.3. SYSTEM CONCEPTUALIZATION AND MODEL FORMULATION

13.3.1. Overview

Mathematical modelling of equilibrium speciation in a particular system provides a succinct description of chemical behavior. Ideally, a model should yield thermodynamic activities for all chemical entities present in a system over the considered range of concentrations for all components. Success in achieving this goal depends on how much is known about the total content of each reactive component, the reaction mechanisms among these components, the chemical potentials for all species, and the kinetics of all reactions. Obviously, total success is rarely approached. Depending on the spatial and temporal boundaries chosen, the chemistry of natural aqueous environments can be considered quite simple, well characterized, and highly predictable or very complex, operationally defined, and only empirically describable. We are at the complex end of the spectrum because we wish to model the speciation of trace metals in aqueous environments containing solid or polymeric ligands of varying heterogeneity.

The approach taken depends on how well reaction mechanisms are known, whether constant reaction energetics can be assumed, and on the degree to which concentrations of the system components (e.g., of polyligand sites) are estimable. Tanford (1961) was able to formulate a model incorporating detailed reaction mechanisms and site–site interactions for cation binding by a relatively simple, well-characterized polyligand, poly(acrylic acid), a synthetic linear polymer having regularly repeating carboxylic acid functional groups capable of trace-metal coordination. A polyligand of intermediate complexity is geothite (FeOOH·H$_2$O), a relatively well-characterized hydrous iron oxide with oxygen electron donor functional groups exposed at the solution–oxide interface. The structure of the functionally active interface is less well known relative to poly(acrylic acid), and the degree of site heterogeneity is greater; nevertheless, the range of site characteristics must be limited by the consistency of its underlying

crystalline structure. The site concentration for goethite cannot be specified directly by mass, as it can for poly(acrylic acid), but it can still be estimated from surface-area and site-density measurements. The most complex polyligand to be considered in this chapter is that of naturally derived soil or sediment. Such a sample is likely to be comprised of a mixture of oxides, clays, and natural organic heteropolymers, each with their own binding characteristics, and the fraction of total surface sites due to each of which is unknown and, quite possibly, unknowable.

To put our model into context we will briefly review the features of previous models. In all instances we explicitly assume equilibrium and therefore do not consider kinetics constraints. The first models considered apply for conditions under which variations due to loading or site heterogeneity would be expected to be minimal; variable binding is considered subsequently. Distinctions are not made between activity and concentration in this review, and charge notation is deleted for simplicity in all equations.

13.3.2. Single-Polyligand, Constant-Binding Energy Systems

The complexation of a trace metal [Me] in a solution containing a single acidic, monomeric ligand [SH] is described conventionally as the mass action equilibrium relationship

$$K = \frac{[\text{SMe}][\text{H}]^z}{[\text{Me}][\text{SH}_z]} \qquad 1$$

where z moles of surface protons [H] are displaced per mole of surface-complexed metal. An alternative expression that explicitly incorporates the total ligand concentration, [S$_T$], in a fractional-binding formulation is

$$\Gamma_{\text{SMe}} = \frac{K[\text{Me}][\text{H}]^{-z}}{1 + K[\text{Me}][\text{H}]^{-z}} \qquad 2$$

where $\Gamma_{\text{SMe}} = [\text{SMe}]/[\text{S}_T]$. Equation 2 is the competitive solution analogue to the familiar Langmuir isotherm. The applicability of a single such expression to polyligand systems requires (1) that site independence, or non-interactivity, holds over the range of solution composition to be modelled and (2) that all sites have the same energy of interaction with both the metal and the proton. If the range of system composition is restricted to allow only a very small fraction of the total sites on any given polyligand assemblage to be complexed with a metal, then behavior in accordance with equations 1 and 2 is frequently observed.

Under the limiting conditions stated, it is also possible to incorporate multiple reaction mechanisms for metal–proton exchange at a presumed single ligand site. For example, Davis and Leckie (1978) modelled trace-metal adsorption by hydrous metal oxides by using a combination of surface proton exchange and surface metal hydrolysis as follows:

$$Me + SH \rightleftharpoons SMe + H \qquad\qquad 3$$

$$Me + H_2O + SH \rightleftharpoons SMeOH + 2H \qquad\qquad 4$$

Experimental observations of increasing proton exchange with increasing pH for certain oxides and correlations of adsorption patterns with known sequences of metal hydrolysis seem to indicate that reactions 3 and 4 are plausible in terms of mechanistic implications; however, other approaches are possible (Schindler, 1981).

In summary, under appropriate conditions, analogies with monomeric dissolved ligands provide useful representations of metal complexation by polyligands in solution. The condition that free-site concentrations exceed those of metal-complexed sites by several orders of magnitude is likely to be realistic for trace metals in many unpolluted environments. This assumes that other nucleophiles that are also able to coordinate with the polyligand do not move the polyligand assemblage into a state of variable binding energy vis-à-vis the metal of interest.

13.3.3. Multiple-Polyligand, Constant-Binding Energy Systems

An extension of the single polyligand, constant energetics approach toward understanding the complexation behavior of natural materials is to consider known mixtures of polyligands under conditions of low degrees of complexation. This perspective has been taken by Honeyman (1984) in examining cation and anion binding by combinations of oxides and a clay wherein

adsorption sites were treated as monomeric, and using surface area and site density as normalization parameters. Others (Oakley et al., 1981; Luoma and Davis, 1983) have also argued that whole-system behavior should be predictable from that of its components, although they did not subject their hypotheses to experimental verification.

The results reported by Honeyman (1984) indicate behavior in accordance with assumed simple additivity to be the exception rather than the rule under conditions of low loading. Depending on system conditions, composite adsorption was found to be either enhanced or decreased relative to strict additivity. Honeyman (1984) concluded that changes in surface properties of the net system resulting from varying degrees of particle–particle interactions (such as coagulation) must be considered if adsorption in mixed oxide systems is to be understood. Such considerations will be necessary in any additivity hypothesis regardless of the approach taken in modelling single-polyligand behavior.

13.3.4. Single-Polyligand, Single-Site, Variable-Energy Systems

A number of authors have extended the single-site model and its inherent chemical bond formation to include effects of electrostatic adsorption and presumed spatial relationships. Davis and Leckie (1978) have employed a triple-layer model of the oxide–electrolyte interface in conjunction with the chemical mechanisms to yield a sophisticated representation of cation–anion binding by these polyligands. Under conditions of a swamping background electrolyte, low surface coverage, and strongly bound metals such as Cu, Cd, and Pb, the electrostatic effect has been found to be negligible relative to chemical bond formation. Stumm et al. (1970) and Schindler (1981) prefer to consider a relatively homogeneous surface-ligand model with state-dependent activity modifications accounting for observed binding variations.

13.3.5. Single-Polyligand, Multiple-Site Systems

The binding of a metal by polyfunctional organic macromolecules has been observed to be inconsistent with the simple models when the degree of complexation is extended beyond some critical point in the case of ligands of known structural homogeneity, or much earlier with polyligands that are known to have functional sites of differing characters (Tanford, 1961). Gamble et al. (1980) and Lytle (1982) have observed and formulated models for this type of behavior in metal and proton binding by fulvic acid, a natural heteropolymer (humic substance). This same variable binding behavior has been theoretically considered and experimentally investigated for oxide systems by Benjamin and Leckie (1981).

A number of approaches can be taken in conceptualizing and mathematically modelling this variable behavior. The one chosen will depend on how much is known about the structure and reaction chemistry of the polyligand. For cases in which differing site types are identifiable, a model with discrete mechanistic aspects is perhaps most appropriate. For cases in which a continuum of behavior is observed and distinctions concerning site structures are not easily made, a model employing more general parameters must be utilized. For systems having more than about three types of sites or for which the differences in binding intensity are too slight to allow resolution of individual contributions, only average, macroscopic characteristics are experimentally accessible, and the model employed must reflect this loss of information.

Many of the polyligands of interest in natural environments (humic substances, oxides, and clays) exhibit a continuously changing macroscopic metal partitioning when the metal:ligand ratio is varied over a wide range. Depending on one's perspective, it is possible to argue that the metal-binding response of these polyligands results from (1) three or less discrete site classes with interactive effects creating variation within a site class or (2) a continuum of essentially heterogeneous sites distributed across the range of metal-binding energy as a result of intrinsic structural differences and unresolvable complex interactions. The model chosen depends on which concept is accepted.

The identifiable-site-class approach has been the predominant one employed by those interested in the metal-complexation characteristics of humic substances. This predisposition derives from structural information indicating the dominance and importance of two bidentate site structures, salicylate (ortho carboxylate/phenolic) and phthalate (ortho carboxylate). Numerous authors employing a wide range of analytical techniques have reported values for either one or two conditional stability constants for binding of a large number of divalent and trivalent cations with various humic materials.

The macroscopic behavior as a function of system composition of a complex mixture of monomeric ligands has been addressed by MacCarthy and Smith (1979) in their stability-surface approach. The distinctive qualitative characteristics of such systems in their approach were limiting average stability parameters at both low and high degrees of complexation with a smooth transition between them. Analogously, nearly constant behavior under conditions of low coverage followed by decreasing average stability was observed by Benjamin and Leckie (1981) for cation–oxide polyligand systems. A decreasing trend with loading followed by a nearly constant region was reported for fulvic acid binding of Cu by Gamble et al. (1980), who derived a conceptually similar model to express proton and Cu binding by fulvic acid. This latter formulation is completely general in that it specifies no functional relationship between the average binding intensity and solution composition. This generality is a strength in that it imposes no arbitrary structure on the site distribution, but it is also a weakness in that there is no simple representation that would allow modelling in terms of macroscopic properties. Another shortcoming shared by most models is the simple stoichiometric representations chosen for both overall and microscopic reactions. The single-proton exchange described by Gamble et al. (1980) is, for example, not consistent with our observations of non-integer proton-exchange stoichiometry for Cu, Cd, and Pb binding by humic acid.

A generic approach, which is at once both an extension of the weighted, summed Langmuir model for simple, discrete multicomponent systems and which also considers the polyligand as comprised of a continuum of site types, is the use of composite-isotherm functions. These functions consider the polyligand to be comprised of a large array of site types, the sites of any specified type being assumed to have the same metal-binding intensity, and all sites being assumed to have the same reaction mechanisms for all solute species. The distribution of total sites among site classes is modelled as a statistical frequency distribution based on metal-binding intensity:

$$\bar{\Gamma}([\text{Me}],[\text{H}]) = \int_0^\infty \Gamma(K,[\text{Me}],[\text{H}])f(K)\,dK \qquad 5$$

where $\bar{\Gamma}([\text{Me}],[\text{H}])$ is the composite isotherm $\Gamma(K,[\text{Me}],[\text{H}])$ is the isotherm assumed operant for each site class, and $f(K)$ is the frequency distribution describing the fraction of total sites present in the site class having a binding constant between K and $K + d$K.

Kinniburgh et al. (1983) have examined several composite-isotherm structures in terms of their ability to model the binding of Ca and Zn by an amorphous hydrous iron oxide. It was found that experimental data were fit very well and to a nearly equal degree by both Tóth and two-site Langmuir isotherms over a wide range of metal:ligand ratios and pH. The Tóth isotherm (Tóth, 1971; Tóth, 1974) has a frequency distribution that is skewed, but near-Gaussian with a low-affinity tail. The individual site-class isotherm function assumed to hold for all site classes and all composite isotherms was a competitive Langmuir model with non-integer proton exchange. The use of composite isotherms such as these seems to offer the best compromise for describing the behavior of complex environmental systems and has, therefore, been selected for our own model.

13.4. GENERALIZED MODEL FOR CATION BINDING IN POLYLIGAND SYSTEMS

The model described in this chapter for cation and proton exchange at polyligand-binding sites was designed to be generally applicable to polyligand systems of widely differing complexity. The ability to describe metal complexation by natural sediments and soils was considered as important as describing that in well-characterized single oxides and synthetic polymers. The chosen approach of summed composite isotherms is highly flexible in its ability to respond to different levels of system specification in terms of macroscopic parameters. As a result of this flexibility, the model necessarily lacks the fundamental detail of other models that could be applied under restricted conditions. The model will be applied to a number of experimental systems involving goethite and humic acid.

The characteristics of composition-dependent macroscopic response considered most important in model development were (1) net metal–polyligand complex formation and (2) the apparent proton exchange accompanying metal complexation. Inclusion of reasonable reaction mechanisms for hydrous metal oxide and humic polyligands was also considered to be important, given their likely environmental significance and differing metal-binding and proton-exchange responses. Mathematical accounting in model development for polyligands of the oxide type is presented in Table 13.1; the parallel development for humic compounds is summarized in Table 13.2.

The oxide reactions shown in equations 2 through 5 in Table 13.1 are those employed for goethite, and include surface-induced metal hydrolysis (Table 13.1, equation 5), as employed by Davis and Leckie (1978),

Table 13.1   Mathematical Formulations of Two-State Isotherm Analogue: Typical Oxide Site

Components

1.	SH, Me, H, H_2O

Polyligand Equilibria

2.	$SH_2 \rightleftarrows SH + H$	K_{a1}
3.	$SH \rightleftarrows S + H$	K_{a2}
4.	$SH + Me \rightleftarrows SMe + H$	K_{m1}
5.	$SH + Me + H_2O \rightleftarrows SMeOH + 2H$	K_{m2}

Polyligand Mass Balance

6. $[S_T] = [S] + [SH] + [SH_2] + [SMe] + [SMeOH]$

Net Stoichiometry

7.	$a(SH_2 \rightleftarrows SH + H)$	K_{a1}^{a}
8.	$b(S + H \rightleftarrows SH)$	K_{a2}^{-b}
9.	$c(SH + Me \rightleftarrows SMe + H)$	K_{m1}^{c}
10.	$(1 - c)(SH + Me + H_2O) \rightleftarrows SMeOH + 2H)$	$K_{m2}^{(1 - c)}$

11. $Me + aSH_2 + bS + (1 - a - b)SH + (1 - c)H_2O \rightleftarrows$
           $cSMe + (1 - c)SMeOH + (2 + a - b - c)H^+$

Mass Action

12. $$U = \frac{[SMe]^{c}[SMeOH]^{(1 - c)}[H]^{(2 + a - b - c)}}{[Me][SH_2]^{a}[S]^{b}[SH]^{(1 - a - b)}}$$

13. $U = K_{a1}^{a} K_{a2}^{-b} K_{m1}^{c} K_{m2}^{(1 - c)}$

Stoichiometric Coefficients

14a. $$a = \frac{[SH_2]}{[SH_2] + [S] + [SH]}$$

14b. $$a = \frac{K_{a1}^{-1}[H]}{K_{a1}^{-1}[H] + K_{a2}[H]^{-1} + 1}$$

15a. $$b = \frac{[S]}{[SH_2] + [S] + [SH]}$$

15b. $$b = \frac{K_{a2}[H]^{-1}}{K_{a1}^{-1}[H] + K_{a2}[H]^{-1} + 1}$$

16a. $$c = \frac{[SMe]}{[SMe] + [SMeOH]}$$

16b. $$c = \frac{[H]}{[H] + K_{m2}K_{m1}^{-1}}$$

Macroscopic Parameters

17.	$[\overline{SMe}] = [SMe] + [SMeOH]$
18.	$[\overline{SH_x}] = [SH_2] + [S] + [SH]$
19.	$[S_T] = [\overline{SMe}] + [\overline{SH_x}]$

Table 13.1.   Mathematical Formulations of Two-State Isotherm Analogue: Typical Oxide Site (continued)

Transform Functions

20a.
$$r = \frac{[\mathrm{SMe}]^{c}[\mathrm{SMeOH}]^{(1-c)}}{[\mathrm{SMe}] + [\mathrm{SMeOH}]}$$

20b.
$$r = \frac{\left(K_{m1}[\mathrm{H}]^{-1}\right)^{c}\left(K_{m2}[\mathrm{H}]^{-2}\right)^{(1-c)}}{K_{m1}[\mathrm{H}]^{-1} + K_{m2}[\mathrm{H}]^{-2}}$$

21a.
$$q = \frac{[\mathrm{SH_2}]^{a}[\mathrm{S}]^{b}[\mathrm{SH}]^{(1-a-b)}}{[\mathrm{SH_2}] + [\mathrm{S}] + [\mathrm{SH}]}$$

21b.
$$q = \frac{\left(K_{a1}^{-1}[\mathrm{H}]\right)^{a}\left(K_{a2}[\mathrm{H}]^{-1}\right)^{b}}{K_{a1}^{-1}[\mathrm{H}] + K_{a2}[\mathrm{H}]^{-1} + 1}$$

Mass-Action Transform

22a.
$$U = \frac{r([\mathrm{SMe}] + [\mathrm{SMeOH}])[\mathrm{H}]^{x}}{q([\mathrm{SH_2}] + [\mathrm{S}] + [\mathrm{SH}])[\mathrm{Me}]}$$

22b.
$$U(qr^{-1}) = \frac{[\,\overline{\mathrm{SMe}}\,][\mathrm{H}]^{x}}{[\,\overline{\mathrm{SH}}_{x}\,][\mathrm{Me}]}$$
$$x = 2 + a - b - c$$

Two-State Isotherm Analogue

23.
$$\Gamma = \frac{[\,\overline{\mathrm{SMe}}\,]}{(\mathrm{ST})} = \frac{U(qr^{-1})[\mathrm{Me}][\mathrm{H}]^{-x}}{1 + U(qr^{-1})[\mathrm{Me}][\mathrm{H}]^{-x}}$$

thereby allowing the modelling of increasing proton exchange with increasing pH. Other surface reactions can be employed as needed to reflect observed proton exchange and the pH dependency of metal binding. Non-integer proton exchange that is relatively constant over the experimental pH range considered in the model necessitates the two independent-site structures shown for humic compounds (Table 13.2, equations 2–7). The net metal-exchange reaction for an oxide (Table 13.1, equation 11), with its attendant mass-action formulation (Table 13.1, equation 12) and intensity parameter (Table 13.1, equation 13), results directly from weighting each of the contributing reactions (Table 13.1, equations 7–10) by the coefficients a, b, and c. A similar treatment of the equilibrium of humic compounds yields the net reaction mass action (Table 13.2, equations 12–14), which incorporates the added factor w, reflecting the presumed proportioning (Table 13.2, equations 8–11) of total sites among the two hypothesized ligand structures.

The stoichiometric coefficients, the form of which are illustrated for the oxide (Table 13.1, equations 14–16), are dictated by the structure of the net stoichiometry, the contributing independent equilibria, the rules governing formulation of valid stoichiometric relationships, and the fact that we have assumed a system at equilibrium. The coefficients are seen to be functions only of [H] and, therefore, independent of metal or polyligand. They can be shown to be equal to the partial concentration responses of each of the species to an infinitesimal perturbation of the extent of reaction at equilibrium. The atom balance of the reactant and the product is maintained for all conditions as required.

It is useful to consider why net stoichiometric formulations are necessary. One could solve for system speciation at equilibrium by standard iterative procedures based on minimization of overall free energy within specified mass-balance constraints. Such a procedure, however, can be faulted on several grounds. First, it yields information that is unverifiable by presently observable data (for example, the terms on the left-hand side of equations 17–19 in Table 13.1). The approach taken here is precisely matched to accessible characteristics and, indeed, the final form has been incorporated into iterative models such as MINEQL + SIGMA and MICROQL (Altmann, 1984) to allow coupling to pertinent solution equilibria. Second, the net proton exchange resulting from metal complexation under conditions of constant pH is given directly by the proton coefficient in the net stoichiometric approach, but it is only obtainable otherwise through successive iterative solutions at varying total metal concentrations. Third, the computation is easier and more rapid. Lastly, a two-state isotherm analogue incorporating all polyligand states is required if the composite-isotherm approach is employed for polyligands with variable binding energy.

The requirement of a two-state isotherm analogue necessitates linking the mass-action expressions (Table 13.1, equation 12; Table 13.2, equation 13) to the mass balances on surface species. For the oxide, the surface species mass balance is reflected in equations 6 and 17–19 (Table 13.1). The mapping of these two independent relationships into a single relationship exhibiting the desired characteristics is accomplished by using the transform functions (Table 13.1, equations 20 and 21). The transform factors q and r are dependent only on [H] and, therefore, after substitution into the mass-action expression (Table 13.1, equation 22a; Table 13.2, equation 15), they can be grouped with the intensity parameter that is also dependent only on [H] (Table 13.1, equation 22b). Rearrangement yields the two-state competitive-isotherm analogue for the oxide (Table 13.1, equation 23) and for the humic material (Table 13.2, equation 16).

Table 13.2.   Summary of Two-State Isotherm Analogue: Typical Humic Acid Site

Components

1. Two types of sites PH and SH, Me, H, H_2O

Equilibria

2. $\quad PH_2 \rightleftharpoons PH + H \quad K_{a1}$
3. $\quad\ \ PH \rightleftharpoons P + H \quad\ K_{a2}$
4. $P + Me \rightleftharpoons PMe \quad K_{m1}$
5. $\quad SH_2 \rightleftharpoons SH + H \quad K_{a3}$
6. $\quad\ \ SH \rightleftharpoons S + H \quad\ K_{a4}$
7. $S + Me \rightleftharpoons SMe \quad K_{m2}$

Mass Balances

8. $[P_T] = [PH_2] + [PH] + [P] + [PMe]$
9. $[S_T] = [SH_2] + [SH] + [S] + [SMe]$
10. $[L_T] = [P_T] + [S_T]$
11. $\quad w = [P_T]/[L_T]$

Net Stoichiometry

12. $Me + w(aPH_2 + bPH + (1 - a - b)P) + (1 - w)(cSH_2 + dSH + (1 - c - d)S)$
$\rightleftharpoons wPMe + (1 - w)SMe + (w(2a + b) + (1 - w)(2c + d))H$

Mass Balance

13. $$U = \frac{[PMe]^w[SMe]^{(1-w)}[H]^x}{[Me]\left([PH_2]^a[PH]^b[P]^{(1-a-b)}\right)^w \left([SH_2]^c[SH]^d[S]^{(1-c-d)}\right)^{(1-w)}}$$
$x = w(2a + b) + (1 - w)(2c + d)$
14. $U = (K_{m1}K_{a1}^a K_{a2}^{a+b})^w (K_{m2}K_{a3}^c K_{a4}^{(c+d)})^{(1-w)}$

Mass-Action Transform

15. $$U = \frac{r([PMe] + [SMe])[H]^x}{[Me]q([PH_2] + [PH] + [P] + [SH_2] + [SH] + [S])}$$

Two-State Isotherm Analogue

16. $$\Gamma = \frac{[\overline{LMe}]}{[L_T]} = \frac{U(qr^{-1})[Me][H]^{-x}}{1 + U(qr^{-1})[Me][H]^{-x}}$$

For any system condition specified by proton and free-metal concentrations, the two-state isotherm analogues give the fractions of total sites in a given site class [as defined by its intensity parameters (Table 13.1, equations 2–5; Table 13.2, equations 2–7)] present as metal complexes. As noted previously, such a model is likely to hold only under conditions of extremely low metal:ligand ratios. The existence of polyligand systems that exhibit binding-response functions consistent with a hypothesized distribution of site-binding intensities and the need to represent such behavior in a mathematically concise fashion have led us to choose a composite isotherm to model these systems. Composite isotherms (equation 5) incorporate both a two-state isotherm, which describes site-class response characteristics, and a distribution function to yield intensity-dependent site-class concentration. The two-state isotherm analogue can be employed in such a mathematical framework.

The form of the composite isotherm can be chosen solely on the basis of goodness-of-fit to experimental data or on the basis of a distribution function for site affinity that reflects some hypothesized structural reality. The Tóth isotherm (Tóth, 1971; Tóth, 1974; Kinniburgh et al., 1983) was selected for use here both because of its superior conformance to experimental observations and because of its inclusion of a continuous and finite (as opposed to discrete or unbounded) frequency distribution. The integrated form of the Tóth isotherm can be expressed as

$$\overline{\Gamma} = \frac{[\overline{MeL}]}{[L_T]} = \frac{U^*[Me][H]^{-x}}{\left\{\left(1 + U^*[Me][H]^{-x}\right)^{\beta}\right\}^{1/\beta}} \qquad 6$$

where U^* is the characteristic value of $U(qr^{-1})$; x is the stoichiometric coefficient for proton exchange; and β is the heterogeneity factor for the frequency distribution of sites. Kinniburgh et al. (1983) expressed this same relationship more conveniently as

$$\overline{\Gamma} = \frac{10^Q}{\left(1 + 10^{Q\beta}\right)^{1/\beta}} \qquad 7$$

where

$$Q = \log U^* + \log[Me] - x \log[H] \qquad 8$$

and where, for an oxide

$$\log U^* = \log(qr^{-1}) + a \log K_{a2}$$

$$- b \log K_{a2} + c \log K_{m1} + (1 - c) \log K_{m2} \qquad 9$$

The purpose of a composite isotherm such as equation 6 is to describe the chemical behavior of a given system. As employed here and by others (Lytle, 1982; Kinniburgh et al., 1983), the composite-isotherm parameters U^* and β are estimated by fitting a selected form of isotherm to experimental data. The methodology here differs from that of others (1) in the more realistic chemical mechanisms included in the two-state isotherm, (2) in the manner of determining the characteristic intensity parameter U^*, and (3) in the modelling of net proton exchange as a function of system composition through the proton equilibria. As regards U^*, it can either be considered (1) as a single entity to be adjusted as needed to fit a particular data set or (2) as a composite of intensity parameters for the elementary reactions assumed for each site class which can be varied. We have chosen the latter approach, as detailed in Tables 13.1 and 13.2, as being more consistent with the idea of an energetically heterogeneous array of site types.

Systems containing multiple polyligands should exhibit an overall behavior predictable from the responses of pure-component polyligands if non-interactivity is assumed to hold and concentrations for each polyligand can be assigned. If the composite-isotherm functions $\overline{\Gamma}_i$ are known for each polyligand i, then the overall metal-binding and proton-exchange responses should be given, respectively, by

$$\overline{\Gamma}_\Sigma = \sum_{i=1}^{n} V_i \overline{\Gamma}_i \qquad 10$$

and

$$X_\Sigma = \sum_{i=1}^{n} V_i X_i \qquad 11$$

Where X_i is the proton-exchange coefficient for polyligand i, and where V_i is the mole fraction of polyligand i as given by

$$V_i = \frac{[L_T]_i}{[L_T]_\Sigma}$$

and $\overline{\Gamma}_i$ is given by equation 6.

The summed composite-isotherm model should exhibit several qualities regarding real-system behavior if it is to be a useful tool. First, the model should be capable of accurately representing the trace metal–proton binding behavior of systems containing single polyligands, both with and without total-site information. For well-characterized single polyligands, the model should yield a total-site concentration in reasonable agreement with that known to be present. The model should also collapse to the single-site, competitive Langmuirian form when fit to the data of a monomeric-ligand system ($\beta = 1$). If single polyligand experimental systems are shown to be well represented by isotherms of the form of equation 6, then proportional summed behavior in accordance with equation 8 should be evidenced.

13.5. MODELLING OF COPPER BINDING IN EXPERIMENTAL SYSTEMS

The binding of Cu, Pb, and Cd by a variety of single, binary, and complex polyligand-containing systems has been determined as a function of total-metal concentration, total-polyligand(s) concentration(s), and pH by batch, metal-into-ligand titration techniques (Altmann, 1984). The procedures employed for polyligand preparation, the experimental protocols utilized, and the study results and interpretation are given by Altmann (1984). Three examples of the system selected from this study to illustrate model application are Cu binding by (1) geothite, a synthetic hydrous iron oxide ($FeOOH \cdot H_2O$), (2) a humic acid sample, and (3) goethite–humic acid mixed systems.

In general, the degree of characterization of the polyligand will determine the number of parameters that must be estimated to fit the model to experimental data. In the most well-known systems, only those parameters comprising U^* and β would need to be estimated from experimental data sets ($[L_T]_i$, $[MeL]_i$, $[Me]_i$, and $[H]_i$). In systems such as soils and sediments for which the total site concentration is unknown, $[L_T]$ would also become a dependent variable. The selection of acidity constants, the proportionality relating the metal-binding constants, and proportioning between the two site classes of humic material was made such that the proton-exchange behavior (as a function of pH) matched the observations as nearly as possible. Once the best attainable proton-exchange response was achieved, the metal-binding constants and β were varied to obtain the best possible match with the observed fractional binding of metal.

The results are shown with the experimental data in Figs. 13.1a and b for fractional binding of Cu by humic acid and the resultant net proton exchange for various conditions of constant pH. The acidity constants utilized for the two types of sites are those reported by Martell and Smith (1977) for phthallic and salicylic acid at 25°C and an ionic strength of 0.1 N. Log intensity parameters also reported by Martell and Smith (1977) for Cu binding are 8.1 and 18.3, respectively, which are reasonably close to the values of 8.4 and 16.4 required by our procedure. Quite good agreement is evident between model and data for metal binding and for the constant proton exchange observed at higher fractional loadings at pHs of 5 and 6.

Similar comparisons of model to data for Cu binding by goethite are shown in Figs. 13.2a and b; again, the model satisfactorily reflects the fractional binding and proton exchange. The acidity constants employed ($\log K_{a1} = -4.9$, $\log K_{a2} = -10.8$) can be compared to those that Yates (1975) ($\log K_{a1} = -4.2$, $\log K_{a2} =$

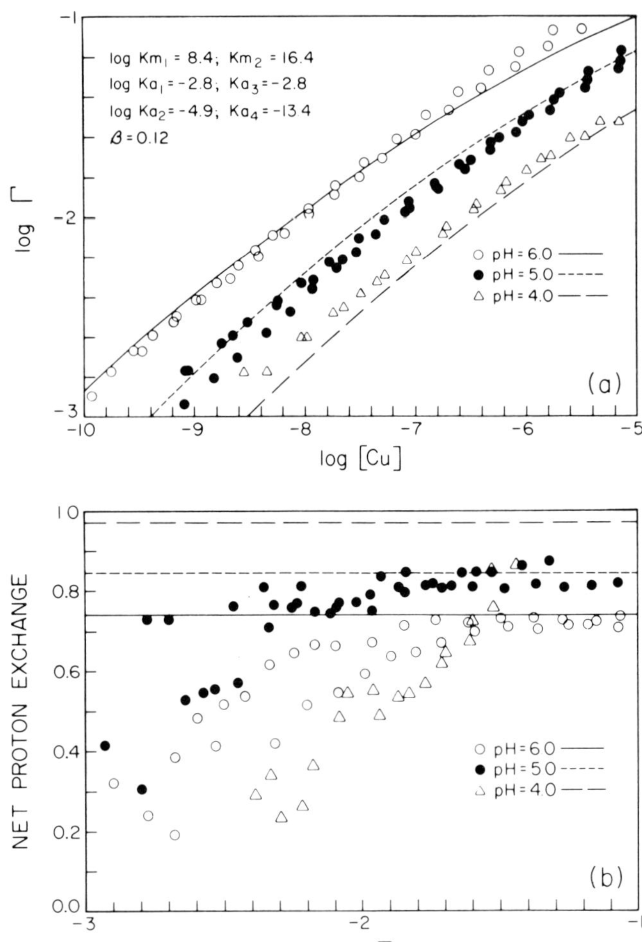

Figure 13.1. (a) Copper-binding isotherms for humic acid as a function of pH, and (b) net proton-exchange stoichiometry for Cu binding by humic acid as a function of pH. All values are for conditions of 25°C and 0.1 N NaOH.

-10.8) determined for goethite of a similar surface area in KNO_3. Two observations can be made regarding the comparison of model with data. First, the requirement of the model that proton stoichiometry be independent of fractional binding is at odds with the experimental trends of increasing exchange with increasing Γ at all pHs. This behavior is consistent with predictions of the electrostatic triple-layer model employed by Davis and Leckie (1978) and results from the tendency of the increasingly positively charged surface to favor the formation of complexes with the singly charged hydrolyzed metal. Second, the data on fractional binding for the goethite–Cu system exhibit a consistent bimodal character with an inflection point occurring in the vicin-

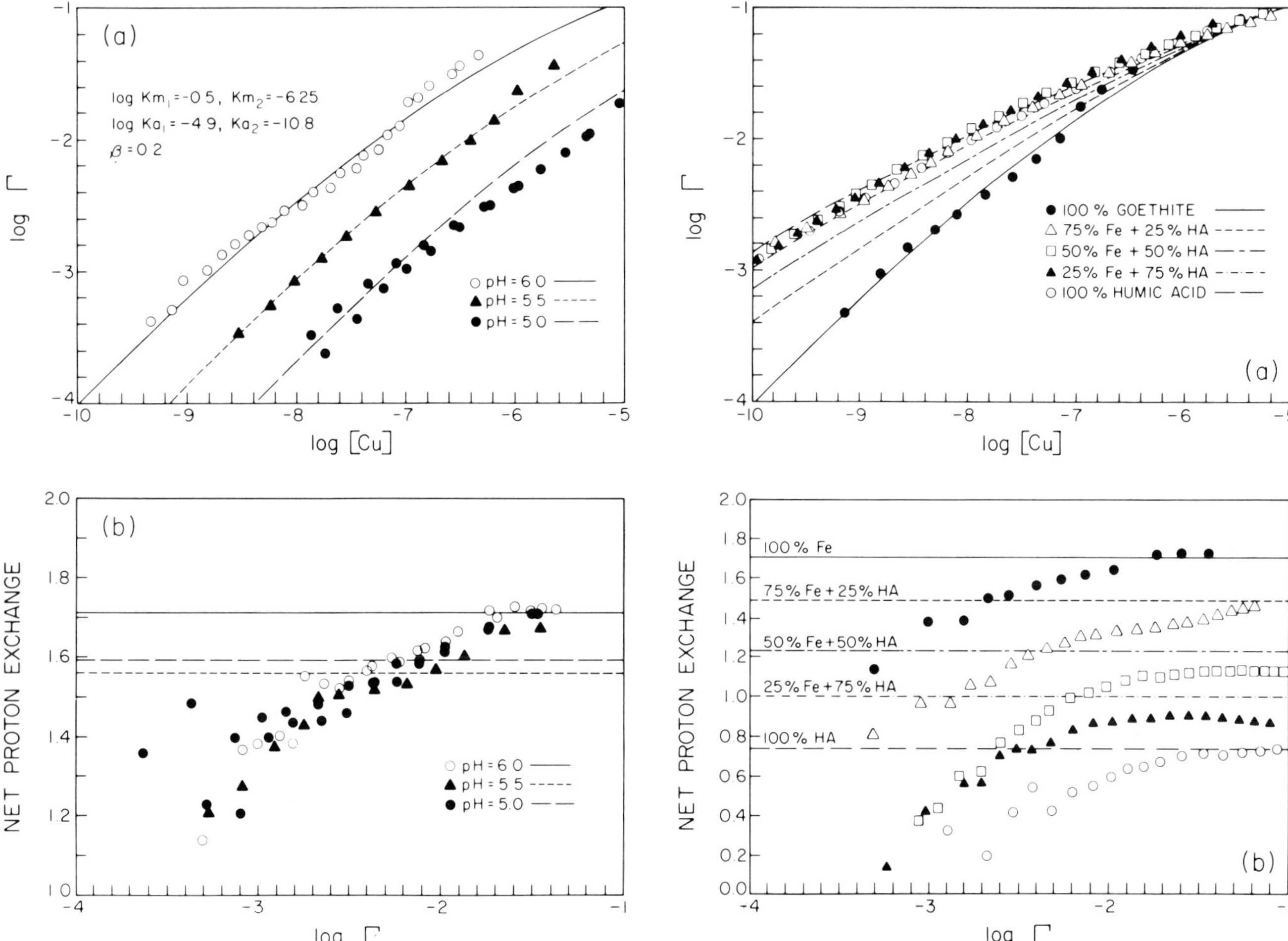

Figure 13.2. (a) Copper-binding isotherms for goethite as a function of pH, and (b) net proton exchange for Cu binding by goethite as a function of pH. All values are for conditions of 25°C and 0.1 N NaOH.

Figure 13.3. (a) Copper-binding isotherms for defined mixed systems containing humic acid and goethite at pH 6, and (b) net proton exchange for Cu binding by defined mixed systems containing goethite and humic acid at pH 6. All values are for conditions of 25°C and 0.1 N NaOH.

ity of log $\Gamma = -2$ (1% surface coverage). Possible explanations for such behavior include (1) different surface planes having unequal metal-binding intensities and differing fractional exposure as indicated by the inflection point and (2) changes occurring in site concentration under inflection-point conditions due to particle–particle interactions or cation bridging. Model refinements including incorporation of electrostatic contributions and inclusion of two-site classes with differing metal-binding intensities yield considerable improvement in correspondence between data and model (Altmann, 1984).

The behavior of mixtures of humic acid and goethite

at a pH of 6 are illustrated in Figs. 13.3a and 13.3b. The proton-exchange data exhibit good qualitative agreement with the strict proportioning required by the additive model. The binding response, however, does not show proportional contribution of the component polyligands but, rather, domination by the humic characteristics. The reproducibility of the mixed systems was excellent and comparable to that evident in the humic-only titrations. In this instance, proton exchange is in accord with polyligand non-interactivity whereas metal binding indicates some interaction such as humic coating of goethite surfaces. Other mixed systems (alumina–humic, rutile–humic, goethite–rutile, and goethite–

alumina) show reasonably proportional behavior for Cu binding and accompanying proton exchange (Altmann, 1984).

13.6. SUMMARY

The composite-isotherm model described in this chapter appears to offer a reasonable means of concisely representing the behavior of polyligand systems, which exhibit characteristics consistent with a hypothesis of variable energy of metal binding. The model incorporates plausible chemical reactions for polyligands of both the oxide and the humic types and yields fractional binding and proton-exchange responses that are generally in good agreement with experimental observations on well-characterized polyligand samples. In most cases, defined mixtures of polyligands are modelled well by assuming simple proportionality based on contributed-site concentration. The model may also be employed to mimic the behavior of complex and ill-defined metal-binding systems (such as soils and sediments) by varying input parameters as appropriate.

ACKNOWLEDGMENTS

This research was supported in part by the National Science Foundation Grants Nos. CME-79-22079 and CME-80-09028.

REFERENCES

Altmann, R. S. 1984. Trace Metal Binding in Heterogeneous, Multi-component Systems: Mathematical and Experimental Modeling. Ph.D. Thesis, Stanford University, Stanford, California, 188 pp.

Anderson, D. M., and F. M. M. Morel. 1978. Copper sensitivity of *Gonyaulax tamarensis*. *Limnology and Oceanography*, **23**, 283–289.

Benjamin, M. M., and J. O. Leckie. 1980. Adsorption of metals at oxide interfaces: effects of the concentrations of adsorbate and competing metals. *In*: Contaminants in Sediments, Vol. 2, R. A. Baker (Ed.). Ann Arbor Science, Ann Arbor, Michigan, pp. 305–322.

Benjamin, M. M., and J. O. Leckie. 1981. Multiple site adsorption of Cd, Cu, Zn, and Pb on amorphous iron oxyhydroxide. *Journal of Colloid Interface Science*, **79**, 209–221.

Brewer, P. G. 1975. Minor elements in seawater. *In*: Chemical Oceanography, 2nd edition, Vol. 1, J. P. Riley and G. Skirrow (Eds.). Academic Press, London, pp. 415–496.

Buffle, J., F. L. Gaeter, and W. Haerdi. 1977. Measurements of complexation properties of humic and fulvic acids in natural waters with lead and copper ion-selective electrode. *Analytical Chemistry*, **49**(2), 216–222.

Cartwright, B., R. H. Merry, and K. G. Tillen. 1977. Heavy metal contamination of soils around a lead smelter at Port Pirie, South Australia. *Australian Journal of Soil Research*, **15**, 69–81.

Davis, J. A., and J. O. Leckie. 1978. Surface ionization and complexation at the oxide/water interface. 2: Surface properties of amorphous iron oxyhydroxide and adsorption of metal ions. *Journal of Colloid Interface Science*, **67**, 90–107.

Dugger, D. L., J. H. Stanton, B. N. Irby, B. L. McConnell, W. W. Cummings, and R. W. Maatman. 1964. The exchange of twenty metal ions with the weakly acidic silanol group of silica gel. *Journal of Physical Chemistry*, **68**, 757–760.

Gamble, D. S., A. W. Underdown, and C. H. Langford. 1980. Copper(II) titration of fulvic acid ligand sites with theoretical, potentiometric, and spectrophotometric analysis. *Analytial Chemistry*, **52**, 1901–1908.

Giesy, J. P., G. L. Leversee, and D. R. Williams. 1977. Effects of naturally occurring aquatic organic fractions on cadmium toxicity to *Simocephalus serrulatus* (Daphnidas) and *Gambusia affinis* (Poeciliidae). *Water Research*, **11**, 1013–1020.

Honeyman, B. D. 1984. Metal and Metalloid Adsorption at the Oxide/Water Interface in Systems Containing Mixtures of Adsorbents. Ph.D. Thesis, Stanford University, Stanford, California, 383 pp.

Hunter, K. A. 1980. Microelectrophoretic behavior of natural surface-active organic matter in coastal seawater. *Limnology and Oceanography*, **25**, 807–822.

Hunter, K. A., and P. S. Liss. 1979. The surface charge of suspended particles in estuarine and coastal waters. *Nature (London)*, **282**, 823–825.

James, R. O., and T. W. Healy. 1972. Adsorption of hydrolyzable metal ions at the oxide/water interface. III: A thermodynamic model of adsorption. *Journal of Colloid Interface Science*, **40**, 65–81.

Jenne, E. A. 1976. Trace element sorption by sediments and soils—sites and processes. *In*: Molybdenum in the Environment, Vol. 2, W. R. Chappel and K. Petersen (Eds.). Marcel Dekker, New York, pp. 425–553.

Kinniburgh, D. G., J. A. Barker, and M. Whitfield. 1983. A comparison of some simple adsorption isotherms for describing divalent cation adsorption by ferrihydrite. *Journal of Colloid Interface Science*, **95**, 370–384.

Lion, L. W., R. S. Altmann, and J. O. Leckie. 1982. Trace-metal adsorption characteristics of estuarine particulate matter: evaluation of contributions of Fe/Mn oxide and organic surface coatings. *Environmental Science and Technology*, **16**, 680–666.

Luoma, S. N., and J. A. Davis. 1983. Requirements for modeling trace metal partitioning in oxidized estuarine sediments. *Marine Chemistry*, **12**, 159–181.

Lytle, C. R. 1982. The Copper Complexation Properties of Dissolved Organic Matter from the Williamson River, Oregon. Ph.D. Thesis, Portland State University, Portland, Oregon, 168 pp.

MacCarthy, P., and G. C. Smith. 1979. Stability surface concept: a quantitative model for complexation in multiligand mixtures. *In*: Chemical Modeling in Aqueous Systems, E. A. Jenne (Ed.).

American Chemical Society Symposium Series, American Chemical Society, Washington, D.C., pp. 201–222.

MacNaughton, M., and R. O. James. 1974. Adsorption of aqueous mercury (II) complexes at the oxide/water interface. *Journal of Colloid Interface Science*, **47**, 431–440.

Mantoura, R. F. C., and J. P. Riley. 1975. The analytical concentration of humic substances from natural waters. *Analytica Chimica Acta*, **76**, 97–106.

Martell, A. E., and R. M. Smith. 1977. Critical Stability Constants, Vol. 3. Plenum Press, New York, 495 pp.

Martin, J. H. 1970. The possible transport of trace metals via moulted copepod exoskeletons. *Limnology and Oceanography*, **15**, 756–760.

Morgan, J. J. 1977. Source functions: fossil fuel combustion products, radionuclides, trace metals and heat. *In*: Global Chemical Cycles and Their Alteration by Man, W. Stumm (Ed.). Dahlen Konferenzen, Berlin, pp. 291–311.

Neihof, R. A., and G. I. Loeb. 1972. The surface charge of particulate matter in seawater. *Limnology and Oceanography*, **17**, 7–16.

Oakley, S. M., P. O. Nelson, and K. J. Williamson. 1981. Model of trace-metal partitioning in marine sediments. *Environmental Science and Technology*, **4**, 474–480.

Saar, R. A., and J. H. Weber. 1980. Comparison of spectrofluorometry and ion-selective electrode potentiometry for determination of complexes between fulvic acid and heavy-metal ions. *Analytical Chemistry*, **52**, 2095–2099.

Schindler, P. W. 1981. Surface complexes at oxide–water interfaces. *In*: Adsorption of Inorganics at Solid–Liquid Interfaces, M. A. Anderson and A. J. Ruin (Eds.). Ann Arbor Science Publishers, Ann Arbor, Michigan, pp. 1–49.

Stevenson, F. J. 1977. Nature of divalent transition metal complexes of humic acids as revealed by a modified potentiometric titration method. *Soil Science*, **123**, 10–16.

Stumm, W., C. P. Huang, and S. R. Jenkins. 1970. Specific chemical interactions affecting the stability of dispersed systems. *Croatica Chimica Acta*, **42**, 223–245.

Suarez, D. L. and D. Langmuir. 1976. Heavy metal relationships in a Pennsylvania soil. *Geochimica et Cosmochimica Acta*, **40**, 589–598.

Sunda, G. W., and R. R. L. Guillard. 1976. The relationship between cupric ion activity and the toxicity of copper to a unicellular algae. *Journal of Marine Research*, **34**, 511–529.

Tanford, C. 1961. Physical Chemistry of Marcromolecules. John Wiley & Sons, New York, 710 pp.

Tóth, J. 1971. State equations of the solid–gas interface layers. *Acta Chimica (Budapest)*, **69**, 311–328.

Tóth, J. 1974. Adsorption of gases on heterogeneous solid surfaces: the energy distribution function corresponding to a new equation for monolayer adsorption. *Acta Chimica (Budapest)*, **82**, 11–21.

Turekian, K. K. 1977. The fate of metals in the oceans. *Geochimica et Cosmochimica Acta*, **41**, 1139–1144.

Turekian, K. K., K. Amitai, and L. Chan. 1973. Trace element trapping in pteropod tests. *Limnology and Oceanography*, **18**, 240–249.

Wood, J. M., and E. D. Goldberg. 1977. Impact of metals on the biosphere. *In*: Global Chemical Cycles and Their Alteration by Man, W. Stumm (Ed.). Dahlen Konferenzen, Berlin, pp. 137–153.

Yates, D. E. 1975. The Structure of the Oxide/Aqueous Electrolyte Interface. Ph.D. Thesis, University of Melbourne, Melbourne, Australia, 246 pp.

Chapter 14

Plutonium and Americium Behavior in Coral Atoll Environments

Victor E. Noshkin, Kai M. Wong, Terrence A. Jokela, James L. Brunk, and Rodney J. Eagle

Lawrence Livermore National Laboratory
University of California
Livermore, California

ABSTRACT

Inventories of $^{239+240}$Pu and ^{241}Am greatly in excess of global fallout levels persist in the benthic environments of Bikini and Enewetak atolls. Quantities of $^{239+240}$Pu and lesser amounts of ^{241}Am are continuously mobilizing into solution from these sedimentary reservoirs. The amount of $^{239+240}$Pu mobilized at any time represents 0.08–0.09% of the sediment inventories to a depth of 16 cm. The mobilized $^{239+240}$Pu has solute-like characteristics, and different valence states coexist in solution; the largest fraction of the soluble Pu is in an oxidized form (V and/or VI). The adsorption of Pu onto sediments is not completely reversible because of changes that occur in the relative amounts of the mixed oxidation states in solution with time. Furthermore, any characteristics of $^{239+240}$Pu described at one location may not necessarily be relevant in describing its behavior elsewhere following mobilization and migration. The relative amounts of ^{241}Am to $^{239+240}$Pu in the sedimentary deposits at Enewetak and Bikini may be altered in the future because of mobilization and radioactive decay. Mobilization of $^{239+240}$Pu is not a process unique to these atolls, and quantities in solution derived from sedimentary deposits can be found at other global sites. These studies in the equatorial Pacific have significance in assessing the long-term behavior of the transuranic elements in any marine environment.

14.1. INTRODUCTION

Many of the questions regarding deep-sea disposal of transuranic elements and other long-lived radionuclides can be partially answered from studies at contaminated, more accessible locations where some of the processes, reactions, and rates that influence the fate of these radionuclides in the marine environment can be identified and evaluated. Reliable information on the behavior of these radionuclides in the ocean is required to improve one's understanding of the pathways that may lead back to humans from future practices involving disposal either onto or into marine sediments.

One important question related to the long-term behavior of transuranic elements in the marine environment is whether the radionuclides, after deposition in the bottom sediments, can return to the water column and eventually reenter food chains remote from the point of origin. At some sites contaminated by global fallout or surface discharges in which high concentrations of Pu are maintained in the overlying water column, mobilization of Pu from sedimentary sources to solution is difficult to demonstrate (Carpenter and Beasley, 1981; Nelson and Lovett, 1981). At sites where Pu was introduced in sufficient quantities directly onto sedimentary materials or where present inputs to sediments are small, however, mobilization of Pu can be easily identified (Noshkin et al., 1978; Noshkin, 1980; Noshkin and Wong, 1980; Schell et al., 1980; Noshkin et al., 1981; Nelson and Metta, 1983). For example, the bottom sediments in the lagoons at Bikini and Enewetak atolls, the sites previously known as the Pacific Proving Grounds, were contaminated with fission and activation products from nuclear devices tested there between 1946 and 1958 by the United States. Following the last nuclear test at Enewetak in 1958, the residual radionuclides in the lagoon water either settled rapidly to the bottom sediments or remained as dissolved or particulate species in the water and were eventually discharged into the open ocean. If the thesis of no mobilization is accepted, a concentration of dissolved Pu in the water mass of the lagoon during any time subsequent to 1958 should be observed at a concentration equivalent to that from global fallout in the northern equatorial Pacific surface water, the replacement water for the lagoon. Since 1972, a considerable number of filtered water samples from the lagoons at Enewetak and Bikini have been analyzed. Invariably these samples contained concentrations of $^{239+240}$Pu and ^{238}Pu in solution and in association with suspended particulate matter greatly exceeding that of the background levels

from fallout in the northern equatorial Pacific surface water. The mean concentration estimated for Enewetak and Bikini lagoons has been relatively constant for at least the last 10 y. It appears that a steady-state condition has been established for $^{239+240}$Pu partitioning from the sedimentary reservoirs at the atolls to solution (Noshkin and Wong, 1980).

Results discussed in this chapter address the question of both $^{239+240}$Pu and ^{241}Am mobilization at Bikini and Enewetak atolls and the geochemical properties of Pu in regions of the equatorial Pacific marine environment.

14.2. METHODS

14.2.1. History of Radionuclides in Enewetak and Bikini Atolls

Bikini and Enewetak atolls, shown with other atolls in the Marshall Islands Fig. 14.1, were the sites for 22 and 43 nuclear tests, respectively, conducted by the United States between 1946 and 1958. The testing produced fallout debris that was contaminated with transuranic elements and other radionuclides. A large amount of this labeled material entered the aquatic environment of the test sites. Other atolls to the east of the test sites were contaminated to lesser degrees by fallout.

In the United States, a moratorium on testing began on 31 October 1958, marking the end of all nuclear testing at Enewetak and Bikini atolls. The fallout debris and other post-testing activities produced a very heterogenous distribution of radionuclides in the lagoon sediments. Today, quantities of long-lived fission products such as ^{137}Cs, ^{90}Sr, and ^{155}Eu; activation products such as ^{55}Fe, ^{60}Co, and ^{207}Bi; and transuranic elements such as $^{238,\,239,\,240,\,241}$Pu and ^{241}Am persist in the atolls' environment. The largest inventory of Pu at Enewetak and Bikini is associated with lagoon sediments.

Analysis of grab and core samples from the lagoons defined the areal distribution of $^{239+240}$Pu and ^{241}Am in the sediment. It is estimated that approximately 9.3 and 2.8 TBq, respectively, are associated with the surface 2.5-cm layer at Enewetak, and at Bikini the inventories of $^{239+240}$Pu and ^{241}Am in the surface layer are 11.5 and 6.7 TBq, respectively. The inventories to a depth of 16 cm in the sediment column at Enewetak are estimated to be 44.4 and 17.8 TBq of $^{239+240}$Pu and ^{241}Am, respectively, and at Bikini the respective values are 54.4 and 42.2 TBq (Noshkin, 1980).

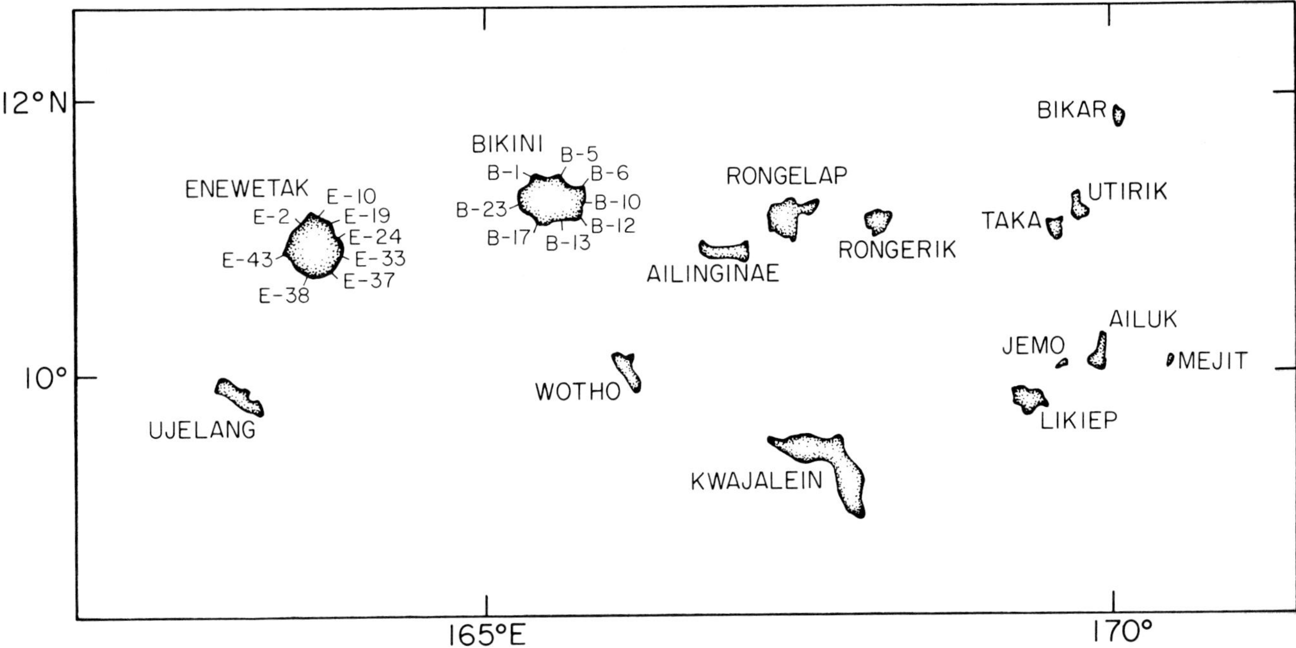

Figure 14.1. Locations of Enewetak, Bikini, and several other atolls in the Marshall Islands.

During the past few years we have been conducting studies at Enewetak and Bikini to better define the physical, chemical, and biological transport mechanisms as well as the fate of the transuranic elements and other long-lived radionuclides in the aquatic environment. Additional studies have been conducted at other atolls in the Marshall Islands and in the equatorial Pacific water outside of the atolls. These radionuclides are studied mainly to evaluate their impact on critical processes essential for the establishment and continuity of life at the atolls and because these studies can provide important data for understanding mobilization and migration processes of transuranic elements.

14.2.2. Experimental Studies

14.2.2a. Seawater-sample collections

Large-volume (50–200 liters) seawater samples obtained for analysis were collected by several methods. Within the lagoon of any atoll, filtered water samples from different depths and locations were collected aboard ship with a pump and hose fitted with an in-line filter. A 1-μm Micro-Wynde II® DCCPY (AMF, Cuno Division, Meriden, Connecticut) filter cartridge was used to remove particulate matter from the majority of the water samples. We have demonstrated that this filter is as efficient in removing suspended particulate matter from the lagoon water as is a 0.2- to 0.3-μm filter (Noshkin et al., 1974; Wong et al., 1980). Therefore, unless otherwise stated, the soluble concentrations discussed in this chapter refer to the quantity passing through a 1-μm filter. Outside the atolls, water samples were collected at various depths by using a large-volume sampler. Plutonium and Am were either preconcentrated from the filtered water in the field by a coprecipitation technique described by Wong et al. (1978) and retained on filters or collected in 60-liter containers and returned to Lawrence Livermore National Laboratory for processing. When measurements of plutonium oxidation states were made in the field, ^{236}Pu (VI) and ^{242}Pu (IV) were used as yield tracers.

14.2.2b. Analysis

Plutonium was separated from the ashed filters, which contained either particulate matter or the preconcentrated soluble fraction of Pu, by the method described by Wong (1971). The ^{241}Am was isolated and purified by using a modification of the procedure described by Bojanowski et al. (1975). Aliquots of ^{242}Pu and ^{243}Am were used as yield tracers except when plutonium oxidation states were determined. Finally, Pu and Am were electroplated onto stainless-steel disks and measured by α-spectrometry. Some samples of sediment and filtered particulate matter from Bikini and Enewetak contained sufficiently high concentrations of [241]Am to be determined directly by γ-spectrometry by using Ge(Li) detectors. Duplicates, blanks, and standards were routinely intermingled and processed along with the samples as part of our quality-assurance program.

14.3. RESULTS AND DISCUSSION

14.3.1. Plutonium Concentrations in Lagoon Seawater

Tables 14.1 and 14.2 show the variation of [239+240]Pu activity in filtered samples of seawater collected from several locations during different seasons at Enewetak and Bikini. Fallout levels of [239+240]Pu in the surface waters of the northern equatorial Pacific have averaged 15 ± 7 μBq liter^{-1} over the last 10 y (Noshkin et al., 1981b). Of this total concentration, $16 \pm 7\%$ is associated with the filterable particulate material. Concentrations of [239+240]Pu greater than background concentrations from global fallout were found in the water sampled from all locations throughout the lagoons. This is a direct indication that [239+240]Pu has been continuously mobilized over the years into solution from the solid phases in these environments.

Ocean water with concentrations of Pu from global fallout flows over the eastern and northern reefs and through the southern channels of the atolls. This water mixes with the labeled lagoon water, producing complex concentration gradients within the lagoons. Significant differences were noted in the concentration of [239+240]Pu with the year during which it was sampled at some stations (Tables 14.1 and 14.2). It is not possible with the data available to discern any consistent trend in the change of concentration with time by comparing the results at any single station. Different seasonal, and even daily, mixing processes influence the concentration at some locations. For example, it has been previously noted that even changes in daily tidal currents affect concentrations of dissolved Pu at specific locations (Noshkin et al., 1974).

One objective was to estimate the fraction of the lagoon inventory transferred to the water column. In order to do so, it was necessary to assume that a reasonable average concentration for the entire lagoon could be determined during periods when sufficient samples were obtained for analysis. Shown in Table 14.3 are the average concentrations of soluble and particulate [239+240]Pu estimated for Enewetak and Bikini lagoons during the periods indicated. The differences in the average concentration of soluble [239+240]Pu determined from samples obtained during the periods indicated are not considered significant, and the assumption is made that the average amount of Pu in the water mass of the lagoon is constant. Future results from the lagoons may, however, show this to be a false premise.

Table 14.1. Concentration of Soluble [239+240]Pu (mBq liter^{-1})[a] in Seawater Samples Collected at Specific Stations in Enewetak Lagoon

Station Designation	October–December 1972[b]	July–August 1974	May 1976	May–June 1982
5-8-4	0.51	—	0.84	0.44
5-8-2	0.67	—	1.15	0.46
5-8-1	1.20	0.77	1.06	0.77
5-7-1	0.25	1.85	0.97	0.57
5-6-6	1.63	1.74	0.77	0.72
5-6-3	0.56	1.22	0.26	0.53
5-6-2	0.27	—	0.05	0.68
5-5-4	2.44	0.08	0.17	1.41
5-5-3	0.48	0.70	0.13	0.37
5-9-4	—	0.32	0.57	0.27
5-9-3	0.36	—	0.85	0.49
5-9-2	0.42	0.26	0.29	0.51

[a] The 1-σ counting error is less than 10% of listed value. A dash indicates that a sample was not taken.

[b] The 1972 Enewetak water samples were unfiltered. All other concentrations were determined in water sample filtered through 1-μm filters.

Table 14.2. Concentration of Soluble $^{239+240}$Pu (mBq liter^{-1})^a in Seawater Samples Collected at Specific Stations in Bikini Lagoon

Station Designation^b	December 1972	January–February 1977	September 1982
B-15 Surface	2.26	1.00	—
B-15 Bottom	2.15	1.11	—
B-19	1.41	1.18	1.41
B-20	1.41	2.04	1.15
B-23 Surface	0.74	1.63	1.33
B-23 Bottom	0.32	0.49	—
B-25	2.56	2.07	—
B-2 Surface	3.44	2.74	—
B-2 Bottom	2.92	3.85	—

^{a}The 1-σ counting error is less than 10% of listed value.

^{b}A dash indicates a sample was not taken.

Since 1972, the average "soluble" $^{239+240}$Pu at Enewetak has been 0.78 mBq liter^{-1}. At Bikini, a mean value of 1.70 mBq liter^{-1} has been determined from the 1972 and 1977 collections. Taking into account the dimensions for each lagoon, these mean concentrations convert to standing inventories of 36 and 48 GBq in the lagoons at Enewetak and Bikini, respectively. These quantities, which exclude $^{239+240}$Pu associated with suspended particles, represent 0.08 and 0.09%, respectively, of the sediment inventories to a depth of 16 cm. The average quantity of $^{239+240}$Pu mobilized and found in solution at any time represents a small fraction of the inventory associated with the major reservoir of the atolls.

14.3.2. Other Evidence of $^{239+240}$Pu Mobilization in the Central Pacific Atolls

Water and sediment samples were collected from regions of several atolls in the Marshall Islands east of Enewetak and Bikini (Fig. 14.1) and from Johnston Atoll, located at 16°44′33″ N, 169°30′59″ W, 1160 km

Table 14.3. The Average Concentrations (mBq liter^{-1}) of $^{239+240}$Pu in Lagoon Water^a

Date of Samples	Number of Samples	Soluble	Particulate	Total
Enewetak Lagoon (Area: 933 km^2; Average Depth: 49 m)				
October–December 1972	35	0.81	0.37	1.18
July 1974	71	0.93	0.70	1.63
May 1976	29	0.59	0.48	1.07
May 1982	23	0.63	—^{b}	—^{b}
Bikini Lagoon (Area: 629 km^2; Average Depth: 45 m)				
December 1972	17	1.55	0.48	2.04
January–February 1977	26	1.81	—^{b}	—^{c}
September 1982^d	5	2.04	—^{b}	—^{c}

^{a}Arithmetic mean values.

^{b}Atoll not sampled.

^{c}Not applicable.

^{d}Incomplete results. The value shown represents an average from only 20% of the total samples obtained.

west–southwest of Hawaii. On 23 July 1962, a missile containing Pu was destroyed while still on the launch pad at Johnston Atoll. The aborted missile dispersed quantities of Pu into the surrounding shallow reef. A variety of samples have recently been collected from the marine environment of Johnston Atoll to assess the concentrations of Pu.

Average concentrations of $^{239+240}$Pu in the lagoon water and surface sediments from several of the northern atolls in the Marshall Islands are presented in Table 14.4, along with the range in concentrations detected at 13 stations sampled at Johnston Atoll. Samples of filtered lagoon water from several of the Marshall Islands contaminated by intermediate-range fallout and from Johnston atoll contained quantities of $^{239+240}$Pu in excess of the mean fallout level in the northern equatorial surface water ouside the lagoon (Table 14.4). At Rongelap, the higher concentrations in the lagoon water noted in 1978 were verified in 1981, when a more extensive sampling program was conducted; therefore, mobilization of $^{239+240}$Pu from the sediments to the water column was evident at some of these atolls as well.

It has been noted that in shallow basins such as the Great Lakes and the Irish Sea a large fraction of introduced Pu becomes attached to sedimentary particles and is transferred rapidly to the bottom sediments (Edgington, 1981). One might, therefore, anticipate observing substantial reductions in the concentration of Pu in the

water of the shallow lagoons of the lesser-contaminated atolls such as Ailuk, Likiep, Utirik, or Wotho. The exhange rate of the lagoon water at these atolls is not accurately known, but we estimated the rate to be ≥ 40 d. This period is sufficiently long to permit most (95%) of the Pu from global fallout entering the lagoon to be effectively scavenged by the suspended materials during passage of water across the shallow reef. The mean concentration of $^{239+240}$Pu within these lagoons (Table 14.4) is not, however, distinguishable from the concentration from the average global fallout in the surface water outside the lagoon. This suggests that an equilibrium has been established at these locations between stabilized forms of fallout $^{239+240}$Pu in the surface seawater and in the sediments.

14.3.3. Oxidation States of Mobilized Plutonium in Lagoon Water

Samples of water from Bikini and Enewetak were collected and immediately processed aboard ship to determine the oxidation state(s) of the mobilized Pu in solution. Results from these analyses for both lagoons are shown in Tables 14.5 and 14.6, along with several replicate results in Table 14.6. The reproducibility among replicate analyses is considered satisfactory. The procedure used to separate reduced $^{239+240}$Pu (III and IV) from oxidized $^{239+240}$Pu (V and VI) was that described by Lovett and Nelson (1981), modified as described by Wong et al. (1985) for our use with water

Table 14.4. The Mean Water and Sediment Concentrations of $^{239+240}$Pu at Several Atolls in the Marshall Islands and Johnston Atoll[a]

| | | | Concentration of $^{239+240}$Pu in Water (μBq liter^{-1}) | | |
Atoll	Year Collected	Surface[b] Sediment (Bq kg^{-1})	Soluble	Particulate	Total
Ailuk	1978	0.30	21	1.5	23
Likiep	1978	0.29	21	8.9	30
Utirik	1978	0.59	10	5.2	15
Wotho	1978	0.35	16	1.1	17
Rongelap	1978	10.32	48	49.0	98
Rongelap	1981	—	42	—	—
Rongerik	1978	3.03	44	5.6	50
Johnston	1980	0.13–61.05	15–574	3.7–326	—
Northern equatorial Pacific surface water	1972–1982	—	—	—	15 ± 7

[a] A dash indicates sample not analyzed.
[b] Surface sediment refers to the top 3-cm layer.

Table 14.5. The Concentration (mBq liter^{-1}) of $^{239+240}$Pu Oxidation States in Filtered Seawater at Bikini Atoll

Station Number	Date	Depth	Reduced (III and IV)	Oxidized (V and VI)	Total	Percent Oxidized
B-1	2/81	Surface	0.33 ± 0.01	0.43 ± 0.02	0.76 ± 0.03	57 ± 3
B-1	2/81	10 m	0.49 ± 0.02	0.44 ± 0.03	0.93 ± 0.04	46 ± 4
B-1	2/81	25 m	0.49 ± 0.02	0.60 ± 0.06	1.08 ± 0.12	55 ± 7
B-5 Reef	2/81	Surface	0.98 ± 0.08	2.45 ± 0.23	3.43 ± 0.31	71 ± 10
B-6 Reef	2/81	Surface	0.21 ± 0.01	1.15 ± 0.04	1.37 ± 0.07	85 ± 6
B-20 Reef	2/81	Surface	0.14 ± 0.01	0.72 ± 0.07	0.86 ± 0.10	84 ± 12
B-1 Reef	2/81	Surface	0.21 ± 0.01	0.23 ± 0.02	0.44 ± 0.03	52 ± 4
B-19	9/82	Surface	0.22 ± 0.40	1.18 ± 0.08	1.40 ± 0.13	84 ± 4
B-18	9/82	Surface	< 0.02	1.98 ± 0.14	0.20 ± 0.14	99 ± 7
B-18	9/82	29 m	1.52 ± 0.09	2.89 ± 0.11	4.40 ± 0.22	66 ± 4
B-26	9/82	Surface	< 0.02	1.79 ± 0.07	1.81 ± 0.07	99 ± 3
B-26	9/82	48 m	0.15 ± 0.02	0.53 ± 0.04	0.64 ± 0.06	80 ± 9

Table 14.6. The concentration (mBq liter^{-1}) of $^{239+240}$Pu Oxidation States in Filtered Seawater at Enewetak Atoll

Station Number	Date	Depth	Reduced (III and IV)	Oxidized (V and VI)	Total	Percent Oxidized
E-24	10/79	Surface	0.17 ± 0.04	0.38 ± 0.03	0.56 ± 0.06	69 ± 9
E-24	10/79	Replicate	0.17 ± 0.04	0.48 ± 0.06	0.65 ± 0.10	74 ± 14
E-24	10/79	10 m	0.28 ± 0.04	0.85 ± 0.05	1.13 ± 0.09	75 ± 8
E-24	10/79	25 m	0.30 ± 0.05	1.01 ± 0.10	1.31 ± 0.13	77 ± 11
E-10	10/79	Surface	0.03 ± 0.01	0.32 ± 0.06	0.36 ± 0.07	91 ± 26
E-10	10/79	Replicate	0.04 ± 0.01	0.24 ± 0.03	0.28 ± 0.04	85 ± 16
5-6-1	10/79	Surface	0.19 ± 0.01	0.74 ± 0.05	0.93 ± 0.07	79 ± 8
E-10	6/82	Surface	0.05 ± 0.01	0.20 ± 0.01	0.24 ± 0.01	80 ± 6
6-2-1	6/82	Surface	0.46 ± 0.03	1.00 ± 0.05	1.45 ± 0.09	69 ± 5
6-2-1	6/82	Replicate	0.46 ± 0.04	0.93 ± 0.07	1.39 ± 0.08	67 ± 6
6-2-1	6/82	Replicate	—	—	1.41 ± 0.04	—
6-2-5	6/82	Surface	0.11 ± 0.01	0.48 ± 0.03	0.59 ± 0.05	82 ± 9
6-2-5	6/82	Replicate	—	—	0.55 ± 0.01	—
6-2-5	6/82	22 m	0.10 ± 0.01	0.40 ± 0.03	0.50 ± 0.04	80 ± 8
6-2-5	6/82	Replicate	—	—	0.47 ± 0.01	—
6-1-3	6/82	Surface	0.09 ± 0.01	0.44 ± 0.03	0.53 ± 0.04	83 ± 8
6-1-3	6/82	Replicate	—	—	0.51 ± 0.01	—
5-31-7	6/82	Surface	0.07 ± 0.01	0.38 ± 0.03	0.45 ± 0.04	85 ± 7
5-31-7	6/82	Replicate	0.06 ± 0.01	0.32 ± 0.02	0.38 ± 0.03	84 ± 7
5-31-7	6/82	Replicate	—	—	0.39 ± 0.01	—

samples larger than 50 liters. Both the reduced and the oxidized forms of $^{239+240}$Pu coexisted in the lagoon water, with oxidized forms generally dominating. On the average, 78% of the total Pu in solution at both atolls was in the oxidized state.

It is difficult to identify any clear regularities or trends in the results, but, in some cases, samples from stations near the reef and near the sediment–water interface contained relatively larger fractions of reduced Pu in solution than was found in mid-lagoon surface samples. Over 90% of the $^{239+240}$Pu found associated with sedimentary or particulate material in the lagoon was in the reduced state (see footnote *b* in Table 14.7); therefore, the near-bottom or reef water would be expected to contain relatively higher concentrations of recently exchanged, reduced Pu. After desorption of reduced and oxidized species, the (III and IV) states disappear from the solution either by oxidation to the higher state or loss by adsorption onto newly exposed

Table 14.7. The concentration of $^{239+240}$Pu Oxidation States Released into Seawater and K_d Values after Mixing Lagoon Sediments from Enewetak off Station E-2[a]

Sample ID	Mixing Time (h)	Sediment[b] Concentration (Bq g^{-1})		Reduced (III and IV) (mBq liter^{-1})		Oxidized (V and VI) (mBq liter^{-1})		$K_d \times 10^5$[c]		
								Apparent	Reduced	Oxidized
				Fine Fraction (< 0.5 mm)						
439/440	3.5	0.74	(2)	1.89	(14)	2.44	(12)	1.7	3.4	0.12
443/444	3.5	0.65	(2)	2.63	(17)	1.78	(17)	1.5	2.4	0.15
399/400	24	0.63	(2)	0.85	(22)	2.00	(9)	2.2	7.1	0.13
411/412	24	0.70	(2)	0.52	(41)	2.48	(14)	2.3	13.0	0.11
427/426	74	0.72	(2)	0.44	(23)	2.96	(8)	2.1	15.6	0.10
				Coarse Fraction (> 0.5 mm)						
379/380	26	0.092	(4)	0.26	(54)	0.48	(26)	1.2	3.4	0.08
383/384	26	0.107	(5)	0.13	(38)	1.04	(13)	0.9	7.9	0.04
391/392	81	0.018	(3)	< 0.01		1.18	(11)	0.9	> 10	0.04

[a]Values in parentheses are the 1-σ counting error expressed as percent of the value listed.

[b]96 ± 2% of the total $^{239+240}$Pu associated with sediments was in the reduced state. There was 3.3–9.2 g of sediment mixed with 0.95 liters of filtered seawater.

[c]Apparent K_d = total concentration of $^{239+240}$Pu in dry sediment (Bq kg^{-1})/total concentration of $^{239+240}$Pu in solution (Bq liter^{-1}). Reduced K_d = 96% of the total concentration in dry sediment/reduced concentration in solution. Oxidized K_d = 4% of the total concentration in dry sediment/oxidized concentration in solution.

surfaces. Laboratory experiments discussed in the next section provide results that support the former possibility. In either case, the oxidized forms of Pu become the dominant species in older water with longer residence times in the lagoons (stations B-18, B-26, and E-10 in Fig. 14.1) and probably persist as a complexed, unreactive species. Since the reduced forms of Pu are ubiquitous in all but a few samples, however, it is possible that the reduced Pu may also be stabilized by complexation to some degree, but that the stability of the complex may change with time.

14.3.4. Exchange of Plutonium between Sediments and Seawater

At any location within Enewetak and Bikini lagoons, bottom sediments consist of different quantities of fine- and coarse-grained carbonate material, shell, foraminifera tests, coral fragments, and *Halimeda* debris. All of these components have different surface areas, and, therefore, different sorptive characteristics. Transuranic elements are associated with all components of the benthic environment (Noshkin, 1980), and their rate of desorption into solution may relate to the composition of the sediment.

To mimic the desorption characteristics of $^{239+240}$Pu

from the sediments to solution observed in the environment, we conducted experiments in the laboratory (or aboard ship) with freshly collected sediments to arrive at a value at a value for the distribution coefficient (K_d) for plutonium. We placed quantities of sedimentary material in plastic containers and added volumes of filtered, equatorial Pacific surface seawater with very low concentrations of dissolved Pu (15μBq liter^{-1}). The containers were shaken over a 3-d period to mix the contents. The water was then separated from the sediments by filtration through 0.2-μm filters. Tracers were added to the solution phase, and Pu was separated and analyzed by the procedures described. Sediments were dried, dissolved, and analyzed for Pu. The range in the apparent K_d values [total $^{239+240}$Pu on sediments (in Bq kg^{-1}) divided by total $^{239+240}$Pu in solution (in Bq liter^{-1})] based on the analyses of 18 different lagoon sediments was between 0.5 and 5.2×10^5, with an average K_d for $^{239+240}$Pu of 2.3×10^5 (Noshkin and Wong, 1980).

With the sediment inventory values and a K_d value of 2.3×10^5 for Pu, one can construct a simple model to predict average concentrations expected in the lagoon water by assuming that the Pu in solution at any time is in equilibrium with that in the sediments. At any time, the amount of Pu in solution is limited by the saturation of the solution under equilibrium conditions. The rate at which water and its dissolved Pu is flushed from the lagoon is balanced by input of the lesser-contaminated oceanic water that is rapidly equilibrated with remobilized Pu from the sediments. Assuming that this steady-state condition can be achieved, the mean Pu inventory and concentration of Pu in lagoon water computed from a K_d of 2.3×10^5 as defined above was 0.04 TBq (0.85 mBq liter^{-1}) at Enewetak and 0.05 TBq (1.51 mBq liter^{-1}) at Bikini. There is good agreement between the average quantities of $^{239+240}$Pu predicted and the average values shown in Table 14.3. The agreement between the calculated and the measured average concentrations, where one would accept a factor of 5 as being good agreement because of the range of K_d values determined, supports the contention of a steady-state condition and demonstrates the general usefulness of this simple model in predicting long-term average concentrations in the lagoon water.

Additional time-dependent experiments and oxidation-state determinations were conducted in the laboratory with contaminated lagoon sediments. Results for the different oxidation states released into seawater from fine- and coarse-sediment fractions obtained from two locations in Enewetak Lagoon are shown in Tables 14.7 and 14.8. Of the total $^{239+240}$Pu associated with the sediments used to generate these results, $96 \pm 2\%$ was in the reduced state. When fresh seawater initially contacted the fine- and coarse-sediment fractions shown in Table 14.7, there was a rapid release of both oxidized and reduced forms of $^{239+240}$Pu into solution. As the contact time with the

sediment increased, the K_d for reduced Pu increased, indicating a disappearance of the reduced forms from solution. Concurrently, the K_d determined for the oxidized forms of Pu decreased slightly, showing that the oxidized forms of $^{239+240}$Pu in solution increased over the same period. This increase of oxidized forms of Pu can be accounted for by assuming that the reduced forms initially mobilized are slowly oxidized when in solution. These results confirm the field observations in which both oxidized and reduced species were initially mobilized to bottom water in the lagoon. The concentration of reduced Pu decreased with time as a result of redox reactions, leaving the oxidized forms of Pu as the dominant dissolved species in solution.

Results in Table 14.8 also show rapid mobilization of both forms into solution, followed, however, by no apparent change in concentration of either species as the contact time with the E-24 sediments increased. After 94 h, the K_d values both for the reduced and for the oxidized forms of $^{239+240}$Pu from the experiment with the fine-sediment fraction were less than the values after 74 h of contact with sediments from station E-2 (Table 14.7). The reasons for the different behavior of the E-2 and E-24 sediments is unknown, and we can only speculate that both mobilized species were stabilized in solution, possibly by complexing agents derived from materials present in the sediments at station E-24 but absent at station E-2. For whatever the reasons, the field measurements suggest that the reduced forms of $^{239+240}$Pu mobilized into solution from the region of station E-24 were capable of remaining in a dissolved state for longer periods of time. Plutonium mobilized

Table 14.8. The $^{239+240}$Pu Oxidation States Released into Seawater and K_d Values after Periods of Mixing with Lagoon Sediments from Enewetak off E-24[a]

Sample ID	Mixing Time (h)	Sediment Concentration (Bq g^{-1})		Reduced (III and IV) (Bq liter^{-1})		Oxidized (V and VI) (Bq liter^{-1})		$K_d \times 10^5$		
								Apparent	Reduced	Oxidized
				Fine Fraction (< 0.5 mm)						
359/360	24	0.53	(3)	0.85	(22)	8.21	(6)	0.6	6.0	0.03
363/364	24	0.54	(2)	2.07	(19)	6.70	(7)	0.6	2.5	0.03
367/364	94	0.63	(3)	1.48	(14)	9.03	(8)	0.6	4.1	0.03
371/372	94	0.63	(3)	1.96	(15)	8.81	(7)	0.6	3.1	0.03
377/620	2880	0.65	(3)	2.15	(15)	8.14	(6)	0.6	2.6	0.03
				Coarse Fraction (> 0.5 mm)						
339/340	23	0.22	(5)	0.56	(19)	1.55	(10)	1.1	3.9	0.06
343/344	23	0.22	(5)	0.56	(21)	1.74	(13)	0.9	3.7	0.05

[a] See footnotes in Table 14.7.

into solution from any region of either lagoon could be expected to behave in a manner between that of the examples in Tables 14.7 and 14.8.

Once the reduced Pu is oxidized and in solution, the process of readsorption of $^{239+240}$Pu onto the sediment particles appears to be determined by an accumulation factor different than the value obtained from the desorptive experiments. Table 14.9 shows two examples in which contaminated water from Enewetak Lagoon was equilibrated with quantities of relatively uncontaminated sediment from Kwajalein Atoll. (The oxidation states of Pu associated with these sediments were not determined.) The first example shows the initial and the final $^{239+240}$Pu concentrations in seawater from station E-24 after mixing with sediments for 22 h. There was no difference noted between the initial and the final concentrations either of reduced or of oxidized forms of Pu in solution. The computed apparent K_d, determined by comparing the final total water concentrations with the sediment concentrations, is significantly less than are the desorption values in Tables 14.7 and 14.8, which were determined after 24 h of mixing contaminated sediment with uncontaminated water.

The second set of results was generated by using aliquots of seawater in contact with sediments for longer periods of time. Unfortunately, the water selected was not as highly labeled with $^{239+240}$Pu as had been anticipated, and therefore the errors associated with the determinations are large. The results, however, show a complete disappearance of reduced $^{239+240}$Pu from solution after 72 h, but it is impossible to assess if this loss resulted from the operation of the redox equilibrium or from adsorption onto the sediment. There was a definite loss of the oxidized state of $^{239+240}$Pu by adsorption

onto the sediment: the concentrations in the sediment increased from 0.030 ± 0.11 to 0.56 ± 0.67 mBq g^{-1} after equilibration. Each of the sediment samples in contact with the water for 24, 72, and 3000 h gained, on the average, 1.1 ± 0.03 mBq, whereas 1.2 ± 0.03 mBq, on the average, was lost from solution. The Pu lost from solution can be accounted for by the gain onto the sediments. The apparent K_d computed in Table 14.9 is the value for total Pu in solution divided by total Pu in the sediments. The adsorption in each timed experiment was rapid in that there was no further change in the concentration of oxidized forms of Pu remaining in solution after 24 h of equilibration.

While it appears possible to predict the desorption of $^{239+240}$Pu by using a K_d of approximately 2.3×10^5, the readsorption process of mobilized Pu resulted in lower accumulation factors and was more complex than predicted by the simple partitioning between solution and solid phases.

14.3.5. Relationships between ^{241}Am and $^{239+240}$Pu

Concentration profiles to a depth of 1000 m in the equatorial Pacific, 27 km west of Bikini (station WBO) indicate that ^{241}Am in the ocean is more strongly associated with particles than is $^{239+240}$Pu (Table 14.10). Some data from the water column 10 km from Johnston Atoll (station JAO) are provided in Table 14.11 to highlight certain features of the results at station WBO. The fact that Pu concentrations were more than ten times higher in the surface waters at station WBO as compared with the global fallout concentrations in the surface layers at station JAO indicates that mobilized

Table 14.9. Equilibration of $^{239+240}$Pu-Labeled Seawater with Lesser-Contaminated Sediments[a]

Contact Time (h)	Sediment Concentration (mBq g^{-1})	Concentration of $^{239+240}$Pu (mBq liter^{-1})			$K_d \times 10^3$
		Reduced	Oxidized	Total	
0	—	0.14 (11)	0.72 (10)	0.86 (11)	—
22	0.17 (21)	0.20 (8)	0.75 (5)	0.96 (6)	0.2
0	—[b]	0.04 (54)	0.32 (19)	0.36 (18)	—
24	0.67 (12)	0.04 (75)	0.19 (27)	0.24 (26)	2.8
72	0.56 (15)	0.03 (40)	0.13 (14)	0.15 (13)	3.7
3000	0.63 (15)	< 0.01	0.16 (19)	0.16 (19)	3.9

[a] Values in parentheses are the 1-σ counting error expressed as percent of the value listed.

[b] Initial concentration in the sediment was 0.30 (37) mBq g^{-1}; 3.84, 2.90, and 3.3 g of sediment were used in the three equilibrium equation experiments, respectively. The volume of seawater used was 1 liter.

Table 14.10. Concentrations of Two Transuranic Elements in the water Column at Station WBO and Inventory to a Depth of 1000 m[a]

Depth (m)	$^{239+240}$Pu (μBq liter^{-1}) Prefiltered	Solution	^{241}Am (μBq liter^{-1}) Prefiltered	Solution	^{241}Am:$^{239+240}$Pu Prefiltered	Solution
Surface	lost	170.0 (4)	lost	lost	—	—
100	1.5 (22)	18.9 (11)	lost	6.3 (45)	—	0.33 (46)
200	7.4 (15)	28.1 (11)	lost	8.1 (26)	—	0.29 (29)
300	17.0 (11)	32.9 (15)	20.0 (47)	7.0 (50)	1.2 (48)	0.21 (52)
400	18.1 (11)	55.5 (9)	22.6 (23)	< 1.0	1.2 (25)	< 0.02
500	7.8 (25)	70.3 (7)	29.2 (20)	4.1 (40)	3.8 (32)	0.06 (40)
600	1.1 (60)	59.2 (7)	14.1 (28)	6.7 (50)	12.0 (65)	0.11 (50)
700	4.4 (30)	59.2 (7)	28.5 (16)	19.2 (37)	6.4 (34)	0.32 (38)
800	4.4 (36)	81.4 (6)	24.1 (25)	13.0 (35)	5.4 (44)	0.16 (36)
900	11.5 (27)	74.0 (9)	26.6 (18)	10.0 (30)	2.3 (32)	0.14 (31)
1000	15.2 (28)	66.6 (7)	26.6 (27)	< 4.0	1.8 (39)	< 0.07
Inventory (MBq km^{-2}) at a depth of 1000 m	8.1	51.8	20.0	8.1	2.4	0.39

[a]Measured in September 1980. Values in parentheses are the 1-σ counting error expressed as percent of the listed value. The K_d ratio of $^{239+240}$Pu:^{241}Am to 1000 m = the ratio of inventories $\dfrac{^{239+240}\text{Pu prefiltered/solution}}{^{241}\text{Am prefiltered/solution}} = 0.06$.

Table 14.11. Transuranic Elements in the Water Column at Station JAO and Inventory to a Depth of 1000 m[a]

Depth (m)	$^{239+240}$Pu (μBq liter^{-1}) Prefiltered	Solution	^{241}Am (μBq liter^{-1}) Prefiltered[b]	Solution	^{241}Am:$^{239+240}$Pu Prefiltered[b]	Solution
Surface	< 4	4.8 (37)	—	< 1	—	< 0.2
100	0.7 (50)	< 4	—	4.1 (50)	—	lost
200	< 0.4	7.0 (26)	—	2.6 (20)	—	0.34 (33)
300	< 4	5.6 (8)	—	10.7 (25)	—	0.26 (26)
400	2.2 (45)	40.3 (9)	—	2.6 (60)	—	0.06 (60)
500	5.6 (23)	38.5 (12)	—	7.4 (37)	—	0.19 (39)
600	2.2 (60)	43.3 (11)	—	8.9 (30)	—	0.20 (30)
700	11.1 (25)	34.1 (9)	—	13.3 (47)	—	0.39 (48)
800	5.6 (28)	40.7 (35)	—	< 7	—	< 0.2
900	9.3 (27)	26.3 (14)	—	6.3 (30)	—	0.24 (33)
1000	8.9 (27)	30.0 (2)	—	7.8 (30)	—	0.24 (32)
Inventory (MBq km^{-2}) at a depth of 1000 m	4.8	29.2	—	6.7	—	0.23

[a]Values in parentheses are the 1-σ counting error expressed as the percent of the listed value.
[b]Samples not analyzed.

forms of $^{239+240}$Pu are passed from the lagoon to the water mass of the northern equatorial Pacific. The concentrations throughout the upper 1000 m at station WBO originate from Bikini and are supplemented by global fallout concentrations typical for this region.

Whereas the $^{239+240}$Pu concentration in solution from all depths sampled at station WBO exceeded that for the particulate materials filtered, the opposite was true for ^{241}Am. The concentration ratios show different but substantial enrichment of ^{241}Am over $^{239+240}$Pu on the particulate fraction throughout the water column. At stations WBO and JAO, 14% of the total Pu inventory in the water column to a depth of 1000 m was associated with particulate matter, whereas 71% of the total [241]Am inventory at station WBO was bound to particles. This partitioning between [241]Am and [239+240]Pu in the ocean is not a feature related to the source term for the radionuclides.

Pentreath et al. (1980) report that the K_d values of [241]Am, determined for particulate material in the surface of the seawater within and outside the Irish Sea, are from 2.5 to 7.8 times greater than are the respective values for [239+240]Pu. Beasley and Cross (1980) show results from the Mediterranean, where an average of 10% of [241]Am from fallout is retained on filters, as compared to only 3.8% of the [239+240]Pu.

Results from our laboratory experiments using different sediments from Enewetak indicate that [241]Am was more strongly associated with lagoon sediments than was [239+240]Pu (Table 14.12). The K_d values for [241]Am were 3.3–18 times greater than were those for [239+240]Pu for the set of samples tested. Our laboratory results suggest there is a lesser, if any, tendency for [241]Am to be mobilized either from sedimentary or from particulate material into solution as a dissolved species.

Table 14.13 compares the [239+240]Pu:[241]Am concentrations in solution and in association with particulate material at Enewetak and Bikini lagoons. Although the concentrations of [241]Am in solution within the lagoons were small, they were nevertheless an order of magnitude greater than were the global fallout concentrations in the surface waters of the northern equatorial Pacific. All results have been corrected for any increase in

[241]Am from [241]Pu decay between the time of collection and separation. Small but measurable amounts of [241]Am were also capable of mobilizing from sedimentary deposits into solution.

The mean activity ratio of [241]Am:[239+240]Pu associated with the particulate material is essentially identical to the mean ratio of the sediment inventory at each atoll (Section 14.2.1). There is a large reduction in this ratio for the radionuclides in solution compared with the values associated with the particulate matter and, by analogy, the sediments. The [241]Am is strongly bound to the solid phases and is therefore less likely to desorb into solution than is Pu.

In the final column of Table 14.13, the prefiltered: solution concentrations of [239+240]Pu and [241]Am, have been converted to the ratio of the K_d for the two transuranic elements. The mean K_d ratio of 0.13 agrees well with the average ratio of 0.11 for values determined in the laboratory experiments (Table 14.12). The value of the mean ratio determined from the inventories in the water column at station WBO (Table 14.10) is 0.06. This is roughly half of the value for the ratio determined from the lagoon water, and it suggests that a larger fraction of the total dissolved Pu found in the water column outside the lagoons is in the oxidized state, as compared to the average (78%) for oxidized forms in solution within the lagoons.

14.3.6. Residence Time of [239+240]Pu and [241]Am in the Lagoons

Results from biological indicators (Noshkin, 1980) and from the direct measurements of the mean concentra-

Table 14.12. Comparative K_d Values for [239+240]Pu and [241]Am[a]

	[239+240]Pu			[241]Am		
ID Number	Solution (mBq liter[-1])	Solid (Bq g[-1])	K_d[b]	Solution (mBq liter[-1])	Solid (Bq g[-1])	K_d
224	2.07 ± 0.07	1.01 ± 0.02	4.8×10^5	0.25 ± 0.10	1.06 ± 0.07	4.3×10^6
226	1.52 ± 0.7	0.79 ± 0.01	5.2×10^5	0.11 ± 0.11	0.63 ± 0.03	5.4×10^6
228	4.04 ± 1.0	1.31 ± 0.01	3.3×10^5	0.11 ± 0.07	0.52 ± 0.03	4.7×10^6
230	3.07 ± 0.07	1.21 ± 0.01	3.9×10^5	0.28 ± 0.12	0.34 ± 0.04	1.3×10^6
236	6.41 ± 0.10	0.73 ± 0.01	1.1×10^5	0.08 ± 0.07	0.16 ± 0.02	2.0×10^6
Mean			$(3.7 \pm 1.6) \times 10^5$			$(3.5 \pm 1.8) \times 10^6$

[a] Determined from grab samples of sediment and seawater collected at Enewetak and transferred to containers and mixed during a two-month storage period; mixed-size range of solids. Seawater filtered through 0.45-μm filter prior to analysis.

$$^b K_d = \frac{\text{(Bq per kg of dry solids)}}{\text{(Bq per kg of water)}}.$$

Table 14.13. The $^{239+240}$Pu and ^{241}Am Concentrations in Seawater and Particulate Matter in Two Lagoons[a]

Location	Date Sampled	$^{239+240}$Pu (mBq liter^{-1}) Prefiltered	Solution	^{241}Am (mBq liter^{-1}) Prefiltered	Solution	^{241}Am:$^{239+240}$Pu Prefiltered	Solution	K_d ($^{239+240}$Pu:^{241}Am)[b]
Enewetak								
Mid-lagoon	8/74	—	1.11 (5)	—	—	—	0.06 (39)	—
Mid-lagoon	4/79	0.16 (5)	—	0.036 (13)	0.067 (39)	0.24 (13)	—	—
Mid-lagoon	4/79	0.13 (5)	—	0.028 (12)	—	0.21 (13)	—	—
Mid-lagoon	10/79	0.12 (5)	—	0.054 (6)	—	0.46 (8)	—	—
Mid-lagoon	10/79	0.19 (3)	—	0.060 (4)	—	0.31 (5)	—	—
E-24 Lagoon	10/79	0.75 (2)	0.84 (2)	0.122 (7)	—	0.16 (7)	0.05 (26)	0.31
E-24 Reef	10/79	2.81 (2)	2.15 (3)	0.322 (9)	0.041 (26)	0.11 (9)	0.02 (25)	0.18
E-10 Reef	10/79	8.95 (3)	0.16 (5)	5.59 (3)	0.041 (26)	0.62 (4)	0.10 (30)	0.16
E-24 Reef	10/79	0.98 (1)	2.33 (2)	0.50 (5)	0.017 (17)	0.51 (5)	0.027 (17)	0.05
E-24 Reef	11/80	—	1.37 (3)	—	0.024 (35)	—	0.018 (35)	—
E-24 Reef	1/81	—	1.63 (2)	—	0.016 (38)	—	0.008 (38)	—
E-24 Reef	7/81	—	1.07 (2)	—	0.024 (20)	—	0.023 (20)	—
Mean		—	—	—	—	0.33 (55)	0.038 (79)	—
Bikini								
B-10	11/78	0.11 (9)	0.28 (4)	0.087 (8)	0.018 (50)	0.78 (9)	0.063 (50)	0.08
B-2	11/78	1.99 (5)	1.78 (2)	1.58 (5)	0.122 (30)	0.79 (7)	0.068 (30)	0.09
B-6	11/78	3.63 (4)	0.85 (3)	3.26 (4)	0.022 (65)	0.90 (6)	0.026 (65)	0.03
B-13	11/78	0.25 (7)	1.45 (2)	0.211 (6)	0.043 (45)	0.82 (9)	0.029 (45)	0.04
Mean		—	—	—	—	0.82 (7)	0.046 (48)	—
Mean		—	—	—	—	—	—	0.13
Equatorial Pacific surface water[c]	6–11/78	1.5×10^{-3} (44)	10.4×10^{-3} (44)	2.2×10^{-3} (51)	4.8×10^{-3} (75)	—	—	—

[a] The 1-σ counting error in parentheses is expressed as percent of the listed value. A dash denotes the absence of data.
[b] K_d ratio $= {}^{239+240}$Pu (prefiltered/solution):^{241}Am (prefiltered/solution).
[c] The average of 18 samples.

tions of the two transuranic elements in the lagoons described in this chapter indicate that the average amount of $^{239+240}$Pu in solution has been constant since at least 1965. Mobilized Pu has come from a sufficiently large reservoir that the amount lost to solution over the years has not substantially depleted the sedimentary inventory. A fraction of the $^{239+240}$Pu and ^{241}Am in the water (Tables 14.3 and 14.13) is associated with particulate matter, some of which is resuspended from the bottom sediments. Very little sedimentary material escapes from the lagoon, and resuspended bottom material probably settles out onto the lagoon floor; therefore, no $^{239+240}$Pu or ^{241}Am associated with particulate matter is assumed to be lost from the lagoon.

Physical circulation data accumulated at Enewetak by Atkinson et al. (1981) indicate that the mean exchange rate for the lagoon water with the ocean is approximately one month. The authors point out, however, that water entering the northern portion of the atoll, where sediments have the highest concentrations of Pu, takes about four months to exit the lagoon. Previously, von Arx (1948) estimated that the water in Bikini Lagoon exchanges seven times annually with the ocean. Rates between 30 and 50 d are substantially faster than the mean exchange rate of (140 d) estimated for $^{239+240}$Pu from radiological data associated with several biological indicators (Noshkin, 1980). Noshkin and Wong (1980) have used this 140-d residence time along with the average inventory of soluble $^{239+240}$Pu (36 GBq) in lagoon water, and they estimate that the soluble inventory represents 0.08% of the inventory in the sediment (to a depth of 16 cm) to calculate a mean life of 435 y (half-residence time of 330 y) for Pu in the Enewetak sediment. Loss of $^{239+240}$Pu by radiological decay was neglected in the above and in subsequent calculations. At Bikini, the computed mean life for $^{239+240}$Pu in the sediment (to a depth of 16 cm) column was 460 y.

The average ^{241}Am:$^{239+240}$Pu ratios of concentrations in solution are 0.038 and 0.046 at Enewetak and Bikini, respectively (Table 14.13). Multiplying these values by the average inventories of $^{239+240}$Pu in solution provides average inventories of 1.3 and 2.2 GBq as the quantities of ^{241}Am in solution at Enewetak and Bikini atolls, respectively. Using a water residence time of 140 d, we calculate that 3.3 and 5.9 GBq of ^{241}Am are lost annually from Enewetak and Bikini. Assuming again that the inventory to a depth of 16 cm in the sediment is the reservoir for the mobilized ^{241}Am, the mean life, neglecting radioactive decay, exceeds 4000 y. Since the mean life of ^{241}Am is 625 y (with a half-life of 433 y), the effective mean lifetime for ^{241}Am in the sedimentary reservoir is 555 y. Taking into account the exchange rates and inventory values discussed in this chapter, ^{241}Am will be depleted in the sedimentary reservoirs of the atolls primarily by radioactive decay, whereas most of the $^{239+240}$Pu will be lost by mobilization and exchange with the open ocean, assuming that the mobilization process continues at the same rate in subsequent years.

The assessments discussed thus far in this chapter were made by using an exchange rate of 140 d and a sedimentary reservoir extending from the sediment–water interface to a depth of 16 cm. The depletion rate of Pu and Am from the lagoons clearly depends on the exchange rate of water between the lagoon and the ocean, and the size and depth of the reservoir from which the mobilized $^{239+240}$Pu or ^{241}Am is derived. The evidence provided by von Arx (1948) and Atkinson et al. (1981), indicating that the exchange rate of water and its dissolved constituents between the lagoon and ocean is more rapid than 140 d, cannot be dismissed. There is some evidence that Pu contained in subsurface sediments is lost to the overlying water by vertical movement of the interstitial water. We do not know, however, the active depth in the sediment column through which Pu can move as a result of advection or diffusion along concentration gradients. In addition, we have not assessed the importance of this mechanism relative to other physical, biological, or chemical disturbances that may result in resuspension of sediment to the water column followed by Pu mobilization into solution. If it is assumed that the water and its dissolved constituents exchange between Enewetak Lagoon and the ocean every 30 d, for example, and that the principal reservoir for Pu is only the surface 2.5-cm layer of the sediment, then the mean and the half-residence times for $^{239+240}$Pu in this reservoir are only 21 and 15 y, respectively, and the mean and half-residence times for ^{241}Am computed for these conditions are 416 and 289 y, respectively.

This computation shows that the mean quantity of plutonium mobilized into solution during 1982 should be half the average amount it was 15 y ago. The ^{241}Am levels would remain essentially unchanged during this period. Our measurements do not agree with this rapid a rate of Pu disappearance from the lagoon. Either the residence time for the lagoon water is substantially longer than 30 d, or the size of the active sedimentary reservoir extends deeper than 2.5 cm. Using an exchange rate of one month at Enewetak and the quantity of Pu contained in the reservoir to a depth of 16 cm, the mean and the half-residence times of $^{239+240}$Pu are 103 and 71 y, respectively, and the mean and half-residence times for ^{241}Am are, respectively, 400 and 277 y. For this example, the measurements of mean lagoon concentrations presented in this chapter lack the degree of precision necessary to distinguish whether this rate is reasonable.

Work to resolve the correct rates of Pu and Am loss from both lagoons is currently a high-priority program. The rate of Pu disappearance from the sediments could be much more rapid than previously estimated.

14.4. CONCLUSIONS

Little reliable information is available to predict the fate of transuranic elements in the deep sea. Only from the accumulation and comparison of results from different sources in the environment can a clearer understanding of the behavior of transuranic elements in the ocean be developed and universal characteristics identified.

Strong association both of $^{239+240}$Pu and of ^{241}Am is found with the sediments, but we found that $^{239+240}$Pu is less firmly bound and is capable of more rapid mobilization from the sediments at Bikini and Enewetak atolls. Small amounts of ^{241}Am are also capable of dissociating from marine sediments. The partitioning of ^{241}Am between sediments and water is controlled by the law of mass action, and the mechanism can be considered reversible. The amount of $^{239+240}$Pu mobilized into solution at the atolls can be reasonably predicted using a K_d of approximately 2.3×10^5 and the mean concentrations in the sediment. This value also provides reasonable estimates of the quantities mobilized into solution at Rongelap Atoll, Johnston Atoll, and from sedimentary deposits at San Clemente Island, California (Noshkin et al., 1981a).

The mobilized $^{239+240}$Pu at Enewetak and Bikini has solute-like characteristics, and different valence states coexist in solution. The largest fraction of the soluble Pu is in an oxidized form (V or VI). Quantities associated with suspended particulate material and sediments are predominantly in the reduced state (III or IV). The sorption–desorption process is not completely reversible because of changes that occur in the relative amounts of the mixed oxidation states in solution with time. The oxidized forms of $^{239+240}$Pu in solution have a lesser tendency to associate with sedimentary or particulate material than does reduced Pu. Complexation after mobilization also affects the resorption rate. Therefore, any characteristics of $^{239+240}$Pu described at a point of reference may not necessarily be relevant in explaining behavior after mobilization and migration into solution.

Water profiles to a depth of 1000 m in the equatorial Pacific show that ^{241}Am:$^{239+240}$Pu ratios are more than two times higher in the particulate phase than in the soluble phase. This separation is mediated by the higher solubility of Pu relative to Am. Since ^{241}Am and $^{239+240}$Pu associated with particulate matter should move vertically in the water column more rapidly than would the species in solution, the rate of vertical transport of ^{241}Am to deep bottom sediments can be assumed to be more rapid than that for $^{239+240}$Pu. As the lagoon studies show, once in the sediment, ^{241}Am will be remobilized into solution at a slower rate than will $^{239+240}$Pu. Some fraction of $^{239+240}$Pu placed on the surface of the seafloor should, in time, disperse into the overlying water mass and migrate from its original site. The ^{241}Am should remain more firmly fixed to the sedimentary material near the point of introduction. The rate of disappearance of the two radionuclides will depend on the physical, biological, and chemical characteristics of the sedimentary deposits and the rate of water movement into and out of the contaminated region. The relative amounts of ^{241}Am to $^{239+240}$Pu in the sedimentary deposits at Enewetak and Bikini will be altered in the future as a result of Pu mobilization and the radiological decay of ^{241}Am.

It is not yet possible to make long-term predictions of the relative quantities remaining in the atolls' sediments. These predictions rely on accurately knowing the size of the sedimentary reservoir capable of continuously supplying plutonium to the overlying water and the mean residence time for the dissolved Pu in the lagoon's water mass, and these parameters are not yet well enough known.

ACKNOWLEDGMENTS

The authors thank Kenneth Marsh, Gale Holladay, and Jack McNabb of Lawrence Livermore National Laboratory for their assistance during phases of the field programs. This work was performed under the auspices of the U.S. Department of Energy, Contract W-7405-Eng-48.

REFERENCES

Atkinson, M., S. V. Smith, and E. D. Stroup. 1981. Circulation in Enewetak Atoll Lagoon. *Limnology and Oceanography*, **26**, 1074–1083.

Beasley, T. M., and F. A. Cross. 1980. A review of biokinetic and biological transport of transuranic radionuclides in the marine environment. *In*: Transuranic Elements in the Environment, W. C. Hanson (Ed.). DOE/TIC-22800, U.S. Department of Energy, Washington, D.C., pp. 524–540.

Bojanowski, R., H. D. Livingston, D. L. Schneider, and D. R. Mann. 1975. A procedure for analysis of americium in marine environmental samples. *In*: Reference Methods of Marine Radioactivity Studies II. Technical Reports Series, No. 169, International Atomic Energy Agency, Vienna, pp. 77–86.

Carpenter, R., and T. M. Beasley. 1981. Plutonium and americium in anoxic marine sediments: evidence against mobilization. *Geochimica et Cosmochimica Acta*, **45**, 1917–1930.

Edgington, D. N. 1981. A review of the persistence of long-lived radionuclides in the marine environment—sediment/water interactions. *In*: Impacts of Radionuclide Releases into the Marine Environment. Proceedings of Symposium (6–10 October 1980), Vienna. International Atomic Energy Agency, Vienna, pp. 67–91.

Lovett, M. B., and D. M. Nelson. 1981. Determination of some oxidation states of plutonium in seawater and associated particulate matter. *In*: Techniques for Identifying Transuranic Speciation in Aquatic Environments. International Atomic Energy Agency, Vienna, pp. 27–35.

Nelson, D. M., and M. B. Lovett. 1981. Measurements of the oxidation state and concentration of plutonium in interstitial waters of the Irish Sea. *In*: Impacts of Radionuclide Releases into the Marine Environment. Proceedings of Symposium (6–10 October 1980), Vienna. International Atomic Energy Agency, Vienna, pp. 105–118.

Nelson, D. M., and D. N. Metta. 1983. The flux of plutonium from sediments to water. *In*: Lake Michigan in Radiological and Environmental Research Division Annual Report, Part 3. ANL-8R-65, Argonne National Laboratory, Argonne, Illinois, pp. 31–36.

Noshkin, V. E. 1980. Transuranium radionuclides in components of the benthic environment of Enewetak Atoll. *In*: Transuranic Elements in the Environment, W. C. Hanson (Ed.). DOE/TIC-22800, U.S. Department of Energy, Washington, D.C., pp. 578–601.

Noshkin, V. E., and K. M. Wong. 1980. Plutonium mobilization from sedimentary sources to solution in the marine environment. *In*: Marine Radioecology. Proceedings of 3rd Nuclear Agency Seminar (1–5 October 1979), Tokyo. Nuclear Energy Agency/Organisation for Economic Co-operation and Development, Paris, pp. 165–178.

Noshkin, V. E., K. M. Wong, R. J. Eagle, and C. Gatrousis. 1974. Transuranics at Pacific Atolls. 1: Concentrations in the Waters at Enewetak and Bikini. UCRL-51612, Lawrence Livermore National Laboratory, Livermore, California, 30 pp.

Noshkin, V. E., K. M. Wong, T. A. Jokela, R. J. Eagle, and J. L. Brunk. 1978. Radionuclides in the Marine Environment near the Farallon Islands. UCRL-52381, Lawrence Livermore National Laboratory, Livermore, California, 17 pp.

Noshkin, V.E., J. L. Brunk, T. A. Jokela, and K. M. Wong. 1981a. ^{238}Pu concentrations in the marine environment at San Clemente Island, *Health Physics*, **40**, 643–659.

Noshkin, V. E., R. J. Eagle, K. M. Wong, and T. A. Jokela. 1981b. Transuranic concentrations in reef and pelagic fish from the Marshall Islands. *In*: Impacts of Radionuclide Releases into the Marine Environment. Proceedings of Symposium (6–10 October 1980), Vienna. International Atomic Energy Agency, Vienna, pp. 293–317.

Pentreath, R. J., D. F. Jefferies, M. B. Lovett, and D. M. Nelson. 1980. The behavior of transuranic and other long-lived radionuclides in the Irish Sea and its relevance to the deep sea disposal of radioactive wastes. *In*: Marine Radioecology. Proceedings of 3rd Nuclear Energy Agency Seminar (1–5 October 1979), Tokyo. Nuclear Energy Agency/Organisation for Economic Cooperation and Development, Paris, pp. 203–221.

Schell, W. R., F. G. Lowman, and E. P. Marshall. 1980. Geochemistry of transuranic elements at Bikini Atoll. *In*: Transuranic Elements in the Environment, W. C. Hanson (Ed.). DOE/TIC-22800, U.S. Department of Energy, Washington, D.C., pp. 541–577.

von Arx, W. W. 1948. The circulation systems of Bikini and Rongelap Lagoons. *Transactions of the American Geophysical Union*, **29**, 861–870.

Wong, K. M. 1971. Radiochemical determinations of plutonium in seawater, sediments and marine organisms. *Analytica Chimica Acta*, **56**, 355–364.

Wong, K. M., G. Brown, and V. Noshkin. 1978. A rapid procedure for Pu separation in large volumes of fresh and saline water by manganese dioxide coprecipitation. *Journal of Radioanalytical Chemistry*, **42**, 7–15.

Wong, K. M., T. A. Jokela, and V. E. Noshkin. 1980. Problems associated with transuranium determination of suspended solids in seawater samples. *In*: Radioelement Analysis Progress and Problems, W. S. Lyon (Ed.). Ann Arbor Press, Ann Arbor, Michigan, pp. 207–214.

Wong, K. M., J. L. Brunk, and T. A. Jokela. 1985. Radiochemical Procedures Used in the Environmental Sciences Division, LLNL, for the Analysis of Selected Radionuclides in the Marine Environment. Lawrence Livermore National Laboratory, Livermore, California, in press.

Chapter 15

The Distribution of ^{239,240}Pu, ^{137}Cs, and ^{55}Fe in Continental Margin Sediments: Relation to Sedimentary Redox Environment

F. L. Sayles and Hugh D. Livingston

Department of Chemistry
Woods Hole Oceanographic Institution
Woods Hole, Massachusetts

ABSTRACT

A series of sediments from the east and west coasts of the United States have been analyzed for fallout nuclides in conjunction with redox-sensitive parameters in the pore waters of the sediments. Depth distributions show a correlation with redox conditions. In particular, depths of penetration of Pu, ^{137}Cs, and ^{55}Fe were greater under more oxidized conditions than in reduced sediment. Modelling of distributions indicates that in a number of sediments dispersion of nuclides at depth is enhanced above that occurring in the uppermost 2–5 cm of sediment. Comparison of ^{239,240}Pu with ^{210}Pb indicates that migration of Pu independent of particle mixing occurs. Modelling of distributions indicates that the occurrence of enhanced migration at depths below 2–5 cm may be widespread. A simple model of the migration of a small soluble fraction of the sedimenting nuclides produces distributions consistent with those observed in the sediments.

15.1. INTRODUCTION

15.1.1. Rationale

Increased utilization of nuclear technologies has brought disposal of radioactive wastes to the forefront of the disposal problems of many industrialized societies. The ocean, in particular the deep-sea floor, is being increasingly considered as a repository for these materials (Hollister et al., 1981). There are many inadequately defined aspects of the behavior of nuclear materials in marine systems. Among these aspects is the physical chemistry of a variety of nuclides in marine sedimentary environments. In the case of many nuclides, effective isolation on or in the seafloor is predicated largely on the presence of irreversible particle–nuclide interactions that trap the nuclides on solid surfaces in the sediments. Present information on the physical chemistry, and particularly the stability of these associations in the range of sedimentary environments that may be encountered, is too limited for rational decision-making with regard to the suitability of the seafloor for disposal.

Any understanding of the behavior of radioisotopes in marine sediment should include a knowledge of the sedimentary environment as it influences the physicochemical behavior of the nuclides of interest. Environmental factors that can influence the chemical behavior of nuclides include the occurrence and specific surface area of detrital oxides, the pH of the pore-water solutions, the quantity and character of organic matter (particulate and dissolved), the anionic composition of the pore water, and the general redox state. This last factor is perhaps the single most important variable. Redox potential influences the valence of a number of important nuclides (e.g., Pu, I, and Tc), strongly affects the amount and character of oxides present in the sediment, is related to anion composition and pH through diagenetic reactions, and probably influences the ligand-binding characteristics of the organic matter present. Numerous studies of radioisotope distributions have made reference to the general redox state of marine sediments in largely qualitative terms such as "reduced," "anoxic," or "oxidized." [For more information, see Sholkovitz (1983).] However, direct attempts to link nuclide distributions to measured redox conditions in marine sediments are virtually absent from the literature. Thus, while there are many reports describing the distribution of a number of nuclides, there is little direct information on the geochemical processes and kinetics that may influence these distributions, a point emphasized by Sholkovitz (1983). Advances in understanding the influence of biogeochemical processes on the distribution of radioactive waste components in marine sediments must include characterization of important redox and chemical parameters.

The research described in this chapter has focused on a characterization of the chemical environment (in-

cluding redox parameters) of sediments in which the distributions of several nuclides were also studied. The studies have included analyses of pore waters, oxide components of the solid phases, and the radioisotopes ^{239,240}Pu, ^{137}Cs, ^{55}Fe, and ^{210}Pb. These nuclides are widely distributed in the marine environment, thus providing an opportunity to study environmental variability under natural conditions (i.e., those that wastes would experience).

One important point should be emphasized in regard to the utilization of fallout-nuclide distributions in the context proposed: the time scales of important processes. Fallout nuclides were first introduced into the environment less than 40 y ago. By contrast, the half-lives of a number of important components of nuclear waste (e.g., ^{237}Np, ^{239}Pu, ^{129}I, and ^{99}Tc) are exceedingly long (10^4–10^7 y). In predicting the fate of such nuclides in the marine environment, it is crucial to identify and characterize not only the processes that are dominant now, but also those of lesser importance and the influence of time on their relative importance.

15.1.2. The Redox Characterization of Marine Sediments

The redox environment of sediments is largely governed by the biologically mediated breakdown of organic carbon. These reactions have been shown to proceed more or less in a stepwise fashion that reflects a decreasing yield of free energy (Bender et al., 1977). A given reaction proceeds until the oxidant is consumed, and the redox potential drops sufficiently to favor the next most-efficient oxidant. As demonstrated by Froelich et al. (1979), the behavior of the reactants (oxidants) and products (reduced species) can be utilized to identify a depth sequence of successively more reduced zones in sediments.

The sequence of redox zones present in the sediments can be most accurately defined on the basis of analysis of the pore waters of the sediments. The pore waters are a far more sensitive indicator of reactions and have a far faster response time than do the solid phases. Consequently, the distributions of the species in the pore waters reflect current conditions (a period of one to a few years) and can be modelled more readily than can the solids. Redox zonation is based on the sequence of oxidants (Froelich et al., 1979): O_2, Mn_s^{4+}, NO_3^-, Fe_s^{3+}, SO_4^{2-}, where the subscript s denotes a solid oxide or solid oxyhydroxide. The zonation as a function of depth is defined on the basis of the assumption of a steady state and the quantity d^2C/dx^2, where C is a soluble product or reactant involved in one of the redox reactions and x is depth (Froelich et al., 1979; Section 15.4).

15.2. METHODS

15.2.1. Sampling and Analyses

Interstitial water samples were collected by *in situ* sampling techniques (Sayles et al., 1976). The sampler employed in these studies was substantially different from that described by Sayles et al. (1976). Samples from the upper 15 cm of sediment were collected from individual probes hydraulically emplaced in the sediment from the base of a tripod frame. This feature permitted location relative to the interface with an accuracy of ± 0.5 cm. Pore-water samples from depths of 30–80 cm were collected from a single "long probe" with an annular sample port. The true depths from these ports depends on the settling of the tripod frame into the sediments and are not accurate to better than 5 cm. Solutions drawn from the sediments through 0.45-μm filters were isolated in Ti cylinders for extraction from the instrument and analysis on board the ship. The sampler also collected bottom-water samples and a core 6.5 cm or 15 cm in diameter for comparison of sediment properties with pore-water profiles.

Sediments were collected by several different types of corers. For the samples taken on the R/V *Oceanus* cruise, a 21-cm sphincter core, as described by Burke (1968) and subsequently modified for use in a tripod frame, was used to obtain sediment for the radiochemical analyses. The 6.5-cm core from the *in situ* pore-water sampler [Woods Hole Interstitial Marine Probe (WHIMP)] was used for the determination of solid-phase chemistry and sediment properties related to pore-water chemistry. The two cores were, of necessity, collected on separate lowerings with the ship being within approximately 500 m (as determined by LORAN-C navigation) for the two lowerings. For the samples taken on the R/V *Wecoma* cruise, the WHIMP was equipped with a 15-cm-diameter corer; all sediment analyses were made on this material and were spatially contiguous with the pore-water samples.

Sediment from the cores was sectioned at 1-cm intervals to a depth of 15 cm and 2-cm intervals below a depth of 15 cm. The outer 1–2 cm was discarded to avoid the mixed sediment near the wall of the corer. Material for analyses was sealed in airtight jars for later water-content determination and analysis. These analyses utilized standard procedures described in the literature and summarized in Table 15.1.

15.2.2. Cruises

Samples were collected primarily on two cruises. The sampling strategy was designed to collect sediment and pore water from as wide a variety of redox environments as possible on each of the cruises. Samples were collected in September 1980 on cruise 86 of the R/V *Oceanus* (designated Oc86) and in September 1981 on cruise 81 of the R/V *Wecoma* (designated W81). The *Oceanus* cruise was a traverse of the continental margin of the eastern coast of the United States and included stations at abyssal depths in the North American Basin (Fig. 15.1a). The *Wecoma* cruise was a transect of the continental margin of the western coast of the United States, also including abyssal depths (Fig. 15.1b). Both transects provided an opportunity to investigate a variety of redox conditions on a fairly limited spatial scale. The transects, in a general sense, sampled a redox gradient that progressed from more oxidized at the greatest water depth to most reduced in more shallow areas. To complement the data on pelagic, oxidized sediments, we have included analyses of three cores from a third cruise, Endeavor 53, which took place during July and August 1980. The station locations for this cruise are included in Fig. 15.1a.

15.3. RESULTS

15.3.1. Redox Conditions

15.3.1a. Pore-water data and present conditions

The progressive utilization of oxidants noted above can be readily seen in the pore-water profiles both of the *Oceanus* and the *Wecoma* cruises (Fig. 15.2). The sequence of reactions was basically the same as that presented in several papers (Bender et al., 1977; Froelich et al., 1979; Emerson et al., 1980) describing the diagenesis of organic carbon. In the uppermost sediments, consumption of O_2 produced NO_3^-, as seen in the well-developed NO_3^- concentration at stations Oc86C, Oc86D, Oc86E, and Oc86G. On the *Wecoma* cruise only station W81C, the deepest, exhibits a maximum, which is unusually weak but rather deep. It is generally believed that biota will utilize NO_3^- as an oxidant only after depletions of O_2 to approximately 2 μM (Cline and Richards, 1972; Froelich et al., 1979). Thus, the onset of NO_3^- utilization may be interpreted as the bottom of the zone containing molecular O_2.

Similarly, Mn^{2+} is oxidized by O_2, and depletion of Mn^{2+} enrichments to ~ 0 at the top of the Mn^{2+} maximum also provided a measure of the bottom of the O_2 zone. With the exception of the more oxidized stations (Oc86C, Oc86D, and W81C), both measures of O_2 exhaustion yielded similar results. In practice, it is often difficult to unambiguously pick the inflection point, $d^2(NO_3^-)/dx^2 = 0$, in the NO_3^- curve. As a con-

sequence, the point at which Mn^{2+} reaches zero will usually be used in this chapter to define the bottom of the uppermost (O_2) zone. At the more oxidized stations (Oc86C and Oc86D), the NO_3^- inflection point was considerably shallower than was the depth at which Mn^{2+} reached zero; here, the bottom of the oxygen-containing zone is defined on the basis of the NO_3^- curve. At station W81C, also considered to be oxidized,

Table 15.1. Summary of Analytical Methods

Factor to be Analyzed	Method	Reference
Sediment Analysis		
Radiochemical components		
^{239,240}Pu	8 *M* nitric acid leach, anion exchange separation followed by α-spectrometry	Livingston et al., 1975
^{137}Cs	8 *M* nitric acid leach, ammonium molybdophosphate collection, cation exchange separation, chloroplatinate precipitation, and anti-coincidence beta counting	Wong et al., 1970
^{55}Fe	8 *M* nitric acid leach, anion exchange separation followed by anti-coincidence X-ray spectrometry	Labeyrie et al., 1975
^{210}Pb	Dissolution in concentrated nitric, perchloric, and hydrofluoric acids; spontaneous deposition of ^{210}Po on silver discs; and α-spectrometry, correction for supported ^{210}Pb by ^{226}Ra (^{222}Rn) measurement	—
Stable components		
Reducible MnO_2	Reducible MnO_2 : hydroxylamine hydrochloride–acetic acid leach	Chester and Hughes, 1967
$CaCO_3$	Manometric analysis; phosphoric acid	—
Porosity	Salt corrected water loss with grain density determined by He pycnometer	—
Pore-Water Analysis		
NO_3^- , NO_2^-	Technicon® Auto-Analyzer® techniques	Atlas et al., 1971
Mn^{2+}, Fe^{2+}	Direct injection flameless atomic absorption spectroscopy with ascorbic acid	Hydes, 1980
SO_4^{2-}	Anion difference chromatography	Sayles and Mangelsdorf, 1976
Alkalinity	Automated Gran Titration	—

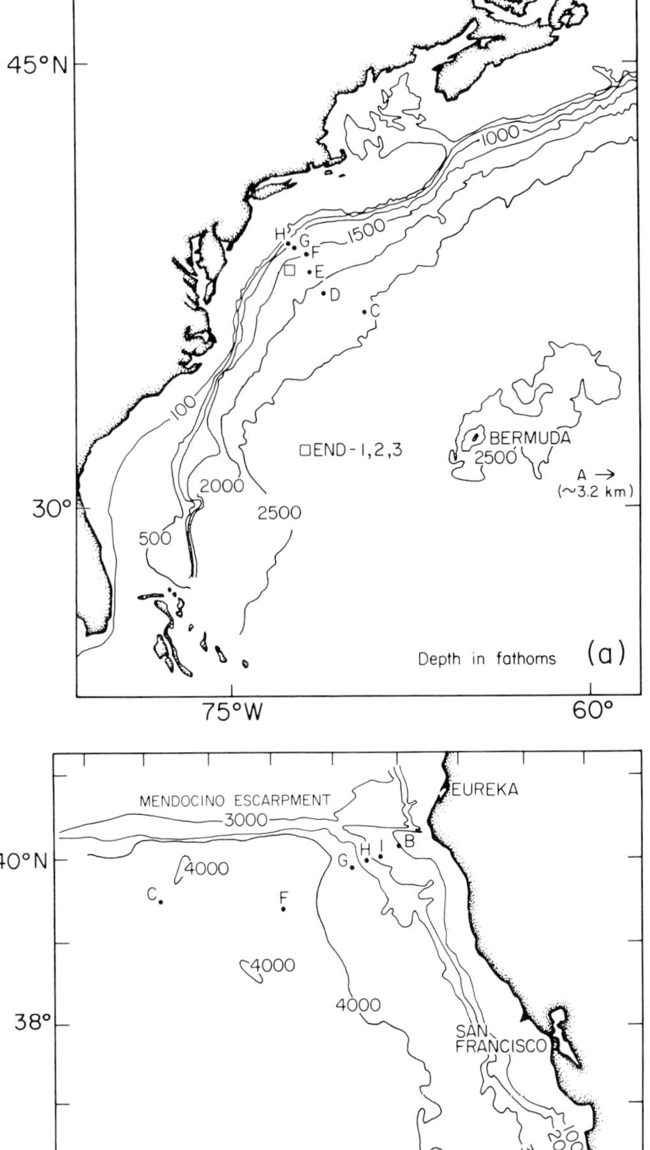

Figure 15.1. Location maps of sampling sites for sediment cores and pore-water samples. (a) *Oceanus* cruise (September 1980) and *Endeavor* cruise (July and August 1980). One fathom = 1.8 m. (b) *Wecoma* cruise (September 1981).

the point at which Mn^{2+} reached zero was well above the NO_3^- inflection point and has been taken as the bottom of the oxygen-containing zone.

It is noteworthy that at virtually all of the stations NO_3^- reduction commenced before MnO_2 reduction began. This is most readily seen in the more oxidized sediments of stations Oc86C, Oc86D, and W81C, where the redox zonation was spread out far more in a spatial sense. At two of these stations, NO_3^- was essentially reduced to blank values prior to the onset of MnO_2 reduction. Most free-energy models predict the onset of MnO_2 reduction at or before that of NO_3^-.

When Mn^{4+} reduction was observed in the pore-water profiles, increases in the Mn^{2+} concentration were rapid. Reduction was observed within the sampled depths at all stations except Oc86C, Oc86D, and W81C (Fig. 15.2.). At station W81C, Mn^{2+} enrichments were seen, but they reflected reduction at depth and diffusion upward to the sampled sediment. At station Oc86D, the Mn^{2+} increase was gradual and also showed extensive diffusive transport. The sharp rise in Mn^{2+} makes it possible to place the onset of Mn^{2+} reduction quite accurately (± 1–2 cm), either by the occurrence of the Mn^{2+} maximum or, as in the case of the sample from station Oc86E, by the observed curvature in the profile $d^2(Mn^{2+})/dx^2 > 0$.

Conditions under which Fe was reduced to the ferrous state were far less common than were those causing Mn reduction. On the *Oceanus* cruise, unambiguous Fe^{2+} enrichment occurred only at station Oc86G. The Fe^{2+} profile appeared to be largely diffusive, with the possibility of some reduction in the interval of 12–15 cm. Alternatively, the whole profile may be diffusive, a hypothesis that we favor because Fe contamination of pore-water samples was found sporadically. We also believe that contamination was the source of the Fe^{2+} found at station Oc86F, for we would not expect reduction in NO_3^--containing solutions, as was the case at this site. Some Fe^{2+} enrichment occurred at three of the *Wecoma* stations. At stations W81G and W81H, reduction occurred at 10–15 cm, and detectable amounts of Fe^{2+} were found at depths greater than 4–6 cm. At station W81I, reduction occurred at $\sim$12 cm, but Fe^{2+} was limited to depths below 10–12 cm. Conditions sufficient for Fe^{3+} reduction did not occur either at station W81C or at station W81F.

The final step in the oxidation sequence considered in this chapter is SO_4^{2-} reduction. While SO_4^{2-} depletions were not uncommon, reduction of SO_4^{2-} in the sampled interval was relatively rare. The SO_4^{2-} data were plotted in Fig. 15.2 only when depletions were greater than 0.3 meq liter^{-1} ($\sim$0.5%). With two excep-

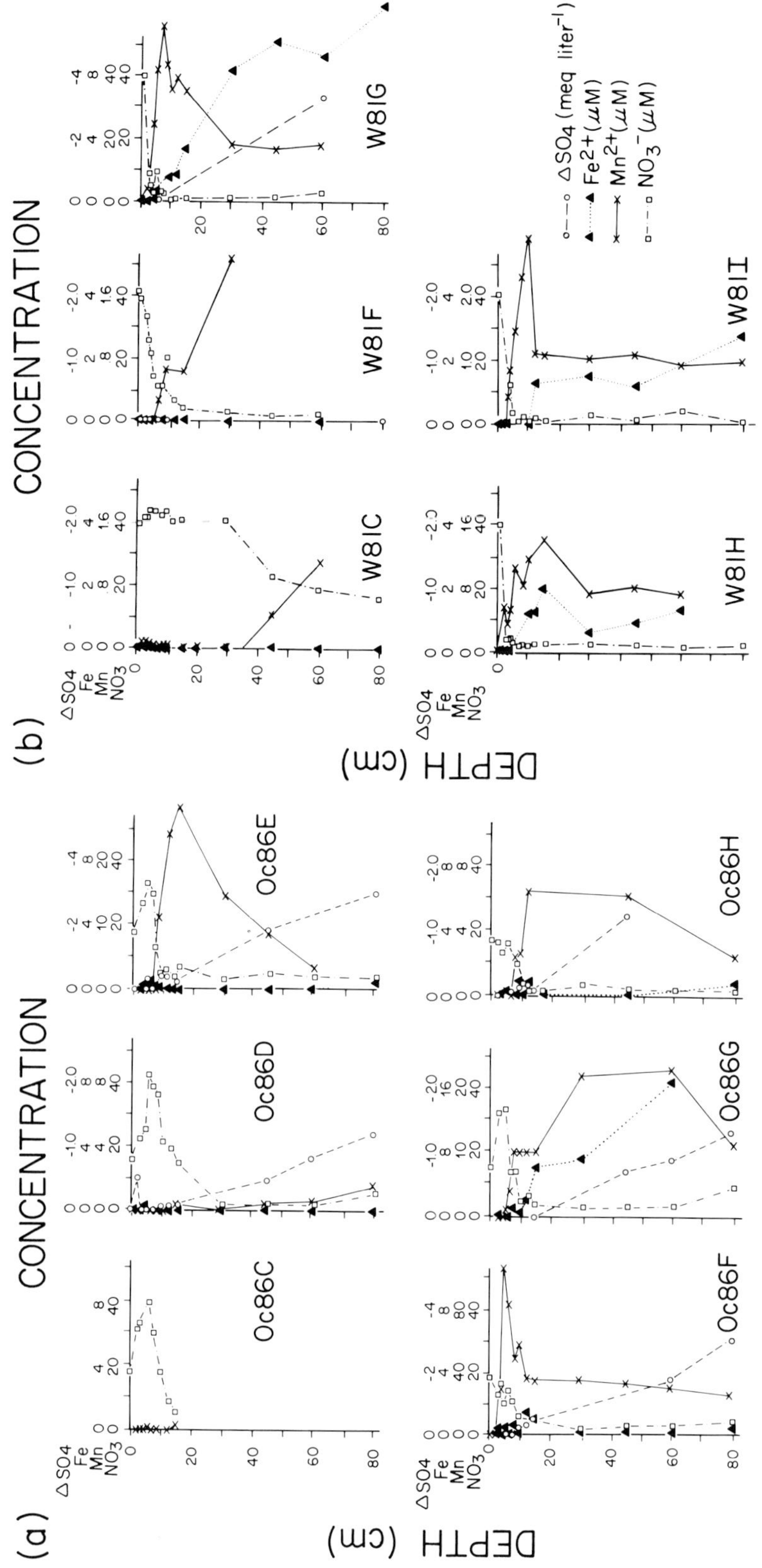

Figure 15.2. Concentration–depth distributions of the oxidants studied in the pore waters of the sediments at stations of the (a) *Oceanus* and (b) *Wecoma* cruises.

tions, the gradients were zero (data not plotted) or linear, requiring the absence of reduction in the sampled interval. Curvature is present in the Oc86F profile and possibly in the Oc86H profile. At both of these stations reduction was limited to depths greater than 15 cm and cannot be accurately placed with the data set available. The depth of the onset of SO_4^{2-} reduction at all stations eliminated it as a major factor in influencing radionuclide distribution in these studies.

From the distribution of various oxidants (electron acceptors) we have defined a set of redox horizons for each of the stations studied (Table 15.2). We have defined specific depths as opposed to zones, in contrast to the approach taken by Froelich et al. (1979). We have done so because it is our purpose to relate radionuclide distributions to species in solution as indicators of chemical environment (dominated by redox reactions) rather than to the reactions occurring within zones. There is often considerable overlap in reaction zones. The depths of the horizons at the various stations, where they can be assigned, are given in Table 15.3.

15.3.1b. Solid-phase chemistry and past redox conditions

The conditions revealed by the pore-water analyses reflect conditons at the time of sampling. The time

Table 15.2. Definition of Redox Horizons[a]

Horizon	General Characteristics/Reactions/Definition
0	The interface
1	Bottom of the zone containing O_2. Defined on the basis of $\dfrac{d^2(NO_3^-)}{dx^2} = 0$ (NO_3^- profile inflection) with $\dfrac{d(NO_3^-)}{dx} < 0$ or $Mn^{2+} \to 0$ above the Mn^{2+} maximum, whichever is shallower.
2	Onset of Mn^{4+} reduction: defined as $\dfrac{d^2(Mn^{2+})}{dx^2} > 0$ at depths above the Mn^{2+} maximum, or the Mn^{2+} maximum if the gradient is very sharp.
3	Occurrence of Fe^{2+} enrichment: detectable Fe^{2+} in pore solutions. The only implication is that conditions sufficient to oxidize Fe^{2+} faster than supply are not met.
4	Onset of Fe^{3+} reduction: $\dfrac{d^2(Fe^{2+})}{dx^2} < 0$

[a] Depth (x) is taken as downwards for the sign of $\dfrac{d^2(C_1)}{dx^2}$.

Table 15.3. Depth (cm) of Redox Horizons

	Oceanus 86 Station							Wecoma 81 Station				
Horizon	A[a]	C[b]	D	E	F	G	H	C	F	G	H	I
1	—	8	8	6	3	5	6	~35	5	3	< 2	25
2	—	> 15	> 80	9	4	8	12	> 60	30	5	6	4
3	—	> 15	> 80	~80	~80	8	~80	> 80	> 80	4	5	11
4	—	> 15	> 80	> 80	> 80	> 80	> 80	> 80	> 80	5–30	12–15	12

[a] No pore-water data collected.

[b] No pore-water samples below 15 cm were collected.

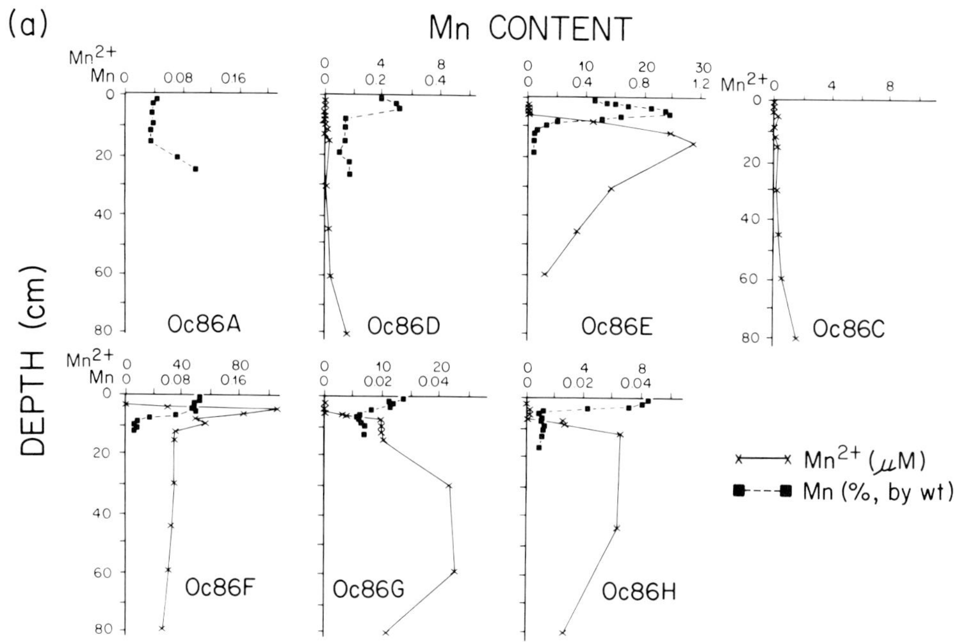

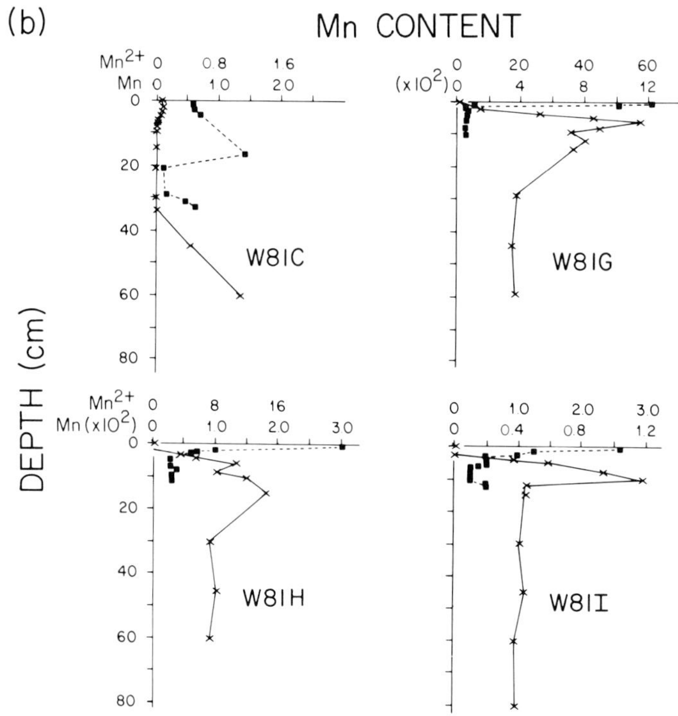

Figure 15.3. Comparisons of depth distributions of Mn^{2+} in pore water and Mn in a solid, reducible phase taken to be a measure of Mn oxyhydroxide content in the (a) *Oceanus* and (b) *Wecoma* cruises.

scales for adjustment of pore-water profiles to changes in sediment-interface conditions to a depth of 25 cm is on the order of a few years. This is short relative to the time radionuclides have been present in the sediments. As a consequence, it is necessary to determine whether the present conditions reflect the long term. This can be done by comparison of the pore-water and the solid-phase profiles for a pore-water component showing evidence of reaction. The most obvious component for such a comparison is Mn^{2+}. It exhibited sharp gradients that should have been reflected in changes in solid-phase Mn if conditions have persisted for periods of time that are long relative to the time since the radionuclides were introduced. The absence of MnO_2 enrichment where precipitation was indicated implies recent (relatively) changes in redox conditions.

The profiles of Mn^{2+} and reducible solid Mn are shown in Fig. 15.3. At nearly all of the stations, the precipitation of Mn^{2+} was reflected in a strong buildup of reducible solid Mn. The spreading of the peaks was due to bioturbation. At two stations (W81C and Oc86D), enrichments of Mn in the solid phase could not reflect current conditions and are therefore relict features. At station W81C, the upper zone of relict MnO_2 was underlain by a second zone of enrichment that appears to reflect the current conditions. At station Oc86D, the zone of enriched MnO_2 must have reflected preexisting conditions during which Mn reduction occurred at a depth of about 8 cm. The existing pore-water profile would produce only weak precipitation at ~60 cm. Calculations of fluxes and depth-integrated MnO_2 enrichment at the stations studied demonstrated that the time scales to establish the observed enrichments were much longer than that since the introduction of radionuclides ($\gg 10^2$ y). Consequently, at all stations but Oc86D the horizons given in Table 15.3 accurately portray the redox depth profiles under which the radionuclide distributions were achieved. At station Oc86D, the case is open since no time scale for the persistence of present conditions can be established.

15.3.1c. Distribution of radionuclides: ^{239,240}Pu, ^{137}Cs, and ^{55}Fe

Depth distributions of the three nuclides, ^{239,240}Pu, ^{137}Cs, and ^{55}Fe, exhibited general patterns reported previously. Most of the profiles for Pu from the samples taken during the *Oceanus* cruise are relatively smooth, exhibiting a roughly exponential decrease with depth (Fig. 15.4a). A slight perturbation of this was seen at station Oc86A and stronger departures at stations Oc86C, Oc86H, and Oc86F. Penetration of Pu was greatest at the more oxidized stations (Oc86A, Oc86D, and Oc86E). Distributions in the sediments from the *Wecoma* cruise were qualitatively different in that concentrations at two of the four stations did not display a maximum at the surface; instead, a subsurface maximum was seen at station W81G, and constant values persisted to a depth of 2–3 cm at station W81H. Overall penetration depths were less than in the *Oceanus* sediments. Distributions of ^{55}Fe were generally similar to that of ^{239,240}Pu. Station W81G, with ^{55}Fe penetration deeper than that of ^{239,240}Pu, appeared to be an exception; however, spectral interference was responsible for high background counts of ^{55}Fe, and the actual values may well be zero. At stations W81C and Oc86D, Pu penetration was deeper. The behavior of ^{137}Cs was generally similar to that of ^{239,240}Pu. Penetration of Cs appeared to be deeper at stations Oc86H and W81I. While Pu penetration appeared to be in excess of that of ^{137}Cs at station W81H, at the low Pu activities found we would not have expected to detect ^{137}Cs.

The gross features of the profiles of the nuclides are largely due to sediment mixing by physical and/or biological processes ("bioturbation") (Glass, 1969; Guinasso and Schink, 1975; Beasley et al., 1982). While we have not measured sedimentation rates from these cores, there is no question that, since the radioisotopes have been introduced, their penetrations have exceeded sediment accumulation (1–5 cm per 10^3 y) by orders of magnitude. The smooth distributions most often seen doubtless reflect both time averaging of discreet mixing events as well as the spatial averaging of our sampling techniques, which permit homogenization of large sediment samples from cores 21 cm in diameter. The sharp perturbations occasionally seen (e.g., at stations Oc86C, Oc86F, and Oc86H) are thought to reflect unusually large inhomogeneities due to the mixing process, involving either slumping or the formation of large, recent burrows (Benninger et al., 1979).

15.3.1d. Relation of fallout-nuclide distributions to redox horizons

At many of the stations, the radionuclides penetrated through several redox zones, as delineated by the horizons defined in Table 15.2 and listed in Table 15.3. The data from Table 15.3 are plotted on the profiles of Pu, ^{137}Cs, and ^{55}Fe (Fig. 15.4). The occurrence of the nuclides in sediments delineated by the various horizons provides an opportunity to investigate the behavior of the isotopes under various redox conditions, the chief objective of this investigation. Behavior under more oxidized conditions should be exhibited by the isotopes

 The Distribution of ^{239,240}Pu, ^{137}Cs, and ^{55}Fe in Continental Margin Sediments

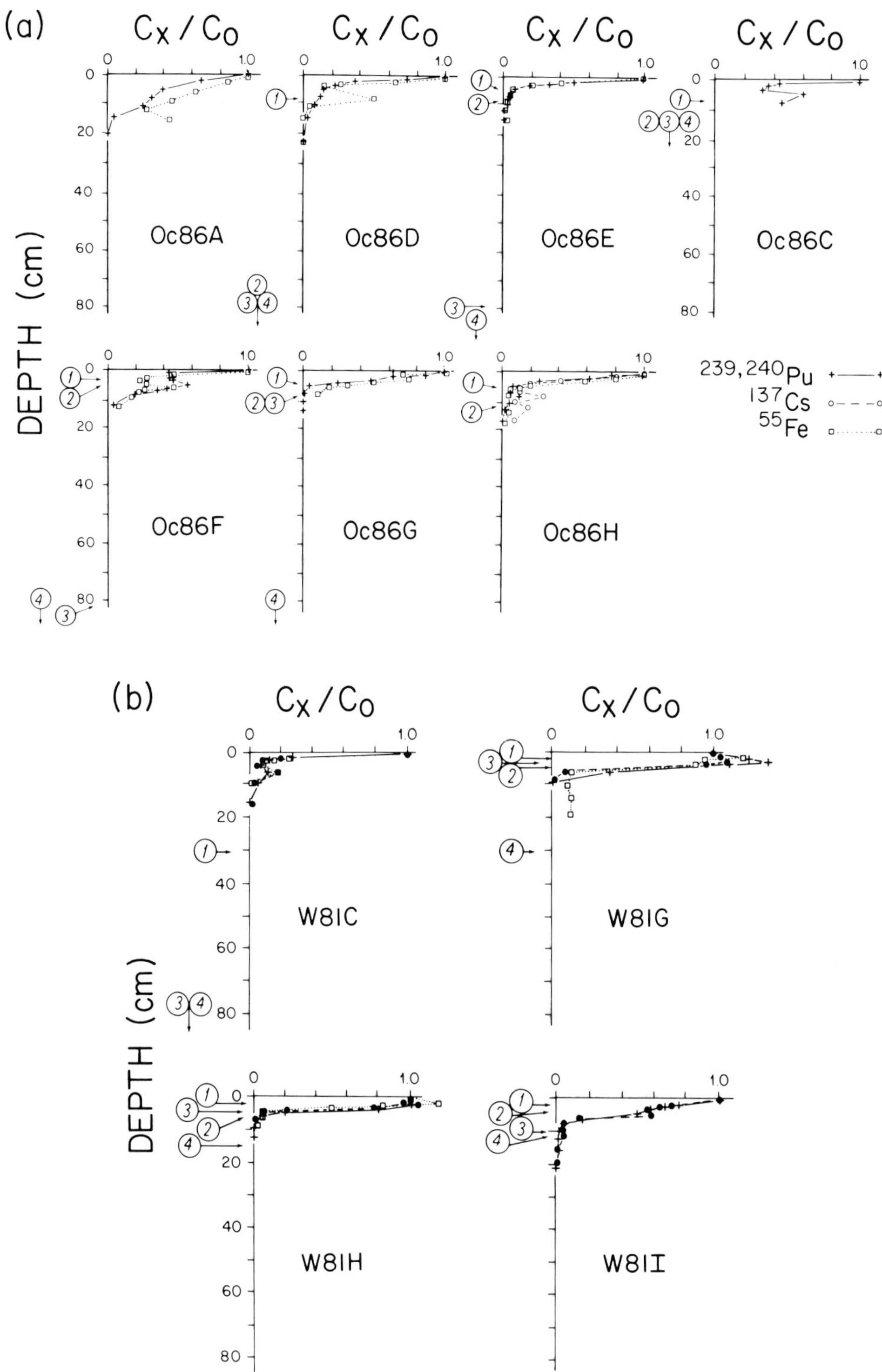

Figure 15.4. Distribution of the radionuclides studied in the sediments sampled on the (a) *Oceanus* and (b) *Wecoma* cruises. Circled numbers indicate the depth of the redox horizons as defined in Table 15.2 and listed in Table 15.3.

at stations Oc86C and Oc86D, and, by analogy, at station Oc86A (for which pore-water data are absent) and at station W81C. Oxygen should have been present throughout the depths at which isotopes were found at these stations, except at Oc86D, where Pu and ^{55}Fe penetrated well below the inferred zero O_2 horizon, but not to the zone containing detectable Mn^{2+}. The isotopes encountered conditions sufficient to reduce Mn^{4+} (horizon 2) at stations Oc86E, Oc86F, and Oc86H, but reduced Fe was not present at the depths penetrated by the radionuclides at these stations. The isotopes reached horizon 3 (where Fe^{2+} was present) but not 4 (where Fe was reduced) only at station Oc86G and possibly at station W81H. The full spectrum of conditions (horizons 1–4) overlapped nuclide penetration depths at station W81I and probably at station Oc86G. At station W81G there appeared to be two zones of Fe reduction, the uppermost possibly reflecting Fe contamination of the pore-water sample.

15.4. PROCESSES AFFECTING RADIOISOTOPE DISTRIBUTIONS

15.4.1. The Utilization of Models to Describe Distributions

As already noted, the gross distribution of the fallout nuclides studied appears to reflect mixing processes. If the isotopes behave entirely as particulate tracers, then the redox environment will have no influence beyond possibly affecting the organisms responsible for the bioturbation. Under these circumstances we would expect the nuclides to mirror organism activity.

Numerous descriptions of particle mixing in deep-sea sediments have been published. The data from these studies form a background for the expectation that nuclide distributions are solely a result of particle mixing. Studies of deep-sea sediment have most commonly assumed a constant mixing coefficient throughout the mixed layer (Goldberg and Koide, 1962; Guinasso and Schink, 1975; Nozaki et al., 1977). Nonconstant mixing rates have also been applied to single-mixed-layer models (Nozaki et al., 1977; Christensen, 1982). Santschi et al. (1980) have described sediment mixing in nearshore sediments as a two-layer process with constant mixing (transfer) coefficients. In deep-sea sediments, mixing coefficients determined by a variety of methods have generally been observed to fall in the range of 10^{-9}–10^{-8} cm^2 s^{-1} (Guinasso and Schink, 1975; Nozaki et al., 1977; Peng et al., 1979); it is noteworthy that the Guinasso and Schink estimates that fall in this range are based on distributions of ^{239,240}Pu. For a majority of the studies

made, a single-mixed-layer model has been applied. Where two mixed layers have been modelled, as was done by Santschi et al. (1980), rates in the deeper layer are lower than in the upper layer by about one order of magnitude.

In attempting to identify processes influencing the distribution of Pu, Cs, and Fe in the sediments sampled, we have applied two contrasting, simple models. A number of more complex and sophisticated models have been applied to the description of depth distributions (e.g., Santschi et al., 1980; Officer and Lynch, 1982). These appropriately utilize time-dependent flux inputs, and have been applied to shallow-water sediment for which the atmospheric fallout function can be used to define the time-dependent input function (Officer and Lynch, 1982). In open-ocean sediments such as those considered in this study, there is virtually no quantitative information on the nature of input functions for the components of global fallout. As a result, we feel that the uncertainties from assuming an input function negate the advantages of this type of model for the analysis of data we have collected. Instead, we attempted to bracket the real conditions by considering both an instantaneous and a continuous input. Sediment-trap data both from the Atlantic and the Pacific (Livingston and Anderson, 1983) demonstrate that present-day fluxes are considerable and argue against the instantaneous-source model. Several of the cores taken on the *Wecoma* cruise, however, exhibited subsurface activity maxima consistent with a decline in the specific activity of sedimenting material. Since both models lead to the same conclusions and clearly illustrate the same qualitative features, we have focused largely on the constant-source model as being more nearly representative of the conditions of input seen today. Such an approach is admittedly an oversimplification and can only be considered as semiquantitative. We believe, however, that despite obvious shortcomings, the model is adequate to demonstrate the qualitative relationships we are seeking to define regarding processes influencing the depth distributions of ^{137}Cs, ^{55}Fe, and, especially, ^{239,240}Pu.

Under conditions of constant supply of sediment with constant activity (Duursma and Hoede, 1967)

$$C_x / C_0 = \sqrt{\pi}\ \mathrm{ierfc}\left[\frac{x}{2\sqrt{Dt}} \right] \qquad\qquad 1$$

where C_0 is the concentration (activity) at the interface, C_x is the concentration at depth x, D is the diffusion coefficient, and t is time. Since C_x/C_0 is known (measured) at any sample depth (as in Fig. 15.4), a value of $f(x)$, or $x/2\sqrt{Dt}$, can be obtained from error function tables; plotting x against $f(x)$ should yield a linear relationship with a slope of $1/2\sqrt{Dt}$.

Data plotted according to the above relationship are presented in Fig. 15.5 for the *Oceanus* stations. Also included are data from three cores collected in the general area of the *Oceanus* cruise transect, about 5° south of station Oc86D in 5400 m of water (Fig. 15.1). In the plots presented, one feature dominates: a break in slope for each of the components plotted at each station. Only ^{55}Fe at station Oc86A did not exhibit this feature. In almost all cases the break in the slope appears to be the result of the intersection of two linear relationships with differing slopes. In some cases the data are scattered, and in one case (Oc86A) the data are distinctly curvilinear rather than appearing as a distinct break in the slope. The data for most stations are, however, sufficiently tightly constrained and suggest that the linear relationships are real and common. The slopes (dy/dx, where x is depth and y is $x/2\sqrt{Dt}$) are consistently smaller at depth than in the near-surface sediments, with station Oc86G being a notable exception in exhibiting an increase in slope at depth. At station Oc86G, dy/dx decreased continually, except at the deepest point. The second persistent feature of note is the lack of separation of the two isotopes $^{239,240}Pu$ and ^{55}Fe along the shallow leg of the plots. There is an apparent separation at depth; however, due to background interferences, we are not certain if this feature is real. In a number of cases the slope for Pu at depth is significantly smaller than is that for ^{55}Fe or ^{137}Cs, but the reverse also occurs.

In several cases the data scatter widely when plotted (Fig. 15.5). Particular examples of this are data from stations Oc86C, Oc86F, and Oc86H, especially at depth. This is a reflection of inversions in the concentration–depth relationship that can also be seen in the C_x/C_0 profiles in Fig. 15.4. We cannot model these inversions with the simple models utilized, so we have focused on stations not exhibiting them because we believe that these features are perturbations of the regular processes that lead to the smooth profiles often seen. We expect that these concentration maxima reflect the occurrence of larger than average mixing events such as slumps or, more likely, burrow infillings that have not yet been homogenized by bioturbation. This then is a filter applied to the data. Whether this interpretation is correct does not affect the interpretation of the ob-

served profiles at the "simpler" stations, but it does arbitrarily reduce the data set.

The observed depth distributions (Fig. 15.5) were qualitatively and quantitatively totally unexpected on the basis of existing concepts of sediment mixing at the water–sediment interface. Existing models define two or three layers in the sediment, with the deepest being unmixed. The upper layer(s) are subject to mixing that is either assumed to be constant over each mixed layer (Santschi et al., 1980) or permitted to vary according to some assumed algorithm (Nozaki et al., 1977; Christensen, 1982). In all cases in which variability of mixing rate is permitted, the mixing rate has been assumed to decrease with increasing depth. This is consistent with the occurrence of biota, which decreases with increasing depth in the sediment. Such models make intuitive sense in that the mixing process starts at the interface and, statistically, must be less frequent at depth. The data in Fig. 15.5 suggest that the rate of dispersion, by whatever means, increased with depth. This follows from the aforementioned model, in which $S^2 = 1/4Dt$ (where S is the slope as plotted, D is the mixing coefficient, and t the time since introduction of the tracer). As a consequence, the decrease in slope observed should reflect an increase in D. Only at station Oc86G was D seen to decrease as previously reported, for instance by Santschi et al. (1980).

If the interpretation of Fig. 15.5 implies enhanced mobility of radionuclides at depths of ~2–5 cm, then determining the origin of this mobility is crucial to an understanding of the chemistry of fallout nuclides in deep oceanic sediments. Presently we can offer three possible explanations of this feature: (1) it is an artifact of the oversimplified input function we have assumed; (2) it is real, and some physical or biological process is more active below the interface than in the upper 2–4 cm; or (3) it is real and reflects a combination of particle mixing and diffusive dispersion of soluble species.

15.4.2. Independent Measures of Particle Mixing: ^{210}Pb

The first two hypotheses mentioned above differ from the third in one important respect: the third hypothesis is predicated on significant dispersion through true solution diffusion. The first two hypotheses, on the other hand, are predicated on dispersal solely by particle mixing, either of a complex input signal or by an undefined depth-dependent mixing process. To test

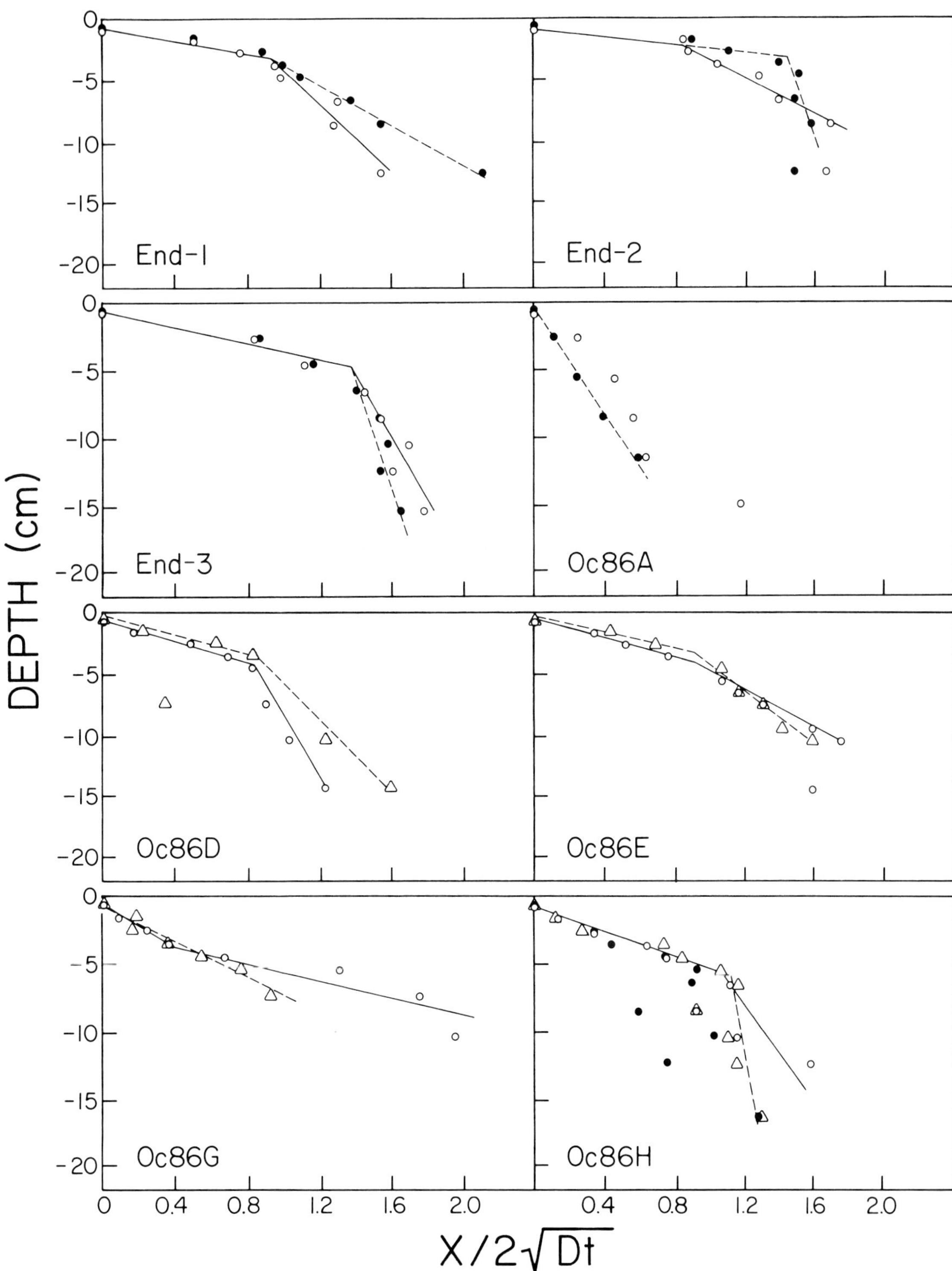

Figure 15.5. Graphic representation of the constant-source model for the *Oceanus* and *Endeavor* stations. Values are for Pu ($\bigcirc$), Cs ($\bullet$), and Fe ($\triangle$). The symbol size is greater than or equal to the sample interval (vertical) and analytical uncertainty of 2 σ (horizontal).

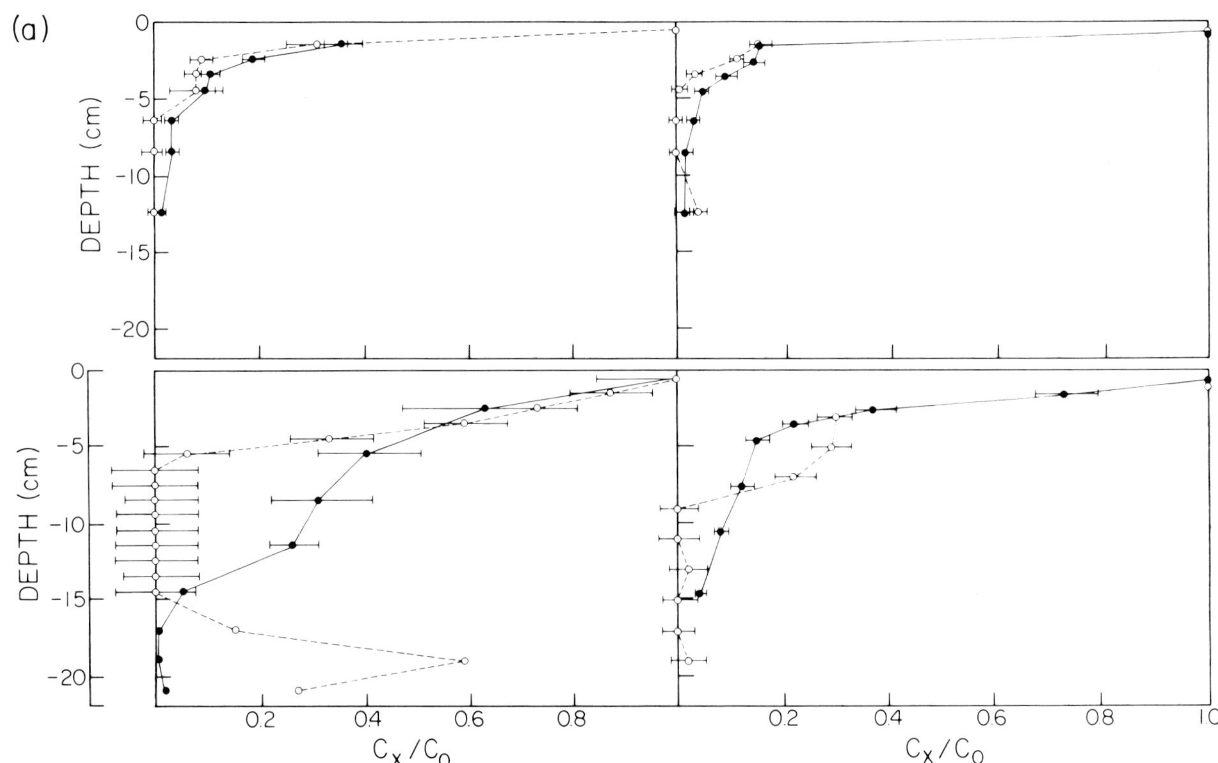

Figure 15.6. (a) Comparison of ^{239,240}Pu and ^{21}Pb distributions with depth in the sediment at two *Endeavor* at two *Oceanus* stations off the eastern coast of the United States. the data are plotted as C_x/C_o with C_o defined as the)- to 1-cm activity. The error bars are 95% confidence intervals.

these two alternatives we sought a particle-mixing indicator that is a steady-state tracer not subject to the vagaries of unknown input. Numerous authors have utilized ^{210}Pb as a tracer of particle mixing in marine sediments from various environments (Nozaki et al., 1977; Benninger et al., 1979; Santschi et al., 1980). The requisite lack of solution mobility is well demonstrated for anoxic sediments [e.g., by Koide et al. (1972)] but is also assumed for oxidized sediment (Robbins, 1978). To provide the needed steady-state particle-tracer information for the *Oceanus* transect, we analyzed several cores for excess ^{210}Pb.

The depth distributions of ^{239,240}Pu and ^{210}Pb are given in Fig. 15.6a for three cores for which ^{210}Pb analyses were made on splits of the same samples utilized for radionuclide analyses. At the remainder of the *Oceanus* stations, ^{210}Pb and Pu analyses were done on cores obtained on separate lowerings. This unfortunately compromises the comparisons for stations

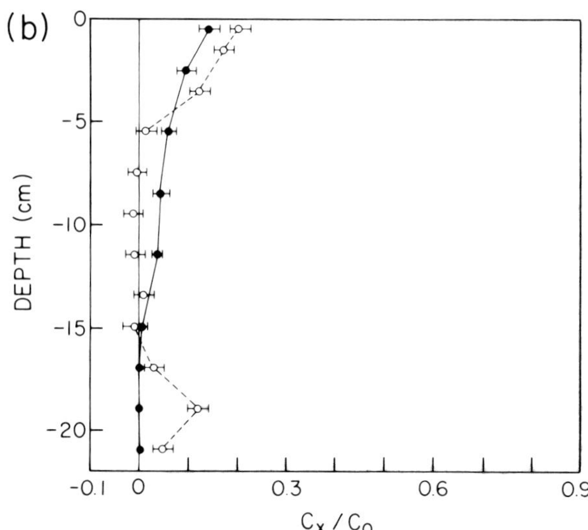

Figure 15.6. (b) Lead-210 and ^{239,240}Pu data for station Oc86A plotted as in Fig. 15.6a but with C/C_o recalculated with C_o for ^{210}Pb and ^{239,240}Pu defined as 27 and 30 dpm g^{-1}, respectively. Error bars are 95% confidence intervals; where no error bar is shown, the symbol size is greater than is the error. Notice that the x-axis scale is -0.1–0.9.

Oc86D through Oc86H since variations between cores occurred on a small spatial scale. In addition, the samples used exclusively for ^{210}Pb determination at these stations were collected in small (6.5 cm in diameter) cores, enhancing the scatter in the data. As a consequence, we will focus our analysis on the three cores for which determinations of all isotopes were completed on the same samples. These are, by coincidence, all of the more-oxidized cores investigated in this study.

The distributions of ^{210}Pb, as compared to ^{239,240}Pu, in each of the three cores (End-1, End-2, and Oc86A) strongly suggest mobility of the latter isotope in excess of that of ^{210}Pb. Distributions in the upper 2–4 cm of sediment were very similar (Fig. 15.6). Below this depth the relative (C_x/C_0) concentrations diverged. In all three cores, ^{239,240}Pu occurred at depths at which no excess ^{210}Pb (determined by a 2-σ confidence interval) was found. At station Oc86A the divergences in concentration were large, and a significant part of the Pu inventory was found in sediments with no detectable ^{210}Pb. This pattern also held for ^{55}Fe at station Oc86A and for ^{137}Cs in the *Endeavor* cores. This pattern is not limited to single depth but, rather, in each case Pu penetrations in sediment horizons devoid of excess ^{210}Pb appeared to persist over intervals ranging from 6 to 10 cm. The largest uncertainty in the determination of excess ^{210}Pb was in the determination of supported ^{210}Pb; we commonly experienced an uncertainty of ± 0.2–0.3 disintegrations per minute (dpm) g^{-1}. With steady-state concentrations of 25–30 dpm g^{-1} of excess ^{210}Pb in the upper 1 cm, excess ^{210}Pb can be determined to < 1 dpm g^{-1} or (since $\lambda = 0.0311$ y^{-1}) an age of $\geqslant 100$ y. The absence of excess ^{210}Pb in these deeper horizons requires that they predate atmospheric testing by a wide margin. The presence of ^{239,240}Pu can only be explained by migration through means other than particle mixing.

The large ^{210}Pb maximum at depth in station Oc86A is difficult to explain on a quantitative basis, but it does not compromise the case for ^{239,240}Pu migration. The ^{210}Pb values between 17 and 21 cm were extremely high for these depths and must represent mechanical injection of surface material by some means. The three ^{239,240}Pu analyses from this interval are not significantly different from zero (at the 2-σ level). If the injection were to have involved surface sediment similar to today's, it would have a ^{239,240}Pu activity 30–50 times that measured. The apparent absence of ^{239,240}Pu and presence of ^{210}Pb in these sediments suggest that the injection of the sediment involved ^{230,240}Pu-free sediment, presumably surface material predating the introduction of Pu.

The amount of ^{210}Pb at depth is difficult to explain. If the sediment represents infilling of a burrow, as suggested by Benninger et al. (1979), then the burrow must be extremely large. The value for C_x/C_0 is 0.60, requiring that the sample analyzed be 60% surface material, an unlikely event in a sample with a cross-sectional area of > 150 cm^2. An origin from before the nuclear age (~ 1 half-life) requires that the ^{210}Pb activity be much higher than that of the present surface material. In short, we do not feel we can adequately explain the extreme C_x/C_0 value of 0.60 with the information presently available.

The relative amount of ^{210}Pb found at depth at station Oc86A may well be an artifact. The ^{210}Pb maximum at 19 cm has a C_x/C_0 value of ~ 0.60. If the sediment at this depth were to predate the nuclear age, as postulated above, then the original excess ^{210}Pb activity would be much higher than that reported for the uppermost section of the core (5.4 dpm g^{-1}). For the other cores from this general area (*Endeavor* 1 and 2), values for the excess ^{210}Pb at the surface were 25–29 dpm g^{-1}, decreasing to 2–3 dpm g^{-1} at a depth of 2–3 cm. Similarly, the ^{239,240}Pu content of the uppermost sections from the *Endeavor* cores was 24–40 dpm g^{-1}, as opposed to 4 at station Oc86A. We believe that the uppermost sediment was missing in this core, probably as a result of overpenetration of the corer. If this were the case, and the true activities of ^{210}Pb and ^{239,240}Pu in the surface section were the average of those observed in the *End*-1 and 2 cores (27 and 30 dpm g^{-1}, respectively), then the C_x/C_0 maximum of ^{210}Pb at 19 cm would be only 0.12.

The occurrence of overpenetration and consequent loss of 1–2 cm of sediment would not, however, change the conclusions drawn regarding ^{239,240}Pu mobility. If true surface values were 27 and 30 dpm g^{-1} for ^{210}Pb and ^{239,240}Pu, respectively, then the scale for Fig. 15.6a would change, but the relative values would not; that is, for the assumed true values of C_0, the C_x/C_0 range as plotted would simply become 0.14 rather than 1.0, and each subdivision would be 0.014 as opposed to 0.10. Figure 15.6b is presented to illustrate this point. The C_0 values of 27 and 30 dpm g^{-1} for ^{210}Pb and ^{239,240}Pu have been used, but the concentration scale has been kept similar to those used in the rest of the figure. The data now fall in the range of 0.2–0.0. The value for Pu remains significantly greater than zero (95% confidence interval plotted) to a depth of at least 10–12 cm, while ^{210}Pb is not significantly different from zero in the interval of 5–6 cm to 14–16 cm. The ^{210}Pb maximum is less pronounced but still well defined, and all of the $^{239,240}Pu$ values in that interval are indistinguishable from zero. The overall slope, when plotted as in Fig. 15.6b, is quite similar to that of the *End*-1 and 2 cores except for the ^{210}Pb maximum at depth.

The occurrence of Pu in ^{210}Pb-free sediment at the *Endeavor* and Oc86A sites leads us to conclude that the apparent increase in dispersal rate seen in Fig. 15.5 is neither an artifact of the model nor a result of rather bizarre particle mixing, but that it reflects migration in solution. Particle-mixing rates in pelagic sediments are commonly on the order of 10^{-9} cm^2 s^{-1} and rarely exceed 10^{-8} cm^2 s^{-1} (Guinasso and Schink, 1975; Nozaki et al., 1977). The ^{210}Pb results reported in this chapter exhibited a similar range. While we do not wish to interpret the data in Fig. 15.5 quantitatively, due to the uncertainty in input assumption, the changes in slope between shallow and deep sediment indicate rate changes of an order of magnitude and more. Thus, what we are ascribing as diffusion in pore solution must be characterized by diffusion coefficients in the range of 10^{-8}–10^{-7} cm^2 s^{-1}. Such values contrast markedly with laboratory values of approximately 10^{-11} cm^2 s^{-1} reported by Schreiner et al. (1981) both for Pu(IV) and for Pu(VI). Since no appreciable migration of Pu in solution would be observed when D is 10^{-11} cm^2 s^{-1}, such values cannot be reconciled with the presence of Pu in ^{210}Pb-free sediment.

15.4.3. Migration and Redox Conditions

The conditions under which Pu migration modes occur are reasonably well documented, but the conditions that prevent solution redistribution are less well established by the data presented in this chapter. Interpretation of the ^{210}Pb–Pu data requires migration in sediments that are oxidized. While we do not have pore-water data for the *Endeavor* cores, we are confident that they would be analogous to stations Oc86C and Oc86D; thus, migration must occur where O_2 is present. Taking enhanced mobility at depth as indicative of the occurrence of migration in solution, the data suggest occurrence to or possibly beyond the point at which Mn^{4+} is reduced (horizon 2 in Fig. 15.4). The data presented in Fig. 15.5 for station Oc86G are distinctly different from those obtained at the other *Oceanus* stations and indicate decreased dispersal at depths below a few centimeters. The environment at this station differed from the other *Oceanus* stations in that readily

measurable Fe^{2+} diffused from depth into the deeper part of the Pu profile, although reduction of Fe^{2+} did not occur in the sampled section. The Fe^{2+} also diffuses to the base of the Pu profiles at three of the *Wecoma* stations (W81G, W81H, W81I). The Pu and ^{137}Cs data are presented in terms of the constant-source model (Fig. 15.7). As a group, the *Wecoma* cores exhibit more scatter than was seen in the *Oceanus* sediments. The plot for the data from station W81H is reasonably linear: there is no enhanced dispersal at depth, nor is there any decrease, an observation that is consistent with the absence of solution migration. The data for station W81I are similar to that for station Oc86G, but the case is not so unambiguously made. There is a possible decrease in dispersal rate with depth, but it depends on how the data are interpreted. The two deepest points exhibited high Pu activity that cannot be explained by a decreasing-rate interpretation. To extract further information from this core we need ^{210}Pb data, particularly to evaluate the origin of the two deepest points. Station W81C, an oxidized station (Fig. 15.2), also has scattered data, but this does not obscure the occurrence of the slope break observed at other oxidized stations.

In summary, these data support the conclusion that under a range of redox conditions, possibly to the point of Mn^{4+} reduction, Pu migration in solution can and does occur in pelagic sediments. The possibility of an important role for Fe^{2+} is suggested by the absence of enhanced migration in the presence of Fe^{2+} in pore waters. This leads us to speculate that Pu migrates in the sampled marine sediments as an oxidized species and that the species is reduced by Fe^{2+} to a far less mobile form.

15.5. A SIMPLE HYPOTHESIS FOR PU DISTRIBUTION

The foregoing description of depth- and redox-dependent distribution of $^{239,240}Pu$ leads us to speculate briefly on the manner in which Pu has achieved the distributions observed in deep-ocean sediments. Such a description must accommodate several factors that stand out in the data. In the uppermost sediments there appears to have been little separation of the fall-out isotopes ^{55}Fe, $^{239,240}Pu$, and ^{137}Cs, as well as, often, ^{210}Pb. At depths below 2–5 cm, there was evidence of substantially increased mobility. In the zone of increased mobility, some separation of various fallout isotopes was observed. In particular, there was evidence of Pu mobility in excess of that of ^{55}Fe under the more oxidized conditions encountered, and, on occasion, Pu

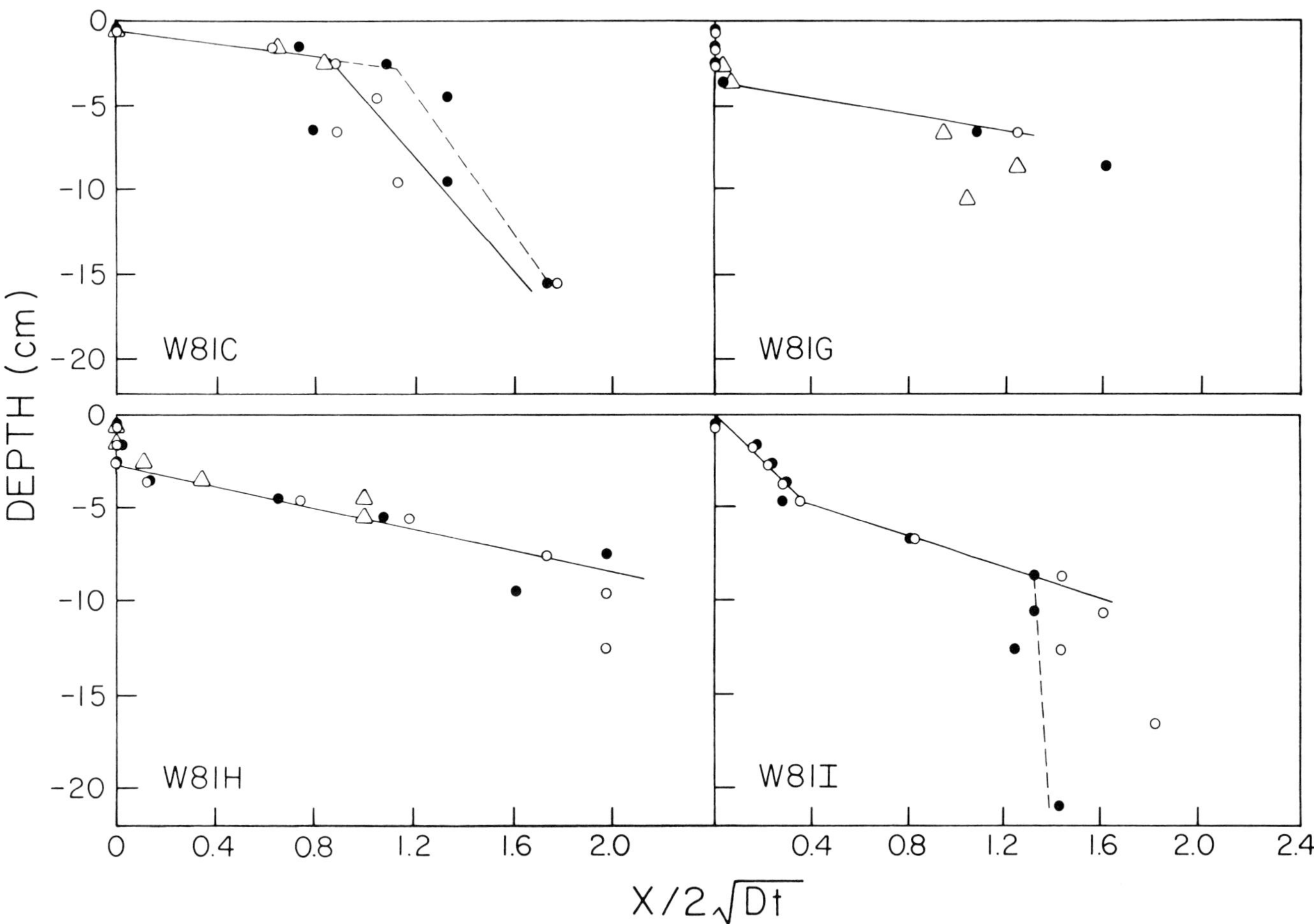

Figure 15.7.   Graphical representation of the constant-source model for the *Wecoma* stations.

mobility appeared to exceed that of ^{137}Cs as well. We speculate that this behavior can be explained on the basis of two pools of Pu, one soluble and one particulate. The soluble Pu would be derived from a fraction of the sedimenting particulate-bound Pu during the early diagenetic reactions that break down the large particles doubtlessly responsible for the flux of Pu (Livingston and Anderson, 1983). We currently have no estimate of the size of such a soluble fraction. We do, however, believe that any soluble fraction that remains in the sediment must be relatively small compared to the particulate-bound pool. This is indicated by the apparent dominance of shallow distributions by sediment mixing. The diffusion of mobile Pu to greater depths, however, leads to the dominance of this species at depths below the bioturbated surface layers, yielding the appearance of higher mobility at depth. We do not envision the mobile Pu as being dissolved, but, more accurately, an exchangeable species with a distribution

coefficient sufficiently low to account for mobility in excess of bioturbation, which is relatively slow. Analysis of sediment samples yielded a sum of the two pools in any sample, dominated in the upper sediment by non-exchangeable, particle-bound Pu and in the lower by exchangeable Pu.

We may quantify this model rather simply in order to illustrate the distributions that result. If we represent the activities of the two forms of Pu by C_m (mobile Pu) and C_p (particulate-bound Pu) and their diffusion coefficients by D_m and D_p, respectively, then

$$C_T^0 = C_m^0 + C_p^0 \qquad \text{2a}$$

and

$$C_m^0 = \alpha C_T^0 \qquad \text{2b}$$

where C_T^0 is the total or observed activity at the interface, and α is the fraction of exchangeable Pu. The constant-source model can be applied to both species, resulting in

$$C_T^x = \alpha C_T^0 \sqrt{\pi} \, \text{ierfc}\left[\frac{x}{2\sqrt{D_m t}} \right]$$

$$+ (1 - \alpha)C_T^0 \sqrt{\pi} \, \text{ierfc}\left[\frac{x}{2\sqrt{D_p t}} \right] \qquad 3$$

where C_T^x is the total measured activity at depth x, and t is the time since introduction of the tracer. The distribution of C_T^x resulting from the application of this relationship and plots of C_T^x according to the simple constant-source equation as used for Fig. 15.5 are presented in Fig. 15.8. The model does result in the depth distributions and the type of slope break observed in the field data. This does not constitute proof of any sort, but does demonstrate that the hypothetical model is, at least, consistent with the observations.

In this chapter we have focused on ^{239,240}Pu and its dispersion because the Pu data are by far the most accurate and sensitive to processes affecting the dispersion of the various nuclides studied. This is not meant to imply anything with regard to the mobilities of ^{137}Cs or ^{55}Fe. Indeed, the data in Figs. 15.5 and 15.7 imply that all three nuclides experience enhanced mobility at depth. Often, any separation of Cs and Pu was marked by the scatter of the data. The first-order feature is the implied enhanced dispersal of nuclides; separation is a second-order feature in the data presented, although it does occur.

15.6. CONCLUSIONS

The determination of the distribution of the fallout nuclides ^{239,240}Pu, ^{137}Cs, and ^{55}Fe in conjunction with various redox horizons showed well-defined differences in the depth of penetration of the nuclides into the sediment. Depth of penetration was deepest at the most oxidized stations, and substantially less in more reduced sediments. Application of a very simple model to describe nuclide distributions with depth in the sediment indicates that dispersion was enhanced at depth relative to that occurring in the uppermost 2–5 cm. Comparison of ^{239,240}Pu distributions with that of ^{210}Pb demonstrates that Pu was distributed independently from ^{210}Pb in the oxidized sediment; Pu penetrates as much as 8–10 cm of ^{210}Pb-free sediment. The separation of Pu and ^{210}Pb is taken to be indicative of the migration of Pu in

sediments, independent of particle mixing. The ^{210}Pb data support the indications of mobility derived from the simple mathematical model. The modelling of ^{239,240}Pu distributions suggests that enhanced migration (at depth) occurred over a wide range of redox conditions, including the reduction of Mn^{4+}. The apparent absence of heightened Pu mobility at depth where Fe^{2+} is observed in pore water suggests that Fe^{2+} may have led to the reduction of the migrating species of Pu to a form that is not mobile. It is speculated that the observed distributions and apparent increase in rate of dispersion at depth can be explained on the basis of conversion of a small fraction of the sedimenting particulate nuclides to an exchangeable form that migrates at rates in excess of particle mixing.

ACKNOWLEDGMENTS

We are indebted to a number of our co-workers for completing the prodigious amount of sampling and analysis involved in this study. Sampling and analysis aboard ship was ably carried out by Will Clarke, Hovey Clifford, Ted Desrosiers (for coring), Wayne Dickinson, Charlie Olson, and Zofia Mlodzinska-Kijowski (for pore-water sampling and analysis). Laboratory analyses were carried out by Lee Goodell, Susan Casso, Bruce Brownawell, David Schneider, Joanne Olmsted, and Lolita Surprenant (for radiochemical work). Metal analyses both for sediments and for pore waters were done by Wayne Dickinson and Scot Birdwhistell. Our participation in the *Wecoma* cruise was at the invitation of G. Ross Heath of Oregon State University and is gratefully acknowleged. For the safe recovery and launching of WHIMP, no mean feat, we are grateful to the officers and deck crew of the *R/V Oceanus* and *R/V Wecoma*.

We also acknowledge the valuable contributions of Vaughan Bowen to this study. He participated in the early planning stages and generously shared his experience and insight regarding both analyses and interpretation of the data as the study progressed.

Finally, we are indebted to Tom O'Connor for his editorial assistance and particularly for calling our attention to an error in the constant-source equations used in the original manuscript.

Financial support has been provided through the Sea Grant Office of the U.S. National Oceanic and Atmospheric Administration under grant NA80AA-00085. Collection and analyses of sediment cores from the *R/V Endeavor* were supported through Sandia Laboratories contract 61-1034. Support for analytical instru-

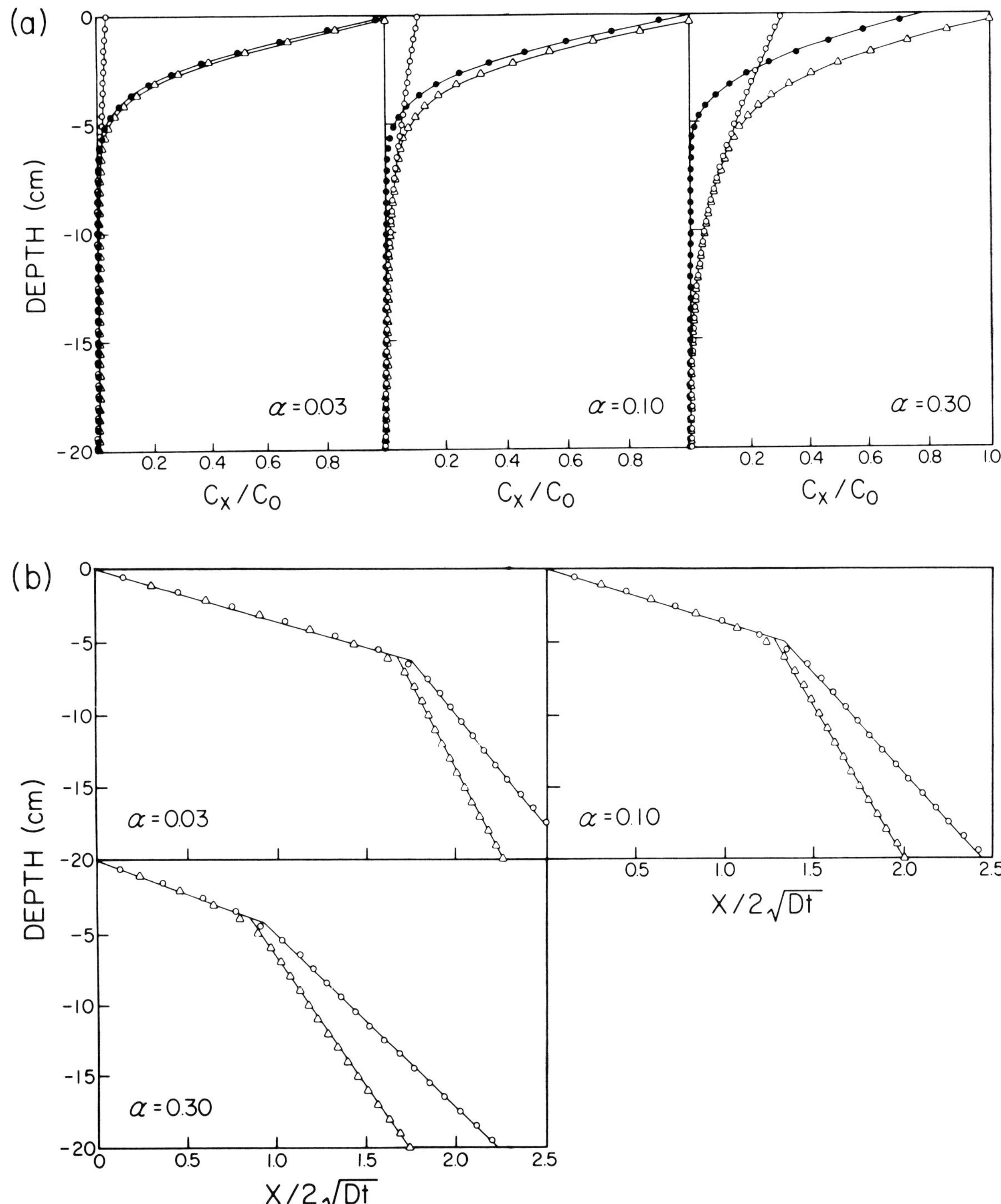

Figure 15.8. (a) Hypothetical depth distributions of two species with different diffusion coefficients and various fractions; α is the total pool of the species. Plots are for $D_m = 5 \times 10^{-8}$ cm^2 s^{-1}, $D_p = 5 \times 10^{-9}$ cm^2 s^{-1}, and $\alpha = 0.03$, 0.10, and 0.30. (b) Graphical representation of the constant-source model, as in Figs. 15.5 and 15.7, for C_T calculated from the data input of Fig. 15.8a, and a second value of $D_m = 1 \times 10^{-7}$ cm^2 s^{-1} to illustrate the substantial influence on the slopes as plotted.

mentation has been provided by the Alcoa Foundation through a grant to the Woods Hole Oceanographic Institution. This chapter is Woods Hole Oceanographic Institution Contribution No. 5755.

REFERENCES

Atlas, E. L., S. W. Hager, L. I. Gordon, and P. K. Park. 1971. A Practical Manual for Use of the Technicon® AutoAnalyzer® in Seawater Nutrient Analyses: Revised. Technical Report 215, Ref. No. 71-22, Department of Oceanography, Oregon State University, Corvallis, 48 pp.

Beasley, T. M., R. Carpenter, and C. D. Jennings. 1982. Plutonium, ^{241}Am and ^{137}Cs ratios, inventories and vertical profiles in Washington and Oregon continental shelf sediments. *Geochimica et Cosmochimica Acta*, **46**, 1931–1946.

Bender, M. L., K. A. Fanning, P. N. Froelich, G. R. Heath, and V. Maynard. 1977. Interstitial nitrate profiles and oxidation of sedimentary organic matter in the eastern equatorial Atlantic. *Science*, **198**, 605–609.

Benninger, L. K., R. C. Aller, J. K. Cochran, and K. K. Turekian. 1979. Effects of biological sediment mixing on the ^{210}Pb chronology and trace metal distribution in a Long Island Sound sediment core. *Earth and Planetary Science Letters*, **43**, 241–259.

Burke, J. C. 1968. A sediment coring device of 21-cm diameter with sphincter core retainer. *Limnology and Oceanography*, **13**, 714–718.

Chester, R., and M. J. Hughes. 1967. A chemical technique for the separation of ferromanganese minerals, carbonate minerals and absorbed trace elements from pelagic sediments. *Chemical Geology*, **2**, 249–262.

Christensen, E. R. 1982. A model for radionuclides in sediments influenced by mixing and compaction. *Journal of Geophysical Research*, **87**, 566–572.

Cline, J. D., and F. A. Richards. 1972. Oxygen deficient conditions and nitrate reduction in the eastern tropical North Pacific Ocean. *Limnology and Oceanography*, **17**, 885–900.

Duursma, E. D., and C. Hoede. 1967. Theoretical, experiment, and field studies concerning molecular diffusion of radiosotopes in sediments and suspended solid particles of the sea. *Netherlands Journal of Sea Research*, **3**, 423–457.

Emerson, S., R. Jahnke, M. Bender, P. Froelich, G. Klinkhammer, C. Bowser, and G. Setlock. 1980. Early diagenesis in sediments from the eastern equatorial Pacific. I: Porewater nutrient and carbonate results. *Earth and Planetary Science Letters*, **49**, 57–80.

Froelich, P. N., G. P. Klinkhammer, M. S. Bender, N. A. Luedtke, G. R. Heath, D. Cullen, P. Dauphin, D. Hammond, B. Hartman, and V. Maynard. 1979. Early oxidation of organic matter in pelagic sediments of the eastern equatorial Atlantic: suboxic diagenesis. *Geochimica et Cosmochimica Acta*, **43**, 1075–1090.

Glass, B. P. 1969. Reworking of deep-sea sediments as indicated by the vertical dispersion of the Australiasian and Ivory Coast microtektite horizons. *Geochimica et Cosmochimica Acta*, **6**, 409–415.

Goldberg, E. D., and M. Koide. 1962. Geochronological studies of deep sea sediments by the thorium–ionium method. *Geochimica et Cosmochimica Acta*, **26**, 417–450.

Guinasso, N. L., and D. R. Schink. 1975. Quantitative estimates of biological mixing rates in abyssal sediments. *Journal of Geophysical Research*, **80**, 3032–3043.

Hollister, C. D., D. R. Anderson, and G. R. Heath. 1981. Subseabed disposal of nuclear wastes. *Science*, **213**, 1321–1326.

Hydes, D. J. 1980. Reduction of matrix effects in carbon furnace atomic absorption spectrometric determination of cobalt, copper, and manganese in seawater using a soluble organic acid. *Analytical Chemistry*, **52**, 959–963.

Koide, M., A. Soutar, and E. D. Goldberg. 1972. Marine geochronology with ^{210}Pb. *Earth and Planetary Science Letters*, **14**, 442–446.

Labeyrie, L. D., H. D. Livingston, and A. G. Gordon. 1975. Measurement of ^{55}Fe from nuclear fallout in marine sediments and seawater. *Nuclear Instruments and Methods*, **128**, 575–580.

Livingston, H. D., and R. F. Anderson. 1983. Large particle transport of plutonium and other fallout radionuclides to the deep ocean. *Nature*, **303**, 228–231.

Livingston, H. D., D. R. Mann, and V. T. Bowen. 1975. Analytical procedures for transuranic elements in seawater and marine sediments. *In*: Analytical Methods in Oceanography, T. R. P. Gibbs (Ed.). Advances in Chemistry Series 147, American Chemical Society, Washington, D.C., pp. 124–138.

Nozaki, Y., J. K. Cochran, and K. K. Turekian. 1977. Radiocarbon and ^{210}Pb distribution in submersible-taken deep-sea cores from project FAMOUS. *Earth and Planetary Science Letters*, **34**, 167–173.

Officer, C. G., and D. R. Lynch. 1982. Interpretation procedures for the determination of sediment parameters from time-dependent flux inputs. *Earth and Planetary Science Letters*, **61**, 55–62.

Peng, T. H., W. S. Broecker, and W. H. Berger. 1979. Rates of benthic mixing in deep-sea sediment as determined by radioactive tracers. *Quaternary Research*, **11**, 141–149.

Robbins, J. A. 1978. Geochemical and geophysical applications of radioactive lead. *In*: The Biochemistry of Lead in the Environment, J. O. Nriagu (Ed.). Elsevier Press, Amsterdam, pp. 285–393.

Santschi, P. H., Y. H. Li, J. J. Bell, R. M. Trier, and K. Kawtaluk. 1980. Pu in coastal marine environments. *Earth and Planetary Science Letters*, **51**, 248–265.

Sayles, F. L., and P. C. Mangelsdorf, Jr. 1976. The analysis of SO_4^{2-} in seawater by difference chromatography. *Limnology and Oceanography*, **21**, 899–905.

Sayles, F. L., P. C. Mangelsdorf, Jr., T. R. S. Wilson, and D. N. Hume. 1976. A sampler for the *in situ* collection of marine sedimentary pore waters. *Deep-Sea Research*, **23**, 259–264.

Schreiner, F., S. Fried, and A. Friedman. 1981. A study of the mobility of plutonium and americium in marine sediments. *In*: Subseabed Disposal Program Annual Report, January to December 1980, Vol. II. SAND81-1095/II, Sandia National Laboratories, Albuquerque, New Mexico, pp. 269–304.

Sholkovitz, E. R. 1983. The geochemistry of plutonium in fresh and marine water environments. *Earth Science Reviews*, **19**, 95–161.

Wong, K. M., V. E. Noshkin, and V. T. Bowen. 1970. Radiochemical procedures for the analyses of strontium, antimony, rare earths, cesium and plutonium in seawater samples. *In*: Reference Methods for Marine Radioactivity Studies, Y. Nishiwaki and R. Fukai (Eds.). International Atomic Energy Agency, Vienna, pp. 119–129.

Chapter 16

The Role of Biogenic Debris in the Vertical Transport of Transuranic Wastes in the Sea

Nicholas S. Fisher and Scott W. Fowler

Laboratory of Marine Radioactivity
Musée Océanographique, Monaco

ABSTRACT

The concentration and flux of transuranic elements (Np, Pu, Am, Cm, and Cf) in six species of phytoplankton, three species of crustacean, and one species of gelatinous zooplankton, and their associated metabolic products (molts, discarded houses, and fecal pellets) were studied in laboratory experiments. Concentration factors of all radionuclides except Np were proportional to the surface:volume ratios of the organisms and ranged from ~10 in mucopolysaccharide larvacean houses to 7×10^5 in diatoms. These elements leached from planktonic debris with half-times ranging from 9 d in appendicularian houses to 50 d in euphausiid fecal pellets. In considering these data together with known sinking rates for these planktonic debris, it is shown that the large, more quickly sinking fecal pellets and crustacean molts probably transport these radionuclides to deep water and sediment, whereas more slowly sinking phytoplankton cells, gelatinous houses, and small fecal pellets are probably involved in remineralization of transuranic elements and recycling processes at intermediate depths. A simple model constructed to assess the potential impact of sinking zooplankton fecal material in removing transuranic elements from surface water indicates that residence times of these elements in surface water is largely a function of the zooplankton biomass.

16.1. INTRODUCTION

Numerous publications have identified the importance of suspended particulate matter in the geochemical cycling of metals in the sea (Broecker and Peng, 1982). Suspended particles, including the phytoplankton in surface water, can effectively scavenge many metals from ambient seawater. These particles may introduce metals into marine food chains and/or act as vectors for their vertical transport to deep water and sediment. Evidence suggests that many toxic wastes, both organic and inorganic, are effectively removed from surface water by association with rapidly sinking biogenic debris (Fowler, 1982; Olsen et al., 1982), but the degree to which this phenomenon might occur with the transuranic elements warrants further study. Transuranic elements are introduced into the ocean primarily via atmospheric fallout from weapons testing, release of wastes from accidents with nuclear devices, disposal of low-level radioactive wastes onto the seabed, and delib-

Table 16.1.   Volume:Volume (VCF) and Wet-Weight (WCF) Concentration Factors (10^5)[a]

Organism	Concentration Factor	Isotope				
		^{235}Np[b]	^{237}Pu[b]	^{241}Am[b]	^{242}Cm	^{252}Cf[b]
Phytoplankton						
Thalassiosira	VCF	0*	6.3	6.9	6.4	6.2
pseudonana	WCF	0*	3.8	4.1	3.8	3.7
Dunaliella	VCF	0*	2.2	1.8	1.2	4.1
tertiolecta	WCF	0*	1.6	1.3	0.9	3.0
Emiliania	VCF	0*	1.6	1.1	2.1	3.2
huxleyi (with coccoliths)	WCF	0*	0.6	0.4	0.8	1.2
Emiliania	VCF	0*	2.2	1.0	1.3	3.4
huxleyi (without coccoliths)	WCF	0*	3.3	1.5	2.0	5.2
Oscillatoria	VCF	0*	1.7	0.3	2.6	1.3
woronichinii	WCF	0*	0.4	0.1	0.6	0.3
Tetraselmis	VCF	0*	0.4	0.3		0.9
chuii	WCF	0*	0.8	0.5	n.d.	1.7
Heterocapsa	VCF			3.8	1.2	
pygmaea	WCF	n.d.	n.d.	2.8	0.9	n.d.
Zooplankton						
Euphausiids	VCF			44–360[c]		
	WCF	n.d.	n.d.	40–330[c]	n.d.	n.d.
Oikopleura	VCF			60[d]		
dioica (trunk and tail)	WCF	n.d.	n.d.	60[d]	n.d.	n.d.
Oikopleura	VCF			10[d]		
dioica (house)	WCF	n.d.	n.d.	10[d]	n.d.	n.d.

[a] An asterisk indicates a value not significantly ($p < 0.05$) different from zero; n.d. means not determined.
[b] From Fisher et al., 1983a.
[c] From Fisher et al., 1983b.
[d] From Gorsky et al., 1984.

erate liquid discharges from nuclear fuel-reprocessing facilities (Watters et al., 1980). Globally, fallout represents the largest source of these wastes (Bowen et al., 1971, 1980), although in some localities such as the Irish Sea liquid discharges can far surpass fallout as a source of these pollutants. The interactions of these wastes with marine biota, particularly with the plankton, have received increasing attention. Because the plankton may strongly influence the rates and routes of transport of these elements in the sea, we have examined the biokinetics of uptake and release of five transuranic nuclides in selected species of marine phytoplankton and zooplankton. In addition, we have examined published information on the vertical distributions of Pu and Am, the two best-studied elements, in oceanic water columns and attempted to explain partially these distributions by assessing the vertical transport of these elements in sinking biogenic debris.

16.2.   BIOKINETICS EXPERIMENTS

The biokinetics of uptake and loss of several trans-uranic elements in marine plankton has been investigated in a series of laboratory experiments. The uptake and retention of γ-emitting isotopes of Am, Pu, Cf, Cm, and Np were measured in cultures of six taxonomically diverse species of phytoplankton. Accumulation and retention of ^{241}Am in crustacean and gelatinous zooplankton (euphausiids and appendicularians) were also measured using animals recently collected from the Mediterranean.

16.2.1.   Phytoplankton

Surface water from the Mediterranean was filtered through sterile 0.2-μm Sartorius (Sartorius Filters, Inc., Hayward, California) filters and innoculated with phytoplankton cells at cell densities that were kept constant by keeping the cultures in the dark. The species studied were *Thalassiosira pseudonana* (a diatom), *Dunaliella tertiolecta* (a chlorophyte), *Oscillatoria woronichinii* (a cyanophyte), *Emiliania huxleyi* (a coccolithophore; two clones used, one with and one without coccoliths), *Tetraselmis chuii* (a prasinophyte), and *Heterocapsa*

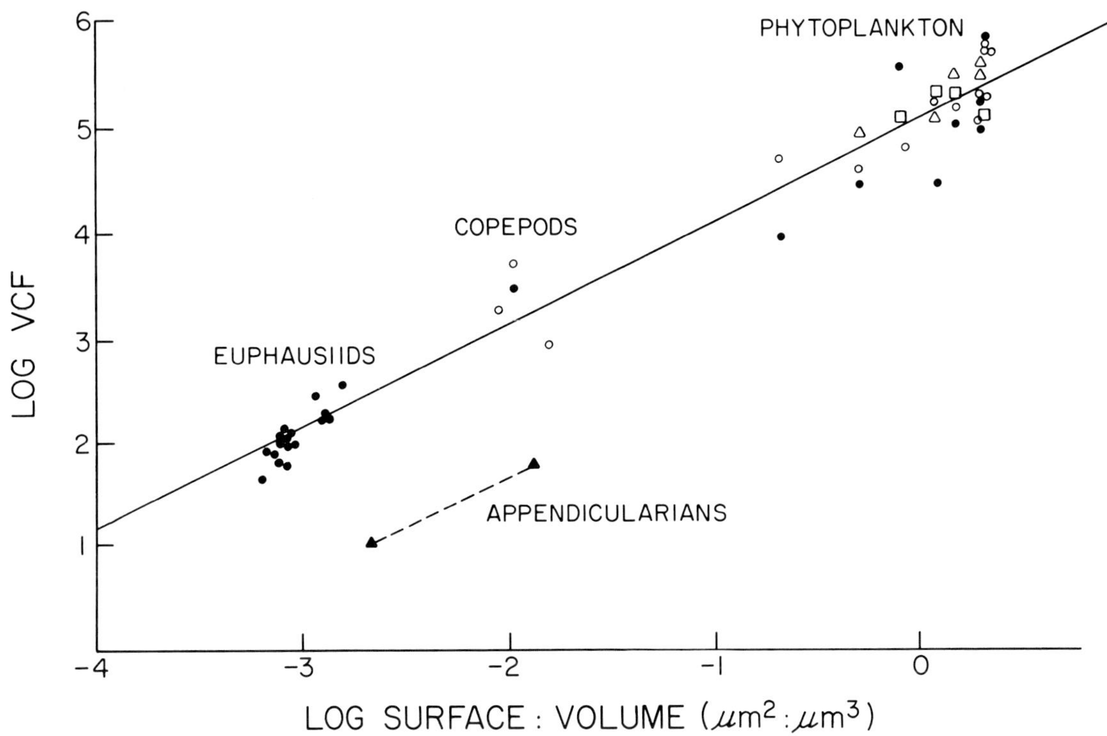

Figure 16.1.   Correlation of log VCF for transuranic elements with log surface; volume ratio of marine plankton. (●) Am, (○) Pu, (△) Cf, (□) Cm, and (▲) Am in appendicularians. Phytoplankton data are from Fisher et al. (1980, 1983a) and from calculations from field data of Fowler et al. (1983) for the dinoflagellate *Ceratium* sp.; copepod data are from field measurements of Fowler et al. (1983); euphausiid data are from Fisher et al. (1983b); and appendicularian data are from Gorsky et al. (1984). For phytoplankton, copepods, and euphausiids: $y = 0.962\,x + 5.110$; $r = 0.981$; $df = 47$ ($p < 0.01$); for appendicularians, $y = 0.962\,x + 3.558$ (where $y = \log$ VCF and $x = \log$ surface:volume).

pygmaea (a dinoflagellate). The cultures, maintained in sterile Erlenmeyer flasks, received either ^{241}Am, ^{242}Cm, or ^{252}Cf in pM (10^{-12} M) concentrations or ^{235}Np or ^{237}Pu in fM (10^{-15} M) concentrations. Details of culturing and experimental conditions are given for ^{235}Np, ^{237}Pu, ^{241}Am, and ^{252}Cf by Fisher et al. (1983a); the same experimental protocol was followed for ^{242}Cm-exposed cells. Uptake rates and retention of radionuclides were determined by radiotracer methods described by Fisher et al. (1983a). Concentration factors (Table 16.1) for each of the five elements and each species were calculated after 96 h, at which time an apparent equilibrium was achieved with respect to isotope partitioning. Concentration factors were calculated as volume:volume and wet-weight concentration factors (VCF and WCF, respectively):

$$\text{VCF} = \frac{\text{atoms of isotope per } \mu\text{m}^3 \text{ of cell volume}}{\text{atoms of isotope per } \mu\text{m}^3 \text{ of medium}} \qquad 1$$

$$\text{WCF} = \frac{\text{atoms of isotope per } 10^{-12}\text{g of cell wet weight}}{\text{atoms of isotope per } \mu\text{m}^3 \text{ of medium}}$$

$$2$$

Surface:volume ratios of phytoplankton (Fig. 16.1) were determined as described by Fisher et al. (1983a). The WCFs for ^{241}Am, ^{237}Pu, ^{252}Cf, and ^{242}Cm (Table 16.1) are of the same order of magnitude as those reported for Am and Pu in particulate matter in natural waters (Gromov, 1976; Krishnaswami et al., 1976a; Santschi et al., 1980; Pentreath et al., 1982). The calculated VCFs from Table 16.1 are presented in Fig. 16.1 as a function of the surface:volume ratios of the organisms. The biological half-life ($tb_{1/2}$) for the loss of ^{241}Am from the diatoms was calculated as 11–13 d (Fig. 16.2) (Fisher et al., 1983a).

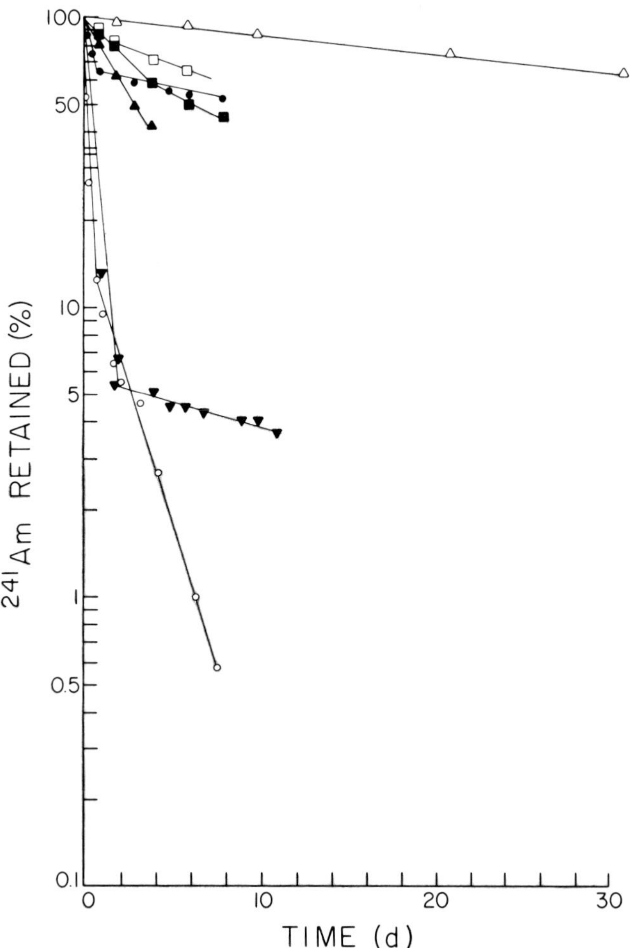

Figure 16.2. Loss of ^{241}Am from organisms or debris resuspended into unspiked seawater. Representative data from (●) diatoms (*Thalassiosira pseudonana*), (○) euphausiids (*Meganyctiphanes norvegica*) after ingestion of labeled *T. pseudonana*, (▼) euphausiids after uptake from water only, (▲) euphausiid molts, (△) euphausiid fecal pellets, (□) appendicularian (*Oikopleura dioica*) houses after feeding on labeled *T. pseudonana*, (■) appendicularian fecal pellets after feeding on labeled *T. pseudonana*. After Fisher et al. (1983a, 1983b) and Gorsky et al. (1984).

16.2.2. Zooplankton

Fisher et al. (1983b) and Gorsky et al. (1984) measured the uptake kinetics of ^{241}Am in the euphausiids *Meganyctiphanes norvegica*, *Euphausia krohni*, and *Nematoscelis megalops*, and in the pelagic tunicate *Oikopleura dioica* (an appendicularian) by using labeled water and labeled food (*T. pseudonana*) as sources of the radioisotope. Figures 16.2 and 16.3 show the kinetics of accumulation and release of ^{241}Am for these species and their particulate metabolic products. The greatest variation in the uptake data for the euphausiids was due to molting during the experiment. The precipitous loss of ^{241}Am from euphausiids following uptake from water occurred during molting, as these molts contained $96 \pm 10\%$ of the ^{241}Am in each animal; the loss of ^{241}Am from animals after feeding was due to defecation. Concentration factors of ^{241}Am in euphausiids varied inversely with the size of the animals and directly with their surface:volume ratios (Fig. 16.1). Volumes of euphausiids were determined by dividing the wet weights by 1.1, and surfaces were computed for each animal, assuming animals consisted of two equal cones [with dimensions related to wet weights as described by Fowler et al. (1971)]. Surface:volume ratios of copepods were computed for spheres of measured average size. Generally, concentration factors for the euphausiids appeared to be about 1000 times lower than for the phytoplankton and about 10 times higher than for the appendicularians (Table 16.1). Concentration factors of Am and Pu in copepods collected from the field and computed from the data reported by Fowler et al. (1983) appear to be 1–2 orders of magnitude lower than those in the phytoplankton (Fig. 16.1). The significant correlation ($r = 0.981$; $p < 0.01$) between the VCF of the transuranic elements and the surface:volume ratios of the organisms [similar to Thomann's (1981) observations] suggests that ^{241}Am adsorbs onto the surfaces of an organism and that, at least for the microcrustaceans and phytoplankton, there is no appreciable difference in the reactivity of their various surfaces for transuranic elements. Thus, the surface reactivity (atoms per μm^2 of surface divided by dissolved atoms per μm^3 of seawater) of all of these organisms was $\sim 10^5$. The mucopolysaccharide surface of the appendicularians appears to be less reactive for these elements, as indicated by the displacement of these organisms below the regression line. Radiotracer studies by Fisher et al. (1980, 1983a) have shown that dead and live algae accumulate equal amounts of ^{237}Pu and ^{241}Am, demonstrating that cellular uptake of these elements is a passive process.

16.3. VERTICAL TRANSPORT

Sholkovitz (1983) has reviewed much of the available data on vertical profiles of transuranic elements in

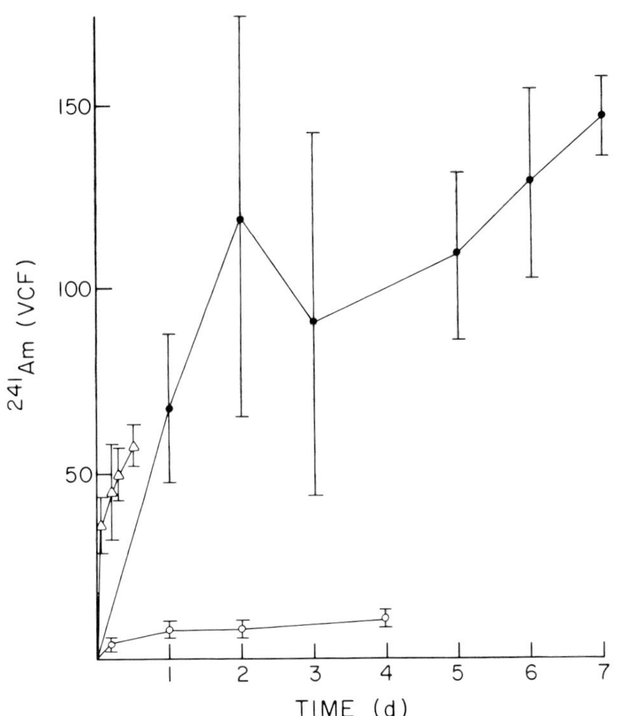

Figure 16.3. Concentration on a VCF basis of ^{241}Am from filtered water by (●) *Meganyctiphanes norvegica*, (△) *Oikopleura dioica* without houses (trunks plus tails only), and (○) empty *O. dioica* houses. Data points are means of 8 replicate animals (●), 20 replicate animals (△), and 15 replicate houses (○) ± one standard deviation. After Fisher et al. (1983b) and Gorsky et al. (1984).

marine water columns. Lowman et al. (1971), Edgington (1981), and Sholkovitz (1983) discussed some of the phenomena probably contributing to these observations, particularly focusing on the importance of sinking and remineralization of particulate matter. In light of new data, summarized in Tables 16.1 and 16.2, we have assessed the potential roles of different types of biogenic debris as vectors for the vertical transport of transuranic elements. In this chapter we attempt to quantify the importance of planktonic debris in the vertical transport of the transuranic elements. The conclusions we draw are generally compatible with those drawn earlier (Lowman et al., 1971; Edgington, 1981; Sholkovitz, 1983), which used different models and a more circumstantial data base than ours.

16.3.1. A Comparison of Vertical Transport in Various Planktonic Debris

Table 16.2 summarizes data on calculated half-times for the leaching of ^{241}Am and the sinking rates for debris from phytoplankton and zooplankton; estimates are also given of the degree to which each kind of debris could transport ^{241}Am to specified depths, in the absence of any degradation. Thus, these depths are upper limits, but the relative importance of the various kinds of biogenic debris in mediating vertical transport of the transuranic elements should conform with the calculations in Table 16.2. Americium-241 was chosen because it is the radionuclide for which most data on concentration and flux are available. Because ^{241}Am and the other reactive transuranic elements behave similarly in phytoplankton (Table 16.1) (Fisher et al., 1983a), it is assumed that those conclusions drawn for Am would generally apply to Pu, Cm, and Cf as well.

Given phytoplankton sinking rates of < 2 m d^{-1} (Bienfang, 1980) and a half-time for ^{241}Am leaching of about 12 d (Fisher et al., 1983a), it seems unlikely that individual, sinking phytoplankton cells would have an appreciable effect on the downward vertical transport of the transuranic elements. This would be particularly true for oligotrophic waters such as the open Mediterranean, in which the number of particles is low and, consequently, the particulate fraction of Pu (4%) and Am (10%) is low (Holm et al., 1980). However, aggregates of phytoplankton cells and other particulate matter including discarded appendicularian houses (Silver and Alldredge, 1981) in the form of "marine snow" are extremely common in the sea and sink much more rapidly [50–100 m d^{-1} (Silver and Alldredge, 1981)] than do individual cells. In water in which the total load of suspended particulate matter is high (e.g., during plankton blooms or in estuaries and in some coastal waters) aggregates of phytoplankton could have an important effect on the redistribution of the transuranic elements. Assuming that (1) marine snow has a constant sinking rate of 50 m d^{-1}, (2) the half-time of ^{241}Am in marine snow is comparable to that in the diatom *T. pseudonana* (12 d), and (3) 50% of the total Am or Pu in surface water is bound to particulate matter (Holm et al., 1980), then 25% of the total Am or Pu in surface water would be transported and released by sinking snow within 12 d, by which time the snow would be at a depth of 600 m. In the absence of any field measure-

Table 16.2. Retention Half-Time of ^{241}Am in Phytoplankton and Zooplankton Debris, Estimates of Debris Sinking Rates, and the Fraction of a Particle's Am That Would Remain on the Particle after it Had Sunk to a Depth of 500 m, 100 m, or 2500 m

Type of Debris	Retention Half-Time of ^{241}Am (d)	Sinking Rate (m d^{-1})	Percentage of ^{241}Am Remaining on Particle at		
			500 m	1000 m	2500 m
Phytoplankton	12[a]	2[d]	≪ 1	≪ 1	≪ 1
Phytoplankton (as marine snow)	12	50[e]	56	31	6
Appendicularian (house)	9[b]	60[b]	53	28	4
Appendicularian (fecal pellets)	8[b]	60[b]	49	24	3
Euphausiid (molts)	2.9[c]	500[f,h]	79	62	30
		1000[f,h]	89	79	55
Euphausiid (fecal pellets)	46[c]	100[g,h]	93	86	69
		300[g,h]	98	95	88

[a] From Fisher et al., 1983a.
[b] From Gorsky et al., 1984.
[c] From Fisher et al., 1983b.
[d] From Bienfang, 1980.
[e] From Silver and Alldredge, 1981.
[f] From Small and Fowler, 1973.
[g] From Fowler and Small, 1972.
[h] A range of slow to moderate sinking rates was selected.

ments, the laboratory-generated data on retention are used; clearly though, the vertical transport of a transuranic element would be linearly related to retention time in the sinking particle, and any variance from the lab-derived values would have a directly proportional effect in the field.

Because there is no appreciable assimilation of transuranic elements by zooplankton (Fig. 16.2), it is evident from the calculations in Table 16.2 that crustacean zooplankton such as euphausiids would act as effective packagers of the transuranic elements, and their debris (fecal pellets and molts) that remain uneaten would have the potential to deliver these elements to very deep water and sediment (Table 16.2). Indeed, the available field data have shown that copepod and euphausiid fecal pellets are highly enriched with Am and Pu relative to the whole animals (Higgo et al., 1977; Fowler et al., 1983), as are euphausiid molts (Higgo et al., 1977, 1980). The greater concentration of Pu and Am in fecal pellets as compared to phytoplankton is probably attributable to the fact that these elements do not associate with proteins in phytoplankton cells but instead

remain on generally unassimilable cell walls (Fisher et al., 1983c), which comprise up to 30% of a cell's dry weight (being species-dependent) (Paasche, 1980), and which are excreted. The resulting fecal pellet, when it is first formed, should have concentrations of transuranic elements comparable to those of the cell walls of the ingested algal food. It is probable that a sinking fecal pellet could scavenge or release transuranic elements in dissolved forms during its descent (Section 16.3.2); however, degradation of the pellet would probably have the overall effect of enhancing the remineralization of the radionuclides, particularly in deeper waters.

Sinking marine snow, which is rich in phytoplankton and discarded appendicularian houses (Silver and Alldredge, 1981), would transport transuranic elements only a few hundred meters. As indicated in Table 16.2, nearly all of the ^{241}Am associated with even a slowly sinking (100 m d^{-1}) euphausiid fecal pellet could reach the sediment at a depth of 2500 m, whereas less than 10% of the ^{241}Am associated with marine snow or with small, slowly sinking copepod fecal material (Small et al., 1979; Hofmann et al., 1981) would reach this depth.

The snow could enrich water less than 1000 m deep, by which time it would have lost approximately three-fourths of its ^{241}Am (Table 16.2). These calculations assume that all particles start to sink at the surface of the ocean and are sinking at constant rates into uncontaminated water; however, as particles sink into colder water, their sinking rates would decline as viscosity increases. More importantly, degradation of biogenic debris may be substantial and would effectively result in remineralization and recycling of the Am in the water column (Hofmann et al., 1981). This process could contribute directly to subsurface maxima of concentrations of transuranic elements in the water column. The depth would depend on the degradation and sinking rates of the sinking material. Degradation rates are temperature-dependent and can range from $\sim$8% d^{-1} loss at 5°C to $\sim$18% d^{-1} loss at 20°C for zooplankton caught in sediment traps (Iturriaga, 1979), whereas Gardner et al. (1983) estimate the decay of organic material in sediment traps in the deep ocean to be on the order of 0.1–1% d^{-1}.

16.3.2. Vertical Profiles and Biogenic Debris

It follows that the degree of vertical redistribution of the transuranic elements by sinking phytoplankton (in whatever form) would increase with the number of suspended cells and, hence, the total suspended reactive surface area. During a bloom, there could be pronounced peaks in the vertical profiles of the concentrations of total (particulate plus dissolved) Am and Pu, whereas in water with few particles, peaks would be less pronounced or undetectable. It is instructive to compare this expectation with the few observations that have been published on the vertical distribution of Am and Pu [see also Sholkovitz (1983)]. In the open Mediterranean, where particulate matter generally contains $\leqslant$10% of the total amount of Am and Pu in surface water (Holm et al., 1980), the vertical distribution of these elements is comparatively homogeneous (Fig. 16.4). Other water columns throughout the Mediterranean have vertical profiles of ^{241}Am and $^{239+240}$Pu similar to those shown in Fig. 16.4 (Fukai et al., 1983), indicating that the profiles in Fig. 16.4 can be considered as representative of the open Mediterranean. In the North Pacific Ocean off the central California coast, where as much as 21 and 32% of $^{239+240}$Pu and ^{241}Am, respectively, are bound to particles in surface water (Fowler et al., 1983), there are pronounced peaks

of both elements at a depth of 250 m (Fig. 16.4). Similarly, Schaule and Patterson (1981) reported subsurface peaks of Pb concentrations at a depth of 400 m in the North Pacific Ocean. Like the transuranic elements, Pb enters open ocean water primarily via the atmosphere (Schaule and Patterson, 1981), is particle-reactive and has no known biological function.

Bowen et al. (1971, 1980) reported pronounced peaks of Pu concentrations in unfiltered water at a depth of about 500 m in different regions of the Pacific and the North Atlantic Oceans (Fig. 16.4) but did not measure partitioning of the Pu between dissolved and particulate phases in these waters. It is noteworthy that this depth does not correspond with the oxygen minimum, the thermocline, or the deep chlorophyll maximum of these waters (Bowen et al., 1980). The load of fine particulate matter at this depth, derived from decaying biogenic material, is unknown. Plutonium and Am would presumably diffuse away from the peak depths, although it would appear from the approximate constancy of the peak over a period of years (Bowen et

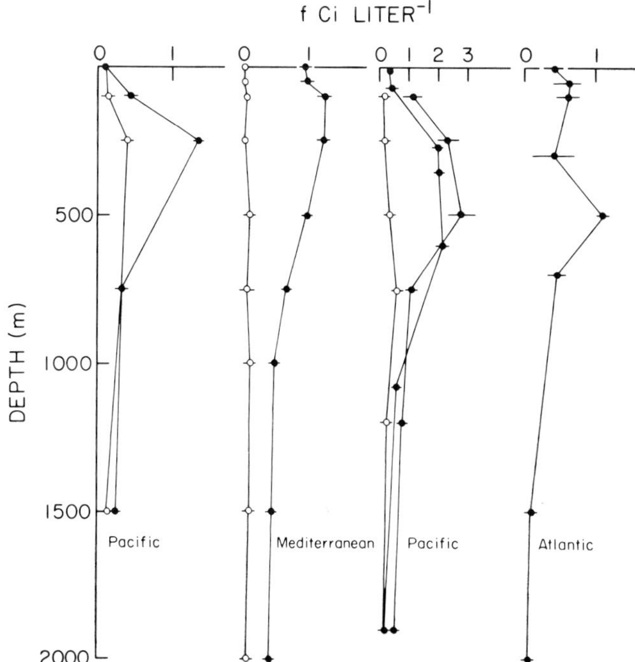

Figure 16.4. Vertical profiles of total ^{241}Am (○) and $^{239+240}$Pu (●) in unfiltered water from the North Pacific (Fowler et al., 1983), the Mediterranean (Fukai et al., 1979), two stations in the Pacific (Bowen et al., 1980; Bowen, 1983), and the Atlantic (Bowen et al., 1971). Note: 1 Ci = 3.7 × 10^{10} Bq.

al., 1980) that inputs balance losses. Also, the high concentrations of transuranic elements in subsurface peaks could serve to enrich particles sinking from surface waters in which these elements are at lower concentrations. Supporting evidence comes from Krishnaswami et al. (1976b), who examined fine particulate matter retained on filters, and Fowler et al. (1983), who analyzed sediment-trap material. Their studies indicated that particles at depths of 700 m and 750 m, respectively, had significantly greater concentrations of transuranic elements than did particles in the surface water.

The great spatial and temporal variability in plankton abundance in the sea makes it difficult to extrapolate from one particular water column to another or to the same water column at a different time. In the absence of extensive field data for various water columns, and considering the time and expense involved in gathering such data, modelling the behavior of transuranic elements on the basis of laboratory experimentation (which is comparatively fast and inexpensive) would seem to be a useful aid in understanding distributions of transuranic elements and processes affecting them. If the models are to have some validity, then more detailed information must be incorporated; however, much of the necessary data is still lacking. It seems a pointless exercise to construct refined models without additional biological data on (1) the abundance in different waters of small, gelatinous zooplankton (e.g., *Oikopleura dioica*), which have been studied little and for which the few data that exist indicate that they can become extremely abundant seasonally in various waters, often dominating the zooplankton, and are efficient packagers of materials (Alldredge and Madin, 1982; Taguchi, 1982); (2) the abundance in different waters of microcrustaceans, considered together with experimental data on the biokinetics of transuranic elements in these animals; (3) a better characterization of the nature and formation of marine snow and its sinking rate and abundance; and (4) rates of degradation and remineralization of marine snow, fecal pellets, and molts in natural waters.

16.3.3. Potential Importance of Zooplankton Fecal Pellets

It is possible to make a crude model of the impact of zooplankton defecation on the rate of removal of transuranic elements from surface water. Table 16.3 presents calculations of the time required to remove 50% of the ^{241}Am from the surface water of a stationary water column. The assumptions are (1) that the ^{241}Am instantly disperses uniformly over the upper 20 m, and there is no ^{241}Am below the upper 20 m; (2) that the WCF of ^{241}Am for suspended particulate matter is 4×10^5 (or 2×10^6, on a dry-weight basis), equivalent to the WCF in diatoms and to the WCFs of transuranic elements in suspended particulate matter in various water columns (Section 16.2.1), and an equilibrium is instantly attained; (3) that the phytoplankton exist only in the upper 20 m, and their biomass remains constant; and (4) that the zooplankton biomass remains constant, they excrete 2.5% of their body weight (dry) per day (Small et al., 1973), and their fecal pellets consist exclusively of perfectly compacted phytoplankton material that sinks at rates > 20 m d^{-1}. Given these assumptions, it can be calculated that half of the total ^{241}Am in the upper 20 m of the water column would be defecated within about 200 d when the phytoplankton and zooplankton biomasses are moderate or low (Table 16.3). At very high zooplankton densities, such as are encountered during blooms, the rate of ^{241}Am removal via defecation is greatly enhanced, so that 40–80 d are necessary to remove half of the total Am. Considering that the residence times of Pu and Am in the upper 80 m of a North Pacific water column have been estimated to be 2.4 y and 4.5 y, respectively (Fowler et al., 1983), 1–3 y for Pu in the Middle Atlantic Bight (50-m depth) (Santschi et al., 1980), and 2.5 y for Pu in Lake Michigan (mean depth of 84 m) (Wahlgren et al., 1980), the estimated removal by zooplankton defecation alone (Table 16.3) could account for a substantial fraction of the total removal of transuranic elements from the surface water. The time required to remove ^{241}Am from the surface layer via fecal pellets (Table 16.3) is inversely proportional to the WCF of ^{241}Am in the hypothetical phytoplankton population and to the defecation rate of the zooplankton. Thus, if the WCF in the phytoplankton is decreased to 1×10^5, a value more typical of the non-diatom groups (Table 16.1), then the corresponding time required to remove Am from the surface water is increased four-fold, whereas a zooplankton defecation rate of 10% of body weight (dry) per day, as can occur with smaller zooplankters (Heyraud, 1979), would accordingly decrease the removal time four-fold. The removal time is therefore dependent on the species composition of the plankton in the surface water, and a general set of calculations cannot apply to all water columns. It is apparent, however, that the standing stock of zooplankton would critically affect residence time of the transuranic elements in surface water.

It is assumed in this chapter that the behavior of

Table 16.3. Hypothetical Data Showing the Potential Influence of Zooplankton Fecal Pellets as a Vehicle for the Vertical Transport of ^{241}Am out of Contaminated Surface Water[a]

Phytoplankton Standing Crop (g m^{-2}, dry wt)	Percentage of Total ^{241}Am on Phytoplankton	Zooplankton Standing Crop (g m^{-2}, dry wt)	Phytoplankton Population Defecated (% d^{-1})	Time Necessary to Remove 50% of Total ^{241}Am from Upper 20 m by Fecal Pellets (d)
0.5	4.8	0.5	2.5	417
0.5	4.8	1.0	5	208
0.5	4.8	5.0	25	42
1.0	9.1	0.5	1.25	440
1.0	9.1	1.0	2.5	220
1.0	9.1	5.0	12.5	44
2.0	16.7	0.5	0.63	479
2.0	16.7	1.0	1.25	240
2.0	16.7	5.0	6.25	48
10.0	50.0	0.5	0.13	800
10.0	50.0	1.0	0.25	400
10.0	50.0	5.0	1.25	80

[a]For each combination of phytoplankton and zooplankton biomass, the fraction of total ^{241}Am (here used as a representative transuranic element) associated with phytoplankton, the fraction of total phytoplankton ingested and defecated by the zooplankton per day, and the dime (d) necessary to remove 50% of the total ^{241}Am from the upper 20 m solely by defecation are given. See text for assumptions made in these calculations.

Am, for which there are the most complete biokinetics data, and of Pu, for which there are the most field data, would be representative of the other highly reactive transuranic elements, such as Cm and Cf, whose concentration factors in phytoplankton are comparable to those of Am and Pu, but for which kinetics data and field data for zooplankton are lacking. Moreover, the same factors that govern vertical transport of transuranic wastes would probably be applicable to a variety of nondegradable organic pollutants (e.g., polychlorinated biphenyls and polynuclear aromatic hydrocarbons) and other anthropogenic contaminants that enter the ocean at the surface. The enrichment of zooplankton fecal material with many toxicants introduced into the oceans by human activity (Fowler, 1982) is consistent with this hypothesis.

16.4. SUMMARY

1. Laboratory experimentation has shown that marine plankton greatly concentrate the transuranic elements ^{237}Pu, ^{241}Am, ^{242}Cm, and ^{252}Cf from seawater but do not appreciably concentrate ^{235}Np. It is proposed that Np would behave essentially conservatively in seawater, whereas the highly reactive elements would concentrate up to 10^6 times in phytoplankton, 100–1000 times in crustacean zooplankton, and 10–100 times in appendicularians.

2. Correlations of concentration factors for transuranic elements with surface:volume ratios for organisms were described by the same equation for phytoplankton, euphausiids, and copepods, suggesting that the uptake process in these organisms proceeds by passive adsorption onto surfaces, and that the surfaces of these organisms have approximately equal affinities for these elements.

3. Following passive accumulation in zooplankters, the transuranic elements are removed either in association with unassimilable material that is defecated or association with exoskeletons that are expended during molting.

4. A model is presented to show the potential of different forms of planktonic debris in the vertical transport of the reactive transuranic elements, for which ^{241}Am is used as a representative radionuclide. The model takes into account sinking rates and abundance of biogenic debris together with estimates of remineralization half-times for ^{241}Am. It is concluded that large zooplankters

such as euphausiids package ^{241}Am into rapidly sinking particles (either fecal pellets or molts), which could transport the ^{241}Am to very deep water and sediment. The phytoplankton and appendicularian houses and fecal material, which are often observed to sink in the form of marine snow in the sea, would be expected to transport the transuranic elements only a few hundred meters, thereby enriching subsurface water with these wastes.

ACKNOWLEDGMENTS

The International Laboratory of Marine Radioactivity operates under a tripartite agreement between the International Atomic Energy Agency, the government of the Principality of Monaco, and the Oceanographic Institute of Monaco. We thank Barry Hargrave and anonymous reviewers for constructive comments on the manuscript.

REFERENCES

Alldredge, A. L., and L. P. Madin. 1982. Pelagic tunicates: unique herbivores in the marine plankton. *Bioscience*, **32**, 655–663.

Bienfang, P. K. 1980. Phytoplankton sinking rates in oligotrophic waters off Hawaii, USA. *Marine Biology*, **61**, 69–77.

Bowen, V. T. 1983. Element specific redistribution processes in the marine water column. *In*: Processes Determining the Input, Behavior and Fate of Radionuclides and Trace Elements in Continental Shelf Environments. U.S. Department of Energy CONF 790382, Washington, D.C., pp. 29–38.

Bowen, V. T., K. M. Wong, and V. E. Noshkin. 1971. Plutonium-239 in and over the Atlantic Ocean. *Journal of Marine Research*, **29**, 1–10.

Bowen, V. T., V. E. Noshkin, H. D. Livingston, and H. L. Volchok. 1980. Fallout radionuclides in the Pacific Ocean: vertical and horizontal distributions, largely from GEOSECS stations. *Earth and Planetary Science Letters*, **49**, 411–434.

Broecker, W. S., and T.-H. Peng. 1982. Tracers in the Sea. Eldigio Press, New York, 690 pp.

Edgington, D. N. 1981. A review of the persistence of long-lived radionuclides in the marine environment—sediment/water interactions. *In*: Impacts of Radionuclide Releases into the Marine Environment. International Atomic Energy Agency, Vienna, pp. 67–91.

Fisher, N. S., B. L. Olson, and V. T. Bowen. 1980. Plutonium uptake by marine phytoplankton in culture. *Limnology and Oceanography*, **25**, 823–839.

Fisher, N. S., P. Bjerregaard, and S. W. Fowler. 1983a. Interactions of marine plankton with transuranic elements. 1. Biokinetics of neptunium, plutonium, americium and californium in phytoplankton. *Limnology and Oceanography*, **28**, 432–447.

Fisher, N. S., P. Bjerregaard, and S. W. Fowler. 1983b. Interactions of marine plankton with transuranic elements. III. Biokinetics of americium in euphausiids. *Marine Biology*, **75**, 261–268.

Fisher, N. S., K. A. Burns, R. D. Cherry, and M. Heyraud. 1983c. Accumulation and cellular distribution of ^{241}Am, ^{210}Po, and ^{210}Pb in two marine algae. *Marine Ecology Progress Series*, **11**, 233–237.

Fowler, S. W. 1982. Biological transfer and transport processes. *In*: Pollutant Transfer and Transport in the Sea, Vol. II, G. Kullenberg (Ed.). CRC Press, Boca Raton, Florida, pp. 1–65.

Fowler, S. W., and L. F. Small. 1972. Sinking rates of euphausiid fecal pellets. *Limnology and Oceanography*, **17**, 293–296.

Fowler, S. W., G. Benayoun, and L. F. Small. 1971. Experimental studies on feeding, growth and assimilation in a Mediterranean euphausiid. *Thalassia Jugoslavica*, **7**, 35–47.

Fowler, S. W., S. Ballestra, J. La Rosa, and R. Fukai. 1983. Vertical transport of particulate-associated plutonium and americium in the northeast Pacific. *Deep-Sea Research*, **30**, 1221–1233.

Fukai, R., E. Holm, and S. Ballestra. 1979. A note on vertical distribution of plutonium and americium in the Mediterranean Sea. *Oceanologica Acta*, **2**, 129–132.

Fukai, R., S. Ballestra, and D. Vas. 1983. Characteristics of vertical transport of transuranic elements through the Mediterranean water column. VIes Journées Etudes Pollutions, Commission Internationale pour l'Exploration Scientifique de la Mer Mediterranée, Monaco, pp. 95–101.

Gardner, W. D., K. R. Hinga, and J. Marra. 1983. Observations on the degradation of biogenic material in the deep ocean with implications on accuracy of sediment trap fluxes. *Journal of Marine Research*, **41**, 195–214.

Gorsky, G., N. S. Fisher, and S. W. Fowler. 1984. Biogenic debris from the pelagic tunicate, *Oikopleura dioica*, and its role in the vertical transport of a transuranium element. *Estuarine and Coastal Shelf Science*, **18**, 13–23.

Gromov, V. V. 1976. Uptake of plutonium and other nuclear wastes by plankton. *Marine Science Communications*, **2**, 227–247.

Heyraud, M. 1979. Food ingestion and digestive transit time in the euphausiid *Meganyctiphanes norvegica* as a function of animal size. *Journal of Plankton Research*, **1**, 301–311.

Higgo, J. J. W., R. D. Cherry, M. Heyraud, and S. W. Fowler. 1977. Rapid removal of plutonium from the oceanic surface layer by zooplankton faecal pellets. *Nature*, **266**, 623–624.

Higgo, J. J. W., R. D. Cherry, M. Heyraud, S. W. Fowler, and T. M. Beasley. 1980. Vertical oceanic transport of alpha-radioactive nuclides by zooplankton fecal pellets. *In*: Natural Radiation Environment III, T. F. Gesell and W. M. Lowder (Eds.). U.S. Department of Energy, Washington, D.C., pp. 502–513.

Hofmann, E. E., J. M. Klinck, and G.-A. Paffenhofer. 1981. Concentrations and vertical fluxes of zooplankton fecal pellets on a continental shelf. *Marine Biology*, **61**, 327–335.

Holm, E., S. Ballestra, R. Fukai, and T. M. Beasley. 1980. Particulate plutonium and americium in Mediterranean surface waters. *Oceanologica Acta*, **3**, 157–160.

Iturriaga, R. 1979. Bacterial activity related to sedimentary particulate matter. *Marine Biology*, **55**, 157–169.

Krishnaswami, S., D. Lal, and B. L. K. Somayajulu. 1976a. Investigations of gram quantities of Atlantic and Pacific surface particulates. *Earth and Planetary Science Letters*, **32**, 403–419.

Krishnaswami, S., D. Lal, B. L. K. Somayajulu, R. F. Weiss, and H. Craig. 1976b. Large-volume in-situ filtration of deep Pacific waters: mineralogical and radioisotope studies. *Earth and Planetary Science Letters*, **32**, 420–429.

Lowman, F. G., T. R. Rice, and F. A. Richards. 1971. Accumulation and redistribution of radionuclides by marine organisms. *In*: Radioactivity in the Marine Environment. National Academy of Sciences, Washington, D.C., pp. 161–199.

Olsen, C. R., N. H. Cutshall, and I. L. Larsen. 1982. Pollutant–particle associations and dynamics in coastal marine environments: a review. *Marine Chemistry*, **11**, 501–533.

Paasche, E. 1980. Silicon. *In*: The Physiological Ecology of Phytoplankton, I. Morris (Ed.). Blackwell, Oxford, U.K., pp. 259–284.

Pentreath, R. J., D. F. Jefferies, J. W. Talbot, M. B. Lovett, and B. R. Harvey. 1982. Transuranic cycling behaviour in marine environment. *In*: Transuranic Cycling Behaviour in the Marine Environment. International Atomic Energy Agency, Vienna, pp. 121–128.

Santschi, P. H., Y. H. Li, J. J. Bell, R. M. Trier, and K. Kawtaluk. 1980. Pu in coastal marine environments. *Earth and Planetary Science Letters*, **51**, 248–265.

Schaule, B. K., and C. C. Patterson. 1981. Lead concentrations in the northeast Pacific: evidence for global anthropogenic perturbations. *Earth and Planetary Science Letters*, **54**, 97–116.

Sholkovitz, E. R. 1983. The geochemistry of plutonium in fresh and marine water environments. *Earth-Science Reviews*, **19**, 95–161.

Silver, M. W., and A. L. Alldredge. 1981. Bathypelagic marine snow: deep-sea algal and detrital community. *Journal of Marine Research*, **39**, 501–503.

Small, L. F., and S. W. Fowler. 1973. Turnover and vertical transport of zinc by the euphausiid *Meganyctiphanes norvegica* in the Ligurian Sea. *Marine Biology*, **18**, 284–290.

Small, L. F., S. W. Fowler, and S. Keckes. 1973. Flux of zinc through a macroplanktonic crustacean. *In*: Radioactive Contamination of the Marine Environment. International Atomic Energy Agency, Vienna, pp. 437–452.

Small, L. F., S. W. Fowler, and M. Y. Unlu. 1979. Sinking rates of natural copepod fecal pellets. *Marine Biology*, **51**, 233–241.

Taguchi, S. 1982. Seasonal study of fecal pellets and discarded houses of appendicularia in a subtropical inlet, Kaneohe Bay, Hawaii. *Estuarine and Coastal Shelf Science*, **14**, 545–555.

Thomann, R. V. 1981. Equilibrium model of fate of microcontaminants in diverse aquatic food chains. *Canadian Journal of Fisheries and Aquatic Sciences*, **38**, 280–296.

Wahlgren, M. A., J. A. Robbins, and D. N. Edgington. 1980. Plutonium in the Great Lakes. *In*: Transuranic Elements in the Environment, W. Hanson and R. A. Watters (Eds.). U.S. Department of Energy, Washington, D.C., pp. 659–683.

Watters, R. A., D. N. Edgington, T. E. Hakonson, W. C. Hanson, M. H. Smith, F. W. Whicker, and R. E. Wildung. 1980. Synthesis of the research literature. *In*: Transuranic Elements in the Environment, W. Hanson and R. A. Watters (Eds.). U.S. Department of Energy, Washington, D.C., pp. 1–44.

Chapter 17

Bioavailability of Technetium, Plutonium, and Americium from Deep-Sea Sediments to the Clam *Venerupis decussata*

Simon R. Aston and Scott W. Fowler

International Laboratory of Marine Radioactivity, IAEA
Musée Océanographique, Monaco

ABSTRACT

Distribution coefficients (K_d) for Tc, Pu, and Am in the sediment, and their accumulation from labeled deep-sea sediments by the clam *Venerupis decussata* were measured in laboratory experiments. The K_d values were on the order of $1, 10^3$, and 10^5 for Tc, Pu, and Am, respectively. Whole-body transfer factors (the radioactivity per gram of animal divided by the radioactivity per gram of sediment) were generally low, ranging from ~ 0.1 for Tc to roughly 5×10^{-3} for Pu and Am. It appeared that Pu was somewhat more available from sediments to the clams than was Am. Technetium was concentrated in the soft parts of bivalves whereas most of the Pu and Am was associated with the shell. The fraction of the two transuranic elements associated with the shell increased when the Pu and Am were taken up directly from seawater. The type of sediment influenced the availability of a radionuclide to the clams; approximately five times more Am was accumulated from Pacific sediments than from Atlantic sediments, despite similar K_d values for the two types of sediment. Sequential leaching experiments revealed that a large fraction ($\sim 60\%$) of Am in the Atlantic sediments was apparently not available for uptake by the clams. These findings underscore the potential importance of sediment-specific geochemical associations of radionuclides in determining the availability of the radionuclide from sediments to the biota.

17.1. INTRODUCTION

The possible use of deep-sea sediment as a repository for high-level radioactive wastes has generated a need to acquire information on the behavior of long-lived radionuclides in these sediments, especially as regards their potential for transfer to the biosphere. Low-level wastes have been, and continue to be, dumped onto the deep-sea floor, further underscoring the importance of understanding radionuclide-transfer processes in the deep ocean. Among the most important radionuclides to be considered are ^{99}Tc, ^{239}Pu, ^{240}Pu, and ^{241}Am. Their abundance in wastes from the nuclear-fuel cycles and their long half-lives ($t_{1/2}$) (2.1×10^5 y for ^{99}Tc, 2.44×10^4 y for ^{239}Pu, 6580 y for ^{240}Pu, and 431 y for ^{241}Am) make it possible for them to be released from containment into deep-sea sediments and subsequently transferred into the biosphere via benthic organisms.

We selected the bivalve *Venerupis decussata* as an infaunal species and used sediments from the vicinity of existing and potential dumpsites in the northeastern Atlantic and the central Pacific oceans. Laboratory experiments were designed to determine the rate and extent of radionuclide transfer to the clams from sediments, the depuration rates, and the distributions of radionuclides in the tissue of the clams. Comparisons with the uptake of Tc, Pu, and Am from seawater by the same species have been made (Fowler et al., 1983; Vangenechten et al., 1983; Aston and Fowler, 1984), and this information, together with some geochemical features of the sediments, have been used to try to explain the uptake routes and sediment-specific features of the bioavailability of these radionuclides.

17.2. METHODS AND MATERIALS

17.2.1. Isotopes and Radioanalyses

Technetium-95m, with a $t_{1/2}$ of 60 d, was supplied as the pertechnetate ion (TcO_4^-) in a solution with a pH of 4. When added to aerated seawater, the pertechnetate ion is extremely stable and is not reduced (Fowler et al., 1981; Beasley et al., 1982). It is also of interest to investigate the important lower oxidation state of Tc, since this may occur in some forms of the nuclear waste. The Tc(IV) state was prepared by the reduction of pertechnetate by using hydrazine sulphate (Fowler et al., 1981). The chemical forms of reduced Tc are not well characterized, so, for convenience, the symbol (IV) will be used here to denote total reduced Tc.

Plutonium-237, with a $t_{1/2}$ of 45.6 d, was obtained in solution with a pH of 3.5. Since Pu may exist in seawater as either the reduced (III + IV) or the oxidized (V + VI) state (Nelson and Lovett, 1978; Aston, 1980), both were used in the experiments. The (III + IV) state was prepared by a reduction–oxidation couple reaction using hydroxylamine hydrochloride and sodium nitrate solutions, and the (V + VI) state was prepared by evaporation to near-dryness with concentrated perchloric acid, followed by dissolution in dilute hydrochloric acid (Fowler et al., 1975).

Americium-241 with a $t_{1/2}$ of 431 y, was supplied in a pH 4 solution as Am^{3+}. Since the introduction of Am into the marine environment in other oxidation states is unlikely considering Am redox equilibria, the Am(III) state alone was used for the experiments.

All radioanalyses were performed by γ spectroscopy with heavily shielded NaI(Tl) scintillation detectors coupled to a pulse-height analyser. The isotopes employed have relatively easily measured photon emissions, while the isotopes of Tc and Pu of interest in nuclear wastes decay by pure β and α emissions, respectively, which render them unsuitable for *in vivo* measurements since radiochemical separations and sacrificing experimental animals are required. Furthermore, the use of isotopes with high specific activity (such as ^{237}Pu and ^{95m}Tc) has the added advantage of more closely approximating the actual atom concentrations of the corresponding radionuclide in the environment (Fowler et al., 1975).

17.2.2. Sediments and Organisms

The experiments were carried out with deep-sea sediments from the vicinity of existing and potential radioactive disposal sites in the northeastern Atlantic and the central Pacific oceans (Table 17.1). In the early experiments performed with Tc, sediment from the actual Nuclear Energy Agency (NEA) Atlantic dumpsite was not available, and experiments were performed with a sediment (SA-1) taken near the NEA site. Clay minerologies were determined by X-ray diffraction.

For the bioavailability experiments, the sediments were labeled with the radionuclides after making a slurry with ~750 g of sediment and 100 ml of seawater with ^{95m}Tc, ^{237}Pu, or ^{241}Am added as required. The activities were in the range of 140 to 3000 Bq g^{-1} (wet weight). After adding the radionuclides, the sediments were periodically stirred and left to equilibrate for one day before being placed into plastic basins and being allowed to settle. The settled sediments were then placed under running, unfiltered seawater with a flow rate of 50 liters h^{-1} and a temperature ranging from 12 to 14°C.

Clams (*Venerupis decussata* L.) were collected near Sete, France. Each of the animals was numbered and acclimated in running seawater for one week prior to the start of the experiments. The clams were not fed during the experiments, and the only sources of available food were particulate organic material in the water and sediments.

17.2.3. Bioaccumulation Experiments

Before introducing the animals into the labeled sediments, these sediments were subsampled to test for

Table 17.1.   Sedimentological Properties of the Deep-Sea Sediments

Sediment and Location	Water Depth (m)	Carbonate Content (%)	Organic Carbon Content (%)	Grain-Size Distribution		Clay Mineralogy[a]
				Size (μm)	Percent of total	
SA-1	3900	42	0.3	> 63	19	Illite
45°14′N				31–63	10	Kaolinite
22°26′W				16–31	1.2	Chlorite
				8–16	8.4	Traces of mixed-layer clays
				4–8	11.2	
				2–4	8.8	
				< 2	41.2	
SA-2	4640	83	0.3	> 63	3.1	Illite
46°02′N				31–63	3.2	Kaolinite
16°55′W				16–31	2.1	Chlorite
				8–16	4.2	
				4–8	16.8	
				2–4	5.3	
				< 2	65	
SP-1	5700	7.6	0.4	> 63	8.7	Illite
31°00′N				31–63	0.3	Kaolinite
159°00′W				16–31	0.6	Chlorite
				8–16	4.5	
				4–8	9.4	
				2–4	18	
				< 2	59	

[a] The dominant clays are listed in descending order of abundance as shown.

homogeneous distribution of the tracer and to measure the distribution coefficient (K_d) values for the radionuclide distribution between solid sediment and pore water. Samples of sediment were centrifuged at 5000 rpm for about 15 min to separate pore water and sediment grains. The pore water was then passed through 0.45-μm Millipore® (Millipore Corp., Bedford, Massachusetts) membrane filters to ensure complete removal of particles. The radioactivity of the two separated fractions was then measured, and the results were used to calculate K_d values [defined as the radioactivity per milliliter of dry sediment divided by the radioactivity per milliliter of pore water (Duursma and Bosch, 1970)]. The quantities of ^{241}Am used in the bioavailability experiments were too low to determine pore-water activities, so 100 g of each sediment was labeled with about ten times as much ^{241}Am activity, and the K_d values were determined as described above.

The clams were first placed directly into the settled sediments and were then periodically removed for radionuclide measurement. The animals were rinsed for 2 h in clean, running seawater to permit defecation of any ingested sediment before radionuclide measurement. In addition, a fine jet of seawater from a squeeze bottle was used to remove sediment particles adhering to the shell. During the ^{241}Am experiments, four clams were subsequently fed with unlabeled phytoplankton for 2 h. Following feeding, the ^{241}Am in the animals and the excreted particulate matter was counted to determine the ^{241}Am content of the clams and the feces. After counting, the clams were immediately returned to the labeled sediments.

To compare radionuclide bioaccumulation from sediments with that from seawater, the clams were exposed to unfiltered, labeled seawater. Fresh radioactive seawater was prepared at least twice weekly to maintain constant seawater activities and to avoid build-up of metabolic wastes that might complex the radionuclide (Fowler et al., 1975). The experimental and radioanalytical techniques used have been described by Fowler et al. (1975, 1981) and Grillo et al. (1981).

After the experiments, some clams were dissected to assess the radionuclide distribution in the tissues. Other animals were placed in unlabeled sediment under running seawater to evaluate the depuration of the radionuclides. The geochemical leaching techniques employed in this study have been described in full by Aston et al. (1984).

17.3. RESULTS AND DISCUSSION

17.3.1. Radionuclide Adsorption onto Sediments

Table 17.2 summarizes the K_d values for radionuclide partitioning between pore water and sediment after one day of labeling the sediments for the bioaccumulation experiments. The transuranic elements showed K_d values far greater than those of Tc: K_d values for both oxidation states of Pu were three orders of magnitude higher than those of Tc. The state of oxidation did not appear to appreciably change the K_d of either Tc or Pu. In the case of Tc, this may be because the (IV) state had been oxidized. Previous studies (Fowler et al., 1981; Beasley et al., 1982) show that approximately 90% of the Tc(IV) in oxygenated seawater is oxidized to pertechnetate within 24 h. The remaining Tc(IV) then appears to be stable and is carried by $Fe(OH)_3$ precipitation. This rapid oxidation would explain the similar K_d values observed for Tc regardless of the oxidation states used in the experiment. Similar information on Pu is not available.

The low K_d values for Tc were paralleled by a fairly rapid release rate from the sediments (Fig. 17.1). The half-time of release was approximately 4–5 d, and there was no difference in loss rates between TcO_4^- and Tc(IV), supporting the hypothesis that the (IV) state had rapidly oxidized during the experiments. The poor adsorption of Tc onto the sediments, as reflected by the

low K_d values and release rates, is confirmed by other studies (Gromov and Spitsyn, 1973; Beasley, 1981; Masson et al., 1981). The loss rates for Tc did not appear to be appreciably influenced by the presence of clams in the sediments since control sediments without animals also lost ^{95m}Tc, although at a slightly faster rate due to the smaller volume of sediment used.

Within experimental errors, the total activities of ^{237}Pu and ^{241}Am in sediment did not change during the course of the experiments, thus reflecting the strong association of these transuranics with sediments. Previous studies of sediments contaminated with fuel-reprocessing wastes also show that Pu and Am tend to stay in the sediment (Beasley and Fowler, 1976; Aston and Stanners, 1981).

17.3.2. Bioaccumulation of Technetium

Transfer factors (expressed as radioactivity per gram of wet animal divided by radioactivity per gram of wet SA-1 sediment or pore water) for whole clams and soft

Table 17.2. $K_d{}^a$ Values for Radionuclide Partitioning between Pore Water and Sediment^b

Radionuclide	Sediment	K_d
^{95m}Tc		
Tc(VII)	SA-1	1.130 ± 0.002
Tc(IV)	SA-1	0.960 ± 0.002
^{237}Pu		
Pu(V + VI)	SA-1	$1.6 \times 10^3 \pm 0.3 \times 10^3$
Pu(III + IV)	SA-1	$1.3 \times 10^3 \pm 0.3 \times 10^3$
^{241}Am		
Am(III)	SA-2	$1.5 \times 10^5 \pm 0.3 \times 10^3$
Am(III)	SP-1	$1.8 \times 10^5 \pm 0.2 \times 10^3$

$^a K_d$ = (radioactivity per milliliter of dry sediment)/(radioactivity per milliliter of pore water).
^{b} After one day of labeling.

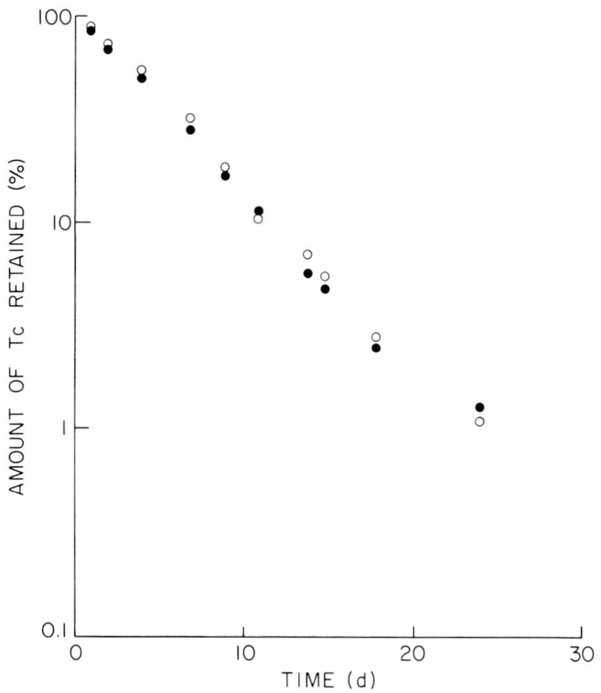

Figure 17.1. Loss of ^{95m}Tc(IV + VII) from labeled surface sediment collected from near the northeastern Atlantic low-level radionuclide dumpsite (SA-1). (●) Tc(VII) and (○) Tc(IV).

Table 17.3. Whole-Body Transfer Factors[a] for 95mTc between Clams and Sediment[b]

Duration of Uptake (d)	Total Sediment		Pore Water[c]	
	Tc(IV)	Tc(VII)	Tc(IV)	Tc(VII)
1	0.09 ± 0.06	0.10 ± 0.03	0.05 ± 0.03	0.05 ± 0.01
4	0.11 ± 0.07	0.14 ± 0.06	0.06 ± 0.04	0.07 ± 0.03
8	0.09 ± 0.05	0.13 ± 0.07	0.08 ± 0.03	0.07 ± 0.04
15	0.12 ± 0.12	0.14 ± 0.07	0.06 ± 0.06	0.07 ± 0.04
24	0.11 ± 0.10	0.13 ± 0.06	0.06 ± 0.05	0.07 ± 0.03
36	0.11 ± 0.11	0.13 ± 0.06	0.05 ± 0.05	0.07 ± 0.03

[a] Transfer factors = (radioactivity per gram wet clam) divided by (radioactivity per gram of wet sediment or pore water).
[b] The temperature of the water was held at 13–14°C. Values are means ±1 σ of 5 individuals weighing between 5.0 and 8.8 g (wet weight).
[c] Analogous to a concentration factor.

Table 17.4. Transfer Factors for 95mTc between Total Soft Tissues of Clams and Sediment[a]

Uptake (d)	Total Sediment		Pore Water[b]	
	IV	VII	IV	VII
1	0.15 ± 0.09	0.17 ± 0.05	0.07 ± 0.04	0.08 ± 0.02
4	0.18 ± 0.12	0.23 ± 0.09	0.09 ± 0.06	0.12 ± 0.05
8	0.14 ± 0.08	0.21 ± 0.12	0.07 ± 0.04	0.11 ± 0.06
15	0.19 ± 0.19	0.21 ± 0.10	0.10 ± 0.09	0.11 ± 0.06
24	0.18 ± 0.16	0.21 ± 0.10	0.09 ± 0.08	0.11 ± 0.05
36	0.18 ± 0.16	0.21 ± 0.10	0.09 ± 0.08	0.11 ± 0.05

[a] Temperature of the water was held at 13–14°C. Values are means ±1 σ of 5 individuals and were computed from results of whole-body radioactivity levels, known fraction distribution of 95mTc between soft parts and shell, and the relative weights of soft parts and shell.
[b] Analogous to a concentration factor.

tissues during bioaccumulation tests are given in Tables 17.3 and 17.4. Since 95mTc losses occurred throughout the experiment, the calculations were based on a running mean of the activities found in the sediment or the pore water. Calculations based on sediment or pore water show that for both oxidation states of Tc the uptake rapidly reached a steady state. Although variable, transfer factors for soft tissues were generally higher (Table 17.4) than were those for whole clams (Table 17.3), indicating that Tc had a higher affinity for the soft tissues than for the shell. Results of the dissections done at the end of the experiment showed that an average of ~20% more 95mTc was taken up by the soft tissue compared to by the shell (Table 17.5). The viscera and gill had the greatest affinity for Tc, and muscle the lowest, which is in accord with previous studies in which bivalves were exposed to 95mTc in water (Fowler et al., 1981; Beasley et al., 1982).

Table 17.5. Transfer Factors for 95mTc between Clam Tissues and Sediment Following a 36-d Exposure to Labeled Sediment[a]

Tissue	Transfer Factors[b]	
	Total Sediments	Pore Water[c]
Shell	0.25 ± 0.14	0.13 ± 0.07
Mantle	0.23 ± 0.09	0.12 ± 0.04
Viscera	0.60 ± 0.30	0.31 ± 0.15
Muscle	0.08 ± 0.03	0.04 ± 0.02
Gill	0.41 ± 0.14	0.22 ± 0.07
Total soft parts	0.30 ± 0.13	0.16 ± 0.07

[a] The temperature of the water was held at 13–14°C. Values are means ±1 σ of 4 individuals.
[b] No difference was noted between Tc(IV) and Tc(VII), so the data have been pooled.
[c] Analogous to a concentration factor.

In general, transfer factors for [95m]Tc (based on sediment or pore water) were low, with the maximum value (0.6) in visceral tissues, including the digestive gland. This has been noted also for other metals taken up by molluscs (Bryan, 1979; Fowler, 1982). A comparison with the uptake of Tc from seawater (Fig. 17.2) shows that the uptake was also low, with whole-body transfer factors of 1 or less; this compares well with results for Tc accumulation by other bivalves (Fowler et al., 1981; Masson et al., 1981; Beasley et al., 1982). As in the case of sediment experiments, no difference between the uptake of Tc(VII) and Tc(IV) from seawater was observed; therefore, the data are combined in Table 17.6 to show tissue transfer factors for [95m]Tc. Uptake from seawater ranges from one to two orders of magnitude greater for the soft tissues than for the shell. Periodic membrane filtration of the seawater indicated a very low fraction of particulate Tc ($\sim$0.5%) throughout the experiment, suggesting that soluble ions were taken up.

Some insight into the pathway of [95m]Tc uptake by the clams can be gained by comparing the Tc distribution among the various tissues after exposure to sediments and to seawater (Table 17.7). When the clams accumulated [95m]Tc from the sediments, $\sim$70% of the accumulated [95m]Tc was found associated with the shell, but only $\sim$11% was accumulated from seawater by the shell. This difference could be partially explained by the observation that clams buried in the sediment extended their siphons into the relatively uncontaminated overlying water. This allowed only limited contact between the contaminated pore water and internal tissues. On the other hand, the shell was in continuous contact with pore water. This hypothesis is supported by the similar transfer factors for shell based on [95m]Tc activities in pore water (Table 17.5) and those in seawater (Table 17.6).

Depuration of [95m]Tc from contaminated clams transferred to clean SA-1 sediment was slow (Fig. 17.3). The curve is best described by a single exponential expression with a biological half-life ($T_{b1/2}$) of 120 d. Dissection of the clams after a 6-month depuration period showed that 56 and 42% of the incorporated [95m]Tc was associated with shell and soft tissues, respectively. This distribution is somewhat different from that noted at the beginning of the depuration phase (Table 17.7) and suggests that loss from shell material may have proceeded more rapidly than that from the soft tissues.

17.3.3. Bioaccumulation of Plutonium

Some preliminary results on the bioaccumulation of [237]Pu(III + IV) and [237]Pu(V + VI) from sediment SA-1 by clams are now available. The complete study of this topic is reported elsewhere (Aston and Fowler, 1984). Transfer factors based on radioactivity per gram wet weight of clams divided by the radioactivity per gram of wet sediment were low for each of the oxidation states. Mean values of the transfer factors were $(6 \pm 0.6) \times 10^{-3}$ and were constant after $\sim$20 d of exposure. Transfer factors calculated relative to Pu activities of pore water are 16 ± 2 for Pu(III + IV) and 10 ± 1 for Pu(V + VI) after 40 d of exposure. The above transfer factors can be compared with [237]Pu from seawater: after 18 d, the transfer factors were 74 ± 5 and 61 ± 1 for [237]Pu(III + IV) and [237]Pu(V + VI), respectively.

Depuration rates for [237]Pu taken up from sediment SA-1 were measured when the clams were transferred to uncontaminated sediment after 40 d (Fig. 17.4). There was no difference in loss rates for the uptake of

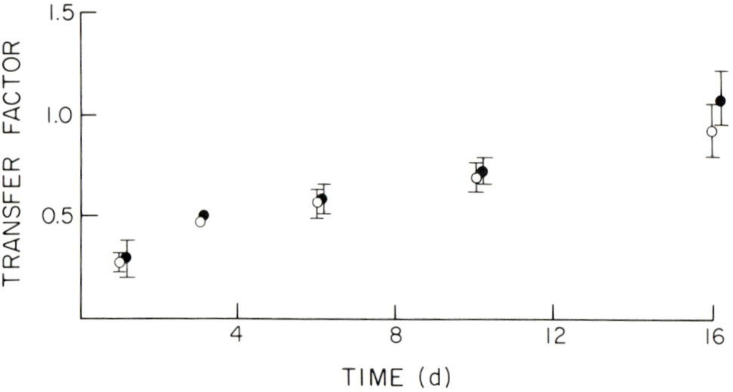

Figure 17.2. Uptake of [95m]Tc(IV) and [95m]Tc(VII) from labeled seawater by clams. Bars represent 1σ for $n = 5$ individuals in each treatment. ($\bullet$) Tc(VII) and ($\circ$) Tc(IV).

Table 17.6. Transfer Factors in Clam Tissues after a 16-d Exposure in Contaminated Seawater[a]

Tissue	Transfer Factor[b]
Shell	0.19 ± 0.4
Mantle	2.6 ± 0.7
Viscera	21.0 ± 5.1
Muscle	1.6 ± 0.2
Gill	4.1 ± 1.4
Total soft parts	3.8 ± 0.8

[a]The temperature of the water was held at 13–14°C. Values are means ± 1 σ of 6 individuals weighing 11 g ± 1 (wet weight).

[b]Tc(IV) and Tc(VII) are treated together.

Table 17.7. Distribution (%) of ^{95m}Tc in Clam tissues Following 36 d of Accumulation from Sediment and 16 d from Seawater[a]

| | | % Total ^{95m}Tc | |
Tissue	% Total (wet weight) ($n = 6$)	Uptake from Sediment ($n = 4$)	Uptake from Seawater ($n = 6$)
Shell	57.5 ± 1.2	69.6 ± 11.5	10.8 ± 2.6
Mantle	4.5 ± 0.7	8.2 ± 2.0	11.4 ± 2.9
Viscera	1.7 ± 0.4	10.8 ± 5.9	32.4 ± 5.7
Muscle	6.7 ± 0.8	2.6 ± 0.9	10.4 ± 2.0
Gill	2.4 ± 0.5	5.4 ± 1.3	9.5 ± 2.6
Pallial fluid[b]	27.3 ± 1.2	3.6 ± 2.5	25.5 ± 2.3
Total soft parts	15.2 ± 1.3	27.0 ± 9.5	63.7 ± 4.6

[a]The $\pm$ denotes standard deviation. Results from experiments with pertechnetate and Tc(IV) are grouped together.
[b]Values determined by difference.

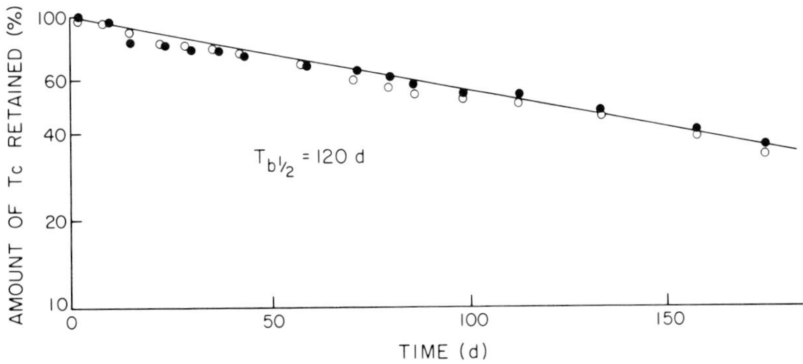

Figure 17.3. Loss of ^{95m}Tc(IV) and ^{95m}Tc(VII) from clams living in unlabeled sediment. (●) Tc(VII) and (○) Tc(IV).

Pu(III + IV) or Pu(V + VI). The loss curve most closely approximated a single exponential function with a $T_{b1/2}$ of 24 d. Depuration following uptake from seawater also indicated that loss was taking place from a single compartment but at a slower rate, as evidenced by a $T_{b1/2}$ of ~50 d (Fig. 17.4). The two-fold difference in depuration rate following exposure to sediment and to water is difficult to explain but may be related to the different exposure times used in the two experiments. Fowler (1982) has shown that exposure can affect subsequent loss rates of heavy metals in marine organisms.

Little has been published on the bioaccumulation of Pu from marine sediments by bivalves. Miramand et al. (1982) reported the uptake of ^{238}Pu(IV) from marine mud by the bivalve *Scrobicularia plana* (da Costa) as a transfer factor (on a wet-weight basis) of ~0.002 after 14 d of exposure. This value is about three times lower than the transfer factors we found for Pu(III + IV) or Pu(V + VI) by *V. decussata* after a comparable period of exposure in sediment from the northeastern Atlantic. The difference may be due to the biological species or to the differences between a deep-sea, carbonate-rich sediment and a muddy, coastal sediment. The importance of sediment site-specificity in relation to geochemical association and bioavailability is discussed in more detail for Am in Section 17.3.4.

17.3.4. Bioaccumulation of Americium

The studies on bioaccumulation of Am differ from those for Tc and Pu in that (1) Am was used in the (III) oxidation state only, and (2) uptake both from northeastern Atlantic (SA-2) and from central Pacific (SP-1) sediments was investigated.

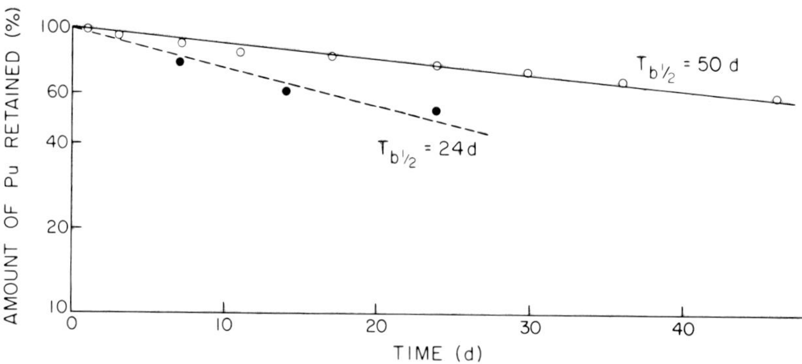

Figure 17.4. Depuration of ^{237}Pu from clams following exposure to (●) labeled sediments and (○) seawater. No differences were noted between the different oxidation states of Pu; therefore, the data have been pooled.

Figure 17.5 illustrates the ^{241}Am transfer factors for uptake from Atlantic and Pacific sediments by the clam as a function of time (calculated as described in Section 17.3.2). Transfer factors for the Atlantic sediment reached a plateau after 2 d, and averaged $(4 \pm 0.2) \times 10^{-3}$. The uptake from Pacific sediment proceeded much more slowly and gave a five-fold larger mean transfer factor of $(20 \pm 0.7) \times 10^{-3}$. If one assumes that bioaccumulation of ^{241}Am results from direct uptake from pore water, the transfer factor for the Atlantic sediment based on pore water alone was 106 ± 5 and was found to be constant after 2 d; for the Pacific sediment the maximum was 780 ± 30 at steady state, which occurred after 25 d.

It is perhaps noteworthy that the whole-body transfer factor for Pu $[(6 \pm 0.6) \times 10^{-3}]$ in Atlantic sediments is ~50% higher than that for ^{241}Am $[(4 \pm 0.2) \times 10^{-3}]$. It may be that this is due to the large difference (10^2) in K_d value between Pu and Am for these sediments; however, any clear-cut interpretation is complicated by the fact that the two deep-sea sediments, while from the same region, differed in their carbonate content (Table 17.1). Nevertheless, our observation of a somewhat greater transfer of Pu for a given sediment is similar to studies with burrowing worms, which took up more Pu than Am from naturally radioactive sediments (Beasley and Fowler, 1976).

In order to assess the distribution of ^{241}Am in the organs of the animals while taking care to avoid the retention of sediment particles in the animals, radioactivity in four clams taken from the labeled Atlantic sediment was measured before and after a 2-h period of feeding on nonradioactive phytoplankton. In three of four animals, total-body radioactivity decreased by 13–24% after feeding, suggesting excretion of some residual sediment (Table 17.8). Virtually all of the ^{241}Am lost from the clams was recovered with the feces onto filter paper. In one animal, loss did not occur, and, furthermore, no feces were observed, although the animal had extruded both siphons and appeared to be in good health.

Following uptake of ^{241}Am from Atlantic and Pacific sediment by the clams for 56 and 43 d, respectively, all remaining animals were dissected in order to examine the distribution of ^{241}Am in the tissues (Table 17.9). The relative distribution of radionuclide was roughly comparable in the animals from both types of sediment. Most of the ^{241}Am was found in the shell (56–70% in SA-2 and 60–75% in the Pacific sediment), whereas most of

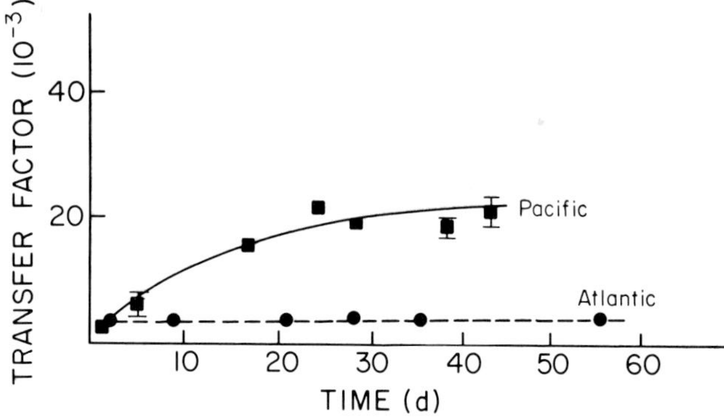

Figure 17.5. Whole-body transfer factors for ^{241}Am as a function of time for clams exposed to labeled Atlantic and Pacific sediments. Error bars represent standard error on the mean for 4 individuals. The curves were fitted by eye.

Table 17.8. Loss of ^{241}Am from Clams after a 2-h Feeding on Unlabeled Phytoplankton[a]

Clam Sample	Before Feeding (cpm $\pm$ 1 σ)	After Feeding (cpm $\pm$ 1 σ)	Lost by Feeding cpm $\pm$ 1 σ	%	Recovered on Filter cpm $\pm$ 1 σ	%
1	4020 $\pm$ 179	4210 $\pm$ 180	$-$ 198 $\pm$ 91	$-$ 5	—	—
2	4920 $\pm$ 180	3750 $\pm$ 170	1167 $\pm$ 93	24	806 $\pm$ 160	16
3	8420 $\pm$ 200	7180 $\pm$ 190	1310 $\pm$ 130	15	1530 $\pm$ 180	18
4	5180 $\pm$ 180	4490 $\pm$ 180	686 $\pm$ 98	13	1050 $\pm$ 160	20

[a]The four animals used in this experiment had accumulated ^{241}Am from Atlantic sediment for 28 d. Data are presented as counts per minute (cmp) $\pm$ 1 σ counting error and as percent of radioactivity before feeding. A dash indicates that the amount was not measurable.

the remaining ^{241}Am range (20–30%) was located in the viscera of clams from both sediments. Only small amounts of ^{241}Am were measured in the pallial fluid, muscles (adductor and foot), mantle and siphons, and the gills.

When the uptake pathway into the same species of clam is from seawater, the fraction associated with the shell increases to 82% (Grillo et al., 1981). The relative bioavailability of ^{241}Am from Atlantic SA-2 and Pacific sediments was reflected in the transfer factors of the radionuclide for the various tissues (Table 17.9). For example, the shells and viscera of the animals in the Pacific sediment had transfer factors about three times higher than the values for animals exposed to the Atlantic sediment SA-2. Except in the case of the gill tissue, transfer factors in the remaining tissues also tended to be higher for the Pacific sediment, although the differences were not significant (2σ) for the two sediment types.

Generally speaking, the bioaccumulation of ^{241}Am, like Pu, from contaminated sediments was low, as evidenced by transfer factors lower than unity for all animals tested in these experiments. Low bioavailability of ^{241}Am from marine sediment seems to be the case for several infaunal species (Beasley and Fowler, 1976; Miramand et al., 1982). It is important to note that about 13–24% of the total radioactivity in the clams in these experiments was excreted during a 2-h period of intense feeding, indicating that ingestion of some labeled sediment had occurred. Because the experimental animals were allowed to void their gut for only 20 min prior to counting, it is possible that some ingested sediment was retained in the gut and, hence, the whole-body transfer factors for clams reported in this chapter may be ~20% higher than in actuality. Nevertheless, the use of these factors is not an unrealistic approach in the ecological sense since animals with full guts are regularly eaten by organisms from higher trophic levels.

The results of this study indicate that the sediment type induced appreciable differences in total ^{241}Am uptake by the animals. The availability of Ag, Co, and Zn from different sediments has been observed by

Table 17.9. Transfer Factors and tissue Distribution (Percent of total Animal Radioactivity) in Clams after 56 and 43 d of Exposure to Contaminated Atlantic and Pacific Sediments[a]

Organ	Atlantic Sediment Transfer Factor (10^{-3})	Atlantic Sediment Total Radioactivity (%)	Pacific Sediment Transfer Factor (10^{-3})	Pacific Sediment Total Radioactivity (%)
Shell	5.1 $\pm$ 0.3	61 $\pm$ 4	19 $\pm$ 1	66 $\pm$ 5
Pallial fluid	0.8 $\pm$ 0.1	4 $\pm$ 1	3.5 $\pm$ 1.5	3 $\pm$ 1
Muscle (foot + adductor)	1.1 $\pm$ 1.1	1 $\pm$ 1	3 $\pm$ 2	0.5 $\pm$ 0.3
Mantle + siphons	4.3 $\pm$ 1.4	5 $\pm$ 1	11 $\pm$ 4	5 $\pm$ 2
Viscera	27 $\pm$ 2	25 $\pm$ 2	80 $\pm$ 14	25 $\pm$ 4
Gills	8 $\pm$ 2	4 $\pm$ 1	2.5 $\pm$ 2.5	0.3 $\pm$ 0.3

[a]Values are means $\pm$ standard error for 9 animals (weight range = 11.0–18.8 g, wet weight) radioanalyzed in 3 groups each with 3 individuals.

Luoma and Jenne (1975). Their results showed that the highest metal availability to the soft parts of the clam *Macoma balthica* is achieved in sediments with the lowest K_d values (those sediments from which most metal was desorbed back into the solution). In our experiments, however, the mean K_d values for [241]Am are approximately 1.5×10^5 in both sediments, and, given the counting errors, cannot be shown to differ from one another.

Sequential leaching experiments performed on these two sediments labeled with [241]Am were therefore used to determine the geochemical associations of this radionuclide. The results show that $\sim$60% of the [241]Am was present in a highly resistant form in the Atlantic sediment, while only $\sim$10% was present in this resistant fraction in the Pacific sediment. The resistant fraction is defined as that Am removed by leaching with hot, concentrated HNO_3 (Aston et al., 1984).

This very important difference between the geochemical associations of [241]Am in the sediments provides an explanation of the biological availability results discussed in this chapter. This difference also underscores the fact that biological availability of radionuclides is, at least for Am, highly dependent on sediment-specific geochemical associations.

To determine the role of the indirect transfer from the pore water to the animals, we calculated the [241]Am transfer factors relative to the pore water (Fig. 17.6), and then we compared these data with the known [241]Am uptake from seawater. Grillo et al. (1981) measured a concentration (or transfer) factor of 350 in the clam *V. decussata* following 24 d of [241]Am uptake from

seawater. No equilibrium was reached within this period, and the concentration factor continued to increase. In our experiments (Fig. 17.6), the same species reached a transfer factor for pore water of $\sim$500 and $\sim$80 for animals in Pacific and in Atlantic sediments, respectively. In the latter sediment, an equilibrium was reached after 2 d, whereas in the former a plateau was achieved only after $\sim$30 d. Thus, the total amount taken up, as well as the uptake rate in our experiments, are clearly different from the data obtained by Grillo et al. (1981) and are dependent on sediment type.

Although accumulation of [241]Am from the sediment by the clams was followed for a considerably long time in these experiments (56 d for Atlantic SA-2 and 43 d for Pacific sediment), little of the element was transferred into the soft tissues (Table 17.9). By far the largest proportion of [241]Am (range between 20 and 30%) was measured in the viscera, but the phytoplankton feeding experiment (Table 17.8) indicated an excretion of about 13–24%. It may be assumed that the bulk of the [241]Am concentration measured in viscera is actually a readily excretable [241]Am fraction such as ingested sediment. Small percentages of total-body [241]Am have been found in the gill, the muscle, and the pallial fluid (Table 17.9). The relative distribution of [241]Am among the tissues of the clams was comparable for both sediments.

17.4. SUMMARY

There were considerable differences between the distribution of Tc and the transuranic elements Pu and Am between deep-sea sediments and their associated pore waters. Technetium was very poorly adsorbed onto the sediments and was easily lost from them; the K_d values for Tc were among the lowest observed for radionuclides of interest in waste disposal that we have examined. The K_d values for both Pu(III $\pm$ IV) and Pu(V + VI) were about three orders of magnitude higher, and the K_d values for Am were as much as five orders of magnitude higher than those for Tc.

Uptake of Tc by all clam tissues (whole body) rapidly reached a steady state, and transfer factors were low. In general, the soft tissues took up more [95m]Tc than did the shell, the highest transfer factor (0.6) being found in the visceral tissues. Technetium uptake from water was also low, with transfer factors of 1 or less, which was in good agreement with other studies. Tissue distributions showed that Tc taken up from seawater was enriched by one to two orders of magnitude for the

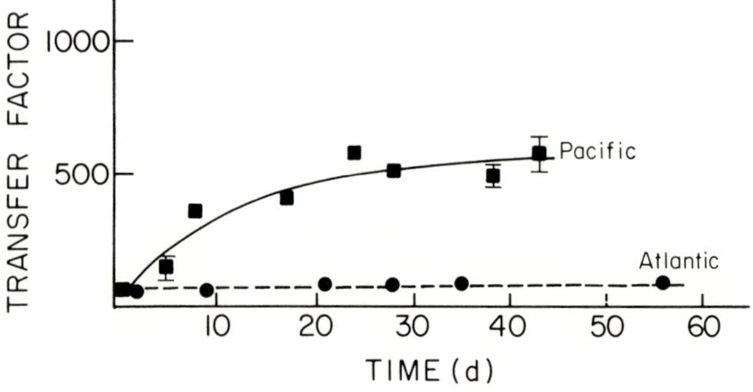

Figure 17.6. Transfer factors for [241]Am based on pore-water concentrations as a function of time in clams exposed to labeled Atlantic and Pacific sediments. Error bars represent standard error on the mean for 4 individuals. The curves were fitted by eye.

soft parts as compared to the shell. The comparison of tissue distributions of ^{95m}Tc taken up by clams in sediment or seawater suggests that the routes of uptake were different, since the shells of those animals living in labeled sediment took up more Tc than did those of animals living in labeled seawater. The depuration of ^{95m}Tc from clams living in labeled sediment had a biological half-life of 120 d, and loss from the shell appeared to be more rapid than that from the soft tissues.

Transfer factors for Pu from the sediment to the clams were very low: about $(6 \pm 0.6) \times 10^{-3}$ on a total sediment–whole-body basis. Transfer factors for the uptake of Pu from seawater were 74 ± 5 for Pu(III + IV) and 61 ± 1 for Pu(V + VI). The biological half-life for the loss of sediment-derived Pu from clams was considerably faster than for Tc, with a mean value of 24 d. The rate is not a function of the oxidation states of the Pu used at the start of the experiments. There seemed to be large differences in the distribution between the soft parts and the shell for ^{237}Pu taken up from both the sediments and the seawater. In contrast to Tc, the greatest fraction of Pu was associated with the shell, regardless of the uptake pathway. Furthermore, the percentage in the shell was greatly increased following exposure to labeled seawater.

Transfer factors of $(4 \pm 0.2) \times 10^{-3}$ for Atlantic sediment SA-2 and $(20 \pm 0.7) \times 10^{-3}$ for the Pacific sediment indicate a five-fold difference in the uptake of ^{241}Am in these two sediments. This difference can be explained by the fact that the Atlantic sediment SA-2 had a higher proportion of strongly held ^{241}Am as compared to the Pacific sediment (Vangenechten et al., 1983). This emphasizes that the characterization of the geochemical associations of radionuclides in sediments is an important prerequisite for predicting their bioavailability.

As with Pu, the shell accumulated a greater fraction of ^{241}Am than did the soft tissues, which took up a relatively small amount. The fraction of Am in the shell increased when the radionuclide was accumulated from seawater. The viscera concentrated more ^{241}Am than did the rest of the soft tissues, but experiments with feeding the clams phytoplankton showed that some of the incorporated radionuclide was readily excreted and may have been due to the ingestion of labeled sediment particles. Comparisons of our results with studies by Grillo et al. (1981) show that the uptake of Am from seawater by these clams is much greater than from sediments.

Our results suggest that Pu in the Atlantic sediment may be somewhat more available ($\sim 50\%$) for uptake by clams than is Am. This may be related to the K_d of Pu being 100 times lower than that of Am in Atlantic deep-sea sediments. While transfer factors for Am and Pu are on the same order of magnitude ($\sim 10^{-3}$), they are 1000 times lower than are those for Tc. All three radionuclides displayed relatively low bioavailability to clams living in labeled sediments as compared to their availability from seawater.

We conclude from these observations that uptake by clams would be a quantitatively unimportant process in transferring weakly bound Tc and the more readily adsorbed Pu and Am from labeled sediments. This conclusion is in agreement with the field studies by Aarkrog et al. (1984), which show that 11 y after an accidental release of Pu to the ocean off Greenland, the entire benthos (including bivalves) contained less than 1% of the Pu in the sediment.

ACKNOWLEDGMENTS

The International Laboratory of Marine Radioactivity operates under a tripartite agreement between the International Atomic Energy Agency, the Government of the Principality of Monaco, and the Oceanographic Institute at Monaco. Support for the present work is gratefully acknowledged. We also thank Roger Wilson (Institute of Oceanographic Sciences, U.K.), Werner Feldt (Isotopenlaboratorium der Bundesforschungsanstalt für Fischerei, F.R.G.), and Dan M. Talbert (Sandia National Laboratories, U.S.A.) for arranging supplies of sediment samples, and N. Toutin (Nice University, France) for X-ray diffraction analyses.

REFERENCES

Aarkrog, A., H. Dahlgaard, K. Nilsson, and E. Holm. 1984. Further studies of plutonium and americium at Thule, Greenland. *Health Physics*, **46**, 29–44.

Aston, S. R. 1980. Evaluation of the chemical forms of plutonium in sea water. *Marine Chemistry*, **8**, 319–325.

Aston, S. R., and S. W. Fowler. 1984. Bioavailability of plutonium(III + IV, V + VI) from a deep-sea sediment to a bivalve mollusc and a polychaete worm. *Journal of Environmental Radioactivity*, **1**, 67–78.

Aston, S. R., and D. A. Stanners. 1981. Pluntonium transport to and deposition and immobility in intertidal sediments from the Irish Sea. *Nature*, **289**, 581–582.

Aston, S. R., J. Gastaud, B. Oregioni, and P. Parsi. 1984. Observations on the adsorption and geochemical associations of technetium, neptunium, plutonium, americium and californium with a deep-sea sediment. *In*: International Symposium on the Behavior of Long-lived Radionuclides in the Marine Environment, C. E. C., Luxembourg, pp. 179–188.

Beasley, T. M. 1981. Biogeochemical Studies of Technetium in Marine and Estuarine Ecosystems. University of Oregon, Corvallis. Unpublished report submitted to the U.S. Department of Energy, 75 pp.

Beasley, T. M., and S. W. Fowler. 1976. Plutonium and americium: uptake from contaminated sediments by the polychaete *Nereis diversicolor*. *Marine Biology*, **38**, 95–100.

Beasley, T. M., J. J. Gonor, and H. V. Lorz. 1982. Technetium: uptake, organ distribution and loss in the mussel *Mytilus californianus* (Conrad) and the oyster *Crassostrea gigas* (Thunberg). *Marine Environmental Research*, **7**, 103–116.

Bryan, G. W. 1979. Bioaccumulation of marine pollutants. *Philosophical Transactions of the Royal Society of London*, **286**, 483–505.

Duursma, E. K., and C. J. Bosch. 1970. Theoretical, experimental and field studies concerning diffusion of radioisotopes in sediments and suspended particles of the sea. Part B: Methods and experiments. *Netherlands Journal of Sea Research*, **4**, 395–469.

Fowler, S. W. 1982. Biological transfer and transport processes. *In*: Pollutant Transfer and Transport in the Sea, G. Kullenburg (Ed.). CRC Press, Boca Raton, Florida, pp. 1–65.

Fowler, S. W., M. Heyraud, and T. M. Beasley. 1975. Experimental studies on plutonium kinetics in marine biota. *In*: Impacts of Nuclear Releases into the Aquatic Environment. International Atomic Energy Agency, Vienna, pp. 157–177.

Fowler, S. W., G. Benayoun, P. Parsi, M. W. A. Essa, and E. H. Shulte. 1981. Experimental studies on the bioavailability of technetium in selected marine organisms. *In*: Impacts of Radionuclide Releases into the Marine Environment. International Atomic Energy Agency, Vienna, pp. 313–339.

Fowler, S. W., S. R. Aston, G. Benayoun, and P. Parsi. 1983. Bioavailability of technetium from artificially labelled north-east Atlantic deep-sea sediments. *Marine Environmental Research*, **8**, 87–100.

Grillo, M. C., J. C. Guary, and S. W. Fowler. 1981. Comparative studies on transuranium nuclide biokinetics in sediment-dwelling invertebrates. *In*: Impacts of Radionuclide Releases into the Marine Environment. International Atomic Energy Agency, Vienna, pp. 273–291.

Gromov, V. V., and V. I. Spitsyn. 1973. Sorption of ^{239}Pu, ^{106}Ru and ^{99}Tc by bottom sediments of the Pacific Ocean. *Radiokhimiya*, **16**, 157–162.

Luoma, S. N., and E. A. Jenne. 1975. The availability of sediment-bound cobalt, silver and zinc to a deposit-feeding clam. *In*: Biological Implications of Metals in the Environment. U.S. Energy Resources Development Agency (ERDA) Symposium Series No. 42, Richland, Washington, pp. 213–230.

Masson, M., G. Aprosi, A. Laniece, P. Guegueniat, and Y. Belot. 1981. Approches experimentales de l'étude des transferts du technetium à des sediments et à des espèces marines benthiques. *In*: Impacts of Radionuclide Releases into the Marine Environment. International Atomic Energy Agency, Vienna, pp. 341–359.

Miramand, P., P. Germain, and H. Camus. 1982. Uptake of americium and plutonium from contaminated sediments by three benthic species: *Arenicola marina, Corophium volutator* and *Scrobicularia plana*. *Marine Ecology—Progress Series*, **7**, 59–65.

Nelson, D. M., and M. B. Lovett. 1978. Oxidation state of plutonium in the Irish Sea. *Nature*, **276**, 599–601.

Vangenechten, J. H. D., S. R. Aston, and S. W. Fowler. 1983. Uptake of americium-241 from two experimentally labelled deep-sea sediments by three benthic species: a bivalve mollusc, a polychaete and an isopod. *Marine Ecology—Progress Series*, **13**, 219–228.

SUBJECT INDEX

AUTHOR INDEX